# VIDEO-BASED SURVEILLANCE SYSTEMS

## *Computer Vision and Distributed Processing*

# VIDEO-BASED SURVEILLANCE SYSTEMS
## *Computer Vision and Distributed Processing*

*edited by*

**Paolo Remagnino**
**Graeme A. Jones**
*Digital Imaging Research Centre*
*Kingston University*

**Nikos Paragios**
*Imaging and Visualization Department*
*Siemens Corporate Research*

**Carlo S. Regazzoni**
*Dept. of Biophysical and Electronic Engineering*
*University of Genova*

SPRINGER SCIENCE+BUSINESS MEDIA, LLC

ISBN 978-1-4613-5301-0 ISBN 978-1-4615-0913-4 (eBook)
DOI 10.1007/978-1-4615-0913-4

**Library of Congress Cataloging-in-Publication Data**

A C.I.P. Catalogue record for this book is available
from the Library of Congress.



*Printed on acid-free paper.*

# Contents

# Preface

Monitoring of public and private sites has increasingly become a very sensitive issue resulting in a patchwork of privacy laws varying from country to country - though all aimed at protecting the privacy of the citizen. It is important to remember, however, that monitoring and visual surveillance capabilities can also be employed to aid the citizen. The focus of current development is primarily aimed at public and corporate safety applications including the monitoring of railway stations, airports, and inaccessible or dangerous environments. Future research effort, however, has already targeted citizen-oriented applications such as monitoring assistants for the aged and infirm, route-planning and congestion-avoidance tools, and a range of environmental monitoring applications.

The latest generation of surveillance systems has eagerly adopted recent technological developments to produce a fully digital pipeline of digital image acquisition, digital data transmission and digital recording. The resultant surveillance products are highly-flexible, capable of generating forensic-quality imagery, and able to exploit existing Internet and wide area network services to provide remote monitoring capability. Despite these advances, however, these systems remain relatively unintelligent - with face recognition and motion detection the only notable exception. Digital technology has not so much changed the nature of surveillance and monitoring operations but rather extended their reach and reliability. In security environments, human operators remain central to the classification, threat assessment, and response planning procedures found in CCTV installations. Nonetheless, the rapidly increasing processing power of affordable, general purpose computing platforms does suggest that the cycle-hungry intellectual property gestating in current academic and industrial research programmes will have a major impact on the next generation of products. In particular, distributed intelligent architectures, explored in the contributions presented in Part Four of this book, are likely to exploit three key recent technological developments - efficient modular image analysis algorithms, light-weight distributed computing methodologies, and intelligent devices (such as the intelligent camera sporting a limited level of onboard processing). Such architectures, in which signal analysis is devolved towards the sensing devices, can exploit the reduced bandwidth requirements of transmitting knowledge rather than pixels.

This book reports on both well-established approaches as well as work in progress in academic and industrial research laboratories on both sides of the Atlantic. It aims to develop and update many of the theoretical problems related to the monitoring of site explored in the two books in the series. A key difference however, has been the invitation of contributions from the key industrial players in the United States where the VSAM and related initiatives have accelerated the solution of outstanding problems as well as highlighted new issues. In Part One, these invited contributions describe the most recent trends in the industrial world, including real-time systems for the monitoring of indoor and outdoor environments, society infrastructures such as subways and motorways, retail stores and aerial surveillance. The first contribution describes a number of component technologies crucial to the Siemens modular approach to building products suited to a range of possible security applications. Modular construction is underpinned by a strong focus on statistical characterisation of algorithms enabling the design, customisation, validation and evaluation of total systems. Embedding aerial surveillance in more effective visualisation and analysis environment is a major focus of surveillance work of the Sarnoff Corporation and described in the second contribution. In particular, mosaicing and the registration of video data with digital elevation data and 3D site models allowing visualisation and interpretation of data from previously unavailable viewpoints. A cooperative camera network is designed by Honeywell in which object colour signatures are propagated between camera neighbours to aid the process of tracking objects in multi-camera environments. Classical tracking algorithms employing these multi-camera fusion techniques are combined with threat assessment algorithms to identify suspicious behaviour in public malls. The final invited contribution from Philips Research describes series of key technologies including a simple step-by-step calibration process for acquiring intrinsic and extrinsic camera parameters, and PTZ tracking algorithms employing image alignment based on estimated global camera motion parameters. These basic technologies underpin an event retrieval system which maps the interactions of tracked events within the scene into complex human behaviours.

Object tracking of vehicles and people is one of the key technologies that will underpin any monitoring applications for in outdoor or public environments. Despite the considerable level and effort which has been expended on this problem of the past decade or so, real-time object tracking that performs reliably in all weather and lighting conditions is a relatively recent achievement. In Part Two, the contributions represent current best practice in the chain of algorithms required to perform robust and accurate real-time tracking including motion detection techniques robust to rapid and frequent lighting changes, the establishment of accurate temporally consistent object trajectories particularly in crowded scenes, and the classification of object types. Part Three represents contributions which attempt to analyse or alarm on the events unfolding in a monitored scheme. For example Koller-Meier and Van Gool learn trajectory manifolds to essentially represent normal behaviours, while the VIGILANT system of Greenhill *et al* attempts to annotate the trajectory tracks to support high-level queries. Alternative pixel-based classification schemes are proposed by Kass *et al* and Sherrah and Gong.

**Dr Paolo Remagnino**
**Dr Graeme A. Jones**

## Acknowledgments

The editors wish to thank all the academic organizations and industrial companies which supported the event. A special thanks goes to the International Association for Pattern Recognition, Associazione Italiana per l'Intelligenza Artificiale, Siemens Corporate Research USA, Sarnoff Corporation USA, Honeywell Laboratories USA, Philips Research USA, University of Genova and Kingston University. The editors are grateful to our colleagues Dr D.Greenhill and Dr N.Monekosso, and all our Ph.D. students Etienne Corvee, John-Paul Renno, Jonathan Rymel and David Thirde who kindly offered their help during the event. A special thank goes to our technicians and MEDIA Services who made all this possible. A great thank you to Professor Graeme Wilkinson for his encouragement and support.

I

# INDUSTRIAL APPLICATIONS

# Chapter 1

# REAL-TIME VIDEO ANALYSIS AT SIEMENS CORPORATE RESEARCH

Real-time Imaging Group *

*Imaging and Visualization Department*

*Siemens Corprorate Research, 755 College Road East, Princeton, NJ 08540*

{nikos,comanici,michael,rameshv}@scr.siemens.com

**Abstract** The proliferation of cheap sensors and increased processing power has made the acquisition and processing of video information more feasible. Real-time video analysis tasks performing object detection, tracking can increasingly be performed efficiently on standard PC's. Smart cameras are being designed that enable on-camera applications to be designed so that they act as intelligent sensors that provide compressed data or meta event information directly. These advances along with major breakthroughs in communication and Internet technology is making possible real-time video monitoring and communication for a variety of application sectors such as: Industrial automation, Transportation, Automotive, Security/Surveillance, and Communications. The real-time imaging group at SCR is focusing on the development of integrated end-to-end solutions for applications requiring object detection, tracking, and action classification / event analysis from video. This paper will present an overview of our research in statistical methods for real-time video surveillance systems. Solutions for subway/highway monitoring for emergency assistance and resource management, real-time tracking technologies for applications in video conferencing, intelligent video communications, and industrial automation, will be highlighted.

**Keywords:** Tracking, Motion Detection, Compression, Surveillance.

*The Real-time Imaging Group is composed of: Visvanathan Ramesh (Project Manager), Dorin Comaniciu, Nikos Paragios, Michael Greiffenhagen, and numerous university collaborators.

# 1. Introduction

We are in an technological era of computing and communications. The availability of cheap sensing and affordable computing power is enabling the development of embedded real-time video analysis systems that can provide compressed data or meta event information directly. These systems are useful for a variety of application sectors such as: Industrial automation, Transportation, Automotive, Security/Surveillance, and Communications. The real-time imaging group at SCR is focusing on the development of integrated end-to-end solutions for applications requiring object detection, tracking, and action classification / event analysis from video. We have developed systems for applications including: subway monitoring, highway monitoring, tunnel monitoring, Intelligent video communication, and Indoor/quasi-outdoor monitoring. These applications share requirements that include: real-time performance, high robustness/accuracy, graceful degradation and self-diagnostic ability, all day, all weather operation, etc. Our approach to building these systems is to follow solid statistical principles while building these complex systems. Statistical methods are utilized at the component level and at the systems integration level in order to fuse coherently the various component modules developed.

This paper is organized as follows. Our view is that the video analysis system is composed of several modules applied in sequence or in iterative fashion. The modules are viewed as statistical estimation techiques and their concatenation essentially corresponds to a compound estimator. A large portion of our research is therefore focused on development of modules based on solid statistical principles. We first describe in this paper applications involving real-time object detection or segmentation, tracking, and video communication. The first application described is an "Intelligent Video Communication System" involving face detection, tracking and log-polar encoding and transmission. This system is useful for mobile video communication, video conferencing, and remote surveillance settings. A second application described involves real-time segmentation of people in subway platforms for the purpose of congestion (crowding) detection in subway platforms.

There is a strong need for validation of these systems before real-world deployment. One way to validate these systems is through systematic testing on a large scale. This often requires extensive tests and is costly. We therefore concentrate on performance characterization of the individual modules as a pre-cursor to the large scale tests. The characterization enables us to identify the theoretical limits of the estimation schemes developed. In addition, we have been focusing on systems identification, i.e. the characterization of performance of complete systems and on the customization of these systems to different installation settings. In the second part of this paper, we will provide our overall design and validation philosophy that draws heavily on systems engineering and statistics and illustrate its use in a dual-camera surveillance system. We conclude with an outlook for future research and open research issues we are pursuing.

## 2. Non-parametric Statistical methods for Detection, Tracking. Efficient Video Communication

In this section we summarize the details of our prototype real-time foveation system for applications such as remote/distributed surveillance, video conferencing, etc. The system uses a Java-based client-server architecture to enable the efficient transmission of events and data of interest from a network of cameras to a central processing unit. A face detection and tracking module that runs at each camera server signals the human presence within an observation area and provides the 2-D face coordinates and its estimated scale to the video transmission module. The captured video is then efficiently represented in log-polar coordinates, with the foveation point centered on the face, and sent to the connected client modules for further processing. The current setting of the system employs active cameras that track the detected person, by switching between smooth pursuit and saccadic movements, as a function of the target presence in the fovea region. The high resolution of the data in the fovea region enables successive application of recognition/re-identification modules on the transmitted video without sacrificing their performance. The system modules are well suited for implementation on the next-generation of Java-based intelligent cameras.

In the following subsections we describe our system for color-based face detection, tracking, log-polar coding and communication.

### 2.1 Color Based Face Detection and Tracking

The module performing the detection and tracking of human faces is based on our recent work, described in [9]. For a more theoretically involved paper, please see [8].

The color model of the human face was obtained from one subject, by computing the mean histogram of the face instances recorded in the morning, afternoon, and at night. Since the experiments were performed in offices with large windows, both outdoor and indoor illumination con-

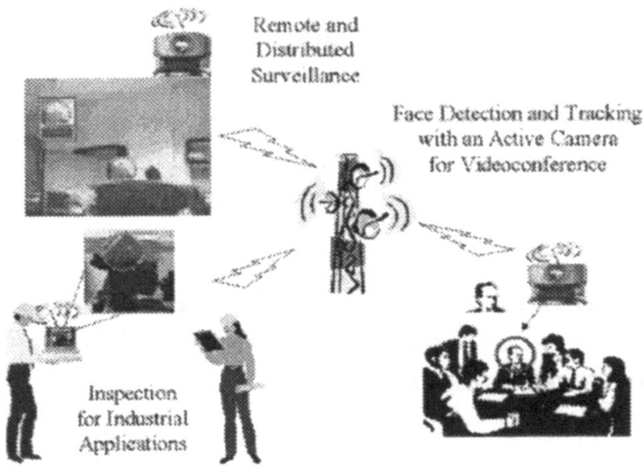


*Figure 1.1.* **Intelligent processing and summarization of video results in very low bit rate. Wireless connections can be employed for the transmission of data of interest to a central processing unit.**

ditions were tested. The histograms were computed in the intensity normalized $RG$ space [16] with $128 \times 128$ bins.

As a dissimilarity measure between the face model and the face candidates, a metric based on the Bhattacharyya coefficient has been employed. Hence, the problem of face localization reduced to a metric minimization, or equivalently to the maximization of the Bhattacharyya coefficient between two color distributions. By including spatial information into the color histograms, we showed, however, that the maximization of the Bhattacharyya coefficient is equivalent to maximizing a density estimate. As a consequence, the gradient ascent mean shift procedure [7] could be employed to guide a fast search for the best face candidate in the neighborhood of a given image location.

The resulting optimization achieves convergence in only a few iterations, being thus well suited for the task of real-time detection and tracking. To adapt to the scale changes of the target the scale invariance property of the Bhattacharyya coefficient is exploited as well as the gradient information on the border of the hypothesized face region.

The detection process involves the mean shift optimization with multiple initializations, each one in a different location of the current image frame. The initialization pattern is determined such that there is at least one initial location in the basin of attraction of a face of typical size, randomly located in the image.

The tracking process involves only optimizations in the neighborhood of the previous face location estimate, being therefore sufficiently fast to run comfortably at the frame rate on a standard PC. As a result, the module that implements the log-polar mapping receives for each frame two vectors representing the estimated position and scale of the currently observed face.

## 2.2 Camera Control

In the case when active cameras are used, the adequate control of the pan, tilt, and zoom is an important phase of the tracking process. The camera should execute fast saccades in response to sudden and large movements of the target while providing a smooth pursuit when the target is quasi-stationary [34, 36]. We implemented this type of control which resembles that of the human visual system. The fovea subimage occupies laterally about 6 degrees of the camera's 50 degrees field of view, at zero zoom.

However, contrary to other tracking systems that suspend the processing of visual information during the saccades movements [19], our visual face tracker is sufficiently robust to deal with the large amount of blurring resulting from camera motion. As a result, the visual tracking is a continuous process that is not interrupted by the servo commands. Note that a standard $RS-232C$ interface is used to communicate with the $SonyEVI-D30$ camera.

## 2.3 Log-Polar Coding and Transmission

Within our architecture each camera is seen as an image server that reports to the client (the central processing unit) whenever a human is present in its field of view. For the transmission of the visual information associated with this event, the data is filtered and sampled in log-polar coordinates, with the foveation point centered on the subject's face.

The log-polar representation of image data is mainly inspired by the primate visual system that reduces the enormous amount of available data through a non-uniformly sampled retina [11]. The field of view is divided into a region of maximal resolution, called the *fovea*, and a region whose resolution decreases towards the extremities, called the *periphery* [20]. The fovea is uniformly sampled, while the periphery is sampled according to a log-polar grid. Generally, the fovea covers the object of interest and the background is represented by the periphery. In [8] we describe the details of the log-polar mapping and the coding procedure. The sampling grid employed to compute the log-polar representation has locally a hexagonal lattice structure ( 1.2), shown to be 13.4% more efficient than rectangular lattices in the case of circularly bandlimited signals [15].

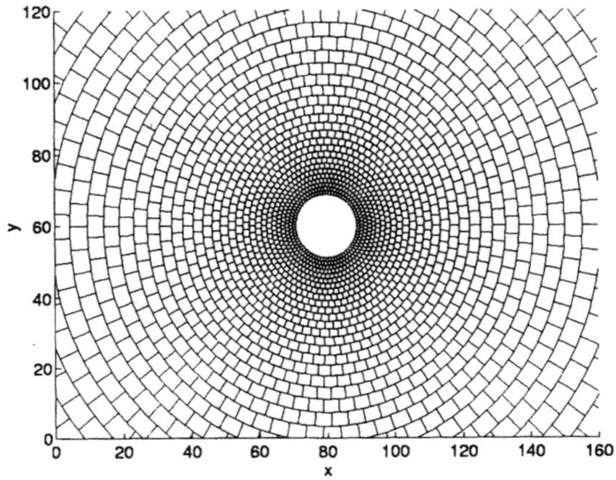


(a)

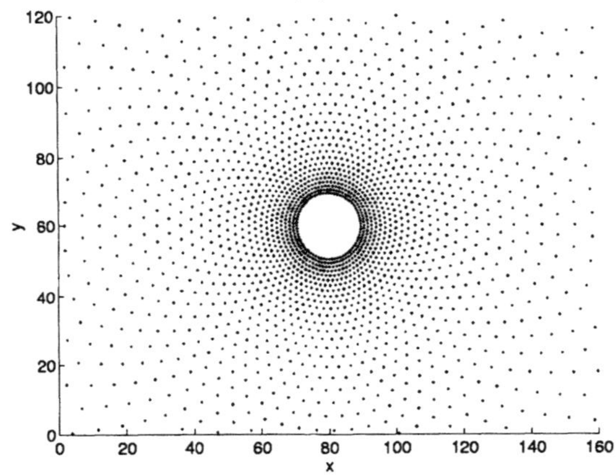


(b)

*Figure 1.2.* (a) Log polar pixels for an image of 160 × 120 pixels. (b) The corresponding sampling grid exhibits a local hexagonal structure.

In our current implementation, each video frame of size 320 × 240 is first converted to the $YCbCr$ color format, the color planes being subsampled to 160 × 120 pixels. Based on the location and scale information of the face that is received from the tracking module, the fovea regions of both the luminance and color planes are transmitted separately. The resulting system maintains approximately constant the transmission bandwidth, while allowing the full resolution transmission of the detected face, independent of the scale. The penalty is paid by the periphery (background) whose quality decreases when the face scale increases.

## 2.4 Decoding

The client performs the inverse mapping (decoding) to derive an approximated replica of the original data. A flexible log-polar representation of the background has been designed to deal in real-time with scale and location changes of the face of interest, while maintaining the required bit rate. Our system allows the high-resolution, quasi-lossless transmission of the fovea region, hence, the successful application of recognition/identification modules at the central unit.

## 2.5 Results

For all the experiments presented in this section we imposed an overall compression ratio of 16 to setup the log-polar coding design parameter (i.e. the number of rings in the coding scheme). Figure 1.3(a) presents the user interface of a camera server, showing the the detected face in the upper-left side of the image. The entire server package, performing, detection, tracking, camera control, and adaptive log-polar representation is well suited for implementation on the next generation of intelligent cameras. Figure 1.3(b) shows the reconstructed image and user interface of the primary client, which is able to control the camera from the distance. Note the face details.

Another pair of original and reconstructed image is presented Figure 1.4(a) and (b), respectively. Again, the face details in the reconstructed image are almost identical to those in the original.

# 3. Markov Random Field Modeling for Real-time Change Detection and Crowd density estimation in Subway Video

In the previous section we described an application involving real-time face detection, log-polar encoding and video communication. The application is of use in video conferencing and remote video surveillance. The system is intended for use in quasi-indoor settings for person to person communication or remote visualization of objects of interest. In this section we describe a system for subway monitoring that uses statistical models for change detection, real-time segmentation to detect areas where persons are located, and the evaluation of the segmentation map to identify crowding/congestion situations [17]. We propose a MRF-based real time approach for crowding/congestion detection in subway platforms. The proposed framework decomposes the corresponding video analysis problem into two steps.

The first step consists of a change detection algorithm that distinguishes the background from the foreground. This is done using discontinuity preserving MRF-based approach where the information from different sources (background subtraction, intensity modeling) is combined with spatial constraints to provide a smooth motion detection map. Two different information sources are used to perform the motion detection. The first is a data attraction and aims at providing a motion detection map that is coherent with the observation set. This coherence is defined through conditional probabilities of three different modules; a difference-based, a color-based and a normalized invariant color-based.

The first detection module is based on the difference frame and aims at capturing dynamic changes (motion) on the observation domain. We assume that density function of the difference frame is composed of two populations. The *static* one contains the background points while the *mobile* one contains the points that belong to moving objects. This assumption leads to a statistical modeling that provides continuous probability density functions for the static as well as the mobile hypothesis [4, 5]. Given the configuration of our system, color image sequences are available. Thus, during the learning phase for the reference frame, some color-based information can be also extracted for each pixel of the image. This information is modeled using independent Gaussian probability density functions that for a given observed intensity value for a given pixel measure the probability of being in the background case or not. This information is used to determine the second data-based detection module. Although the color module captures the background intensity properties, it is very sensitive to **global illumination changes** as well as **to shadows**. To deal with these limitation introduced by the color module, a third data attraction module that is based on the normalization of the RGB color space is used [28].

Finally, to preserve regularity of the detection map, we consider a second-order neighborhood and the set of two pixel cliques. Based on these assumptions we introduce a constraint that aims at finding a detection map that without many discontinuities. This is formulated by penalizing the objective function, when two neighborhood pixels are assigned to different labels (discontinuity) and relieving it for the opposite case.

Then, the obtained change detection map is combined with a geometry module to estimate the congestion of the observed area (platform). This module takes into account the real camera position (soft auto-calibration). To do this, we make use of the fact that the platform is planar and that one can obtain quasi-calibration information by using the images of the trains. The calibration information provides relative heights and widths of persons as a person translates across the scene at various depths from the camera. Based on these relative height and width measures we assign to each pixel a weight (far points are weighted more, while points closer to the camera are assigned smaller weights). Given these geometric factors the crowdedness estimate, is obtained by weighted average over all pixels with the segmentation result.

The minimization of the objective function is a very difficult task, since the corresponding solution space is huge and even the simplest minimization algorithm demands considerable amount of computations. Besides the cost function has many local minima, that are provided as final solutions from the non-optimal minimization algorithms. The most common way to overcome these constraints is by the use of *multi-scale* techniques that reduce to a significant ratio the required cost and perform a smooth operation to the the objective function. The main idea is to solve the problem in many different label spaces (subsets of the finest one). We adopt a sophisticated multi-scale approach which consists on defining a consistent *multi-grid* label model by using detection maps which are constrained to be piecewise constant over smaller and smaller pixel subsets [3]. Finally, fast deterministic relaxation algorithms (ICM [1], HCF [2]) are used for the minimization in each level.

Our current system has been developed and tested on several hours of subway video taken on different days over a period of 6 months [**fig** (1.5)]. Tests indicate that the implemented system meets the user requirements with low false alarm rate and no miss-detection.

As far the future directions are concerned, several theoretical issues have to be dealt with. The selection of the objective function as well as the model parameters is an interesting issue. Also several improvements can be considered for the change detection module since it is not yet clear how significant the MRF approach is compared to other alternatives. Approaches that may just take the difference map and incorporate knowledge of people sizes and geometry to estimate rough counts of people in the frame. Furthermore the system has to be validated on other settings.

## 4. Systems Engineering of Video analysis Systems

In the previous sections we described two applications wherein statistical estimation techniques as components in the system. The first system uses non-parametric statistics for real-time tracking, while the second uses Markov-Random Field models for real-time change segmentation in a subway setting. The systems are indicative of how robust solutions to real-world problems can be achieved through proper application of statistical methods. One major issue in these systems is their validation in real settings and the theoretical characterization of their performance. Our group is active in performance characterization research in the context of video analysis systems. This section will outline our current work on methodologies for performance characterization and illustrate a case study involving a dual camera video surveillance system.

## 4.1 Performance Characterization and Systematic Engineering Methodology

The typical scenario in an industrial research and development unit developing vision systems is that a customer defines a system specification and its requirements. The engineer then translates these requirements to a system design and validates that the system design meets the user-specified requirements. The system requirements in the video analysis setting often involves the specification of the operating conditions, the types of sensors, the accuracy requirements, and the computational requirements to be met by the system. The operating conditions essentially restrict the space of possible inputs by restricting the type of scene geometry, the physical properties such as object material types, and illumination conditions, and object dynamics. The accuracy requirements are usually defined in terms of detection and false alarm rates for objects, while the computational requirement is specified typically by the system response time to an object's presence (e.g. real-time or delayed?). The objective of the vision systems engineer is to then exploit these restrictions (i.e. constraints) and design a system that is optimal in the sense that it meets customer requirements in terms of speed, accuracy and cost. The main problem, however, is that there is no known systematic way for vision systems engineers to go about doing this translation of the system requirements to a detailed design. It is still an art to engineer systems to meet given application specific requirements. There are two fundamental steps in the design process: the choice of the system architecture and the modules for accomplishing the task, and the statistical analysis and validation of the system to check if it meets user requirements.

We have been focusing on the development and refinement of a systematic engineering methodology for the design and analysis of vision systems. We draw upon the methodology described in [82]. Ramesh et al [82] summarize a systems engineering methodology for building vision systems and illustrate performance characterization of a system for building parameter estimation. In this methodology there are two main steps: statistical modeling or performance characterization of component algorithms (component identification) and application domain characterization. Component identification involves the derivation of the deterministic and stochastic behavior of each module. This entails the specification of the ideal model and an error model in the input and relating their parameters to the output ideal model and error model parameters. The essence of the methodology is that each sub-step used in a vision system is treated as an estimator and therefore the estimator's behavior has to be characterized in terms of its distribution as a function of the input samples and error distribution parameters. When a system is composed of multiple estimation steps concatenated together then performance characterization is a daunting task. We will describe some of the issues related to this step in the next section. Application domain characterization is a learning or estimation step wherein the restrictions on the application data relevant to the task at hand are specified in terms of prior distributions of parameters relevant to the algorithm/system representation chosen. These prior distributions can be viewed as specifying the range of possible images for the given application. The average or worst case performance of the system can be determined by combining the Component Identification steps and the Application domain modeling steps.

While [82] discusses systems analysis, there has been limited work on making systems design choices from user requirements. The more recent trend in the community is to emphasize statistical learning methods, more appropriately Bayesian methods for solving computer vision problems (See for example [71]). However, there still exists the problem of choosing the right statistical likelihood model and right priors that suit an application. Even if this were possible, it is still computationally infeasible to satisfy real-time application needs. In the context of video analysis systems, real-time considerations play a big role in the design of video processing systems.

Sequential decomposition of the total task into manageable sub-tasks (with reasonable computational complexity) and the introduction of pruning thresholds, is the common way to tackle the problem. This introduces problems because of the difficulty in approximating the probability distributions of observables at the final step of the system so that Bayesian inference is plausible. This approach to perceptual Bayesian inference has been attempted, (see for example [82], [70]). The idea of gradual pruning of candidate hypotheses to tame the computational complexity of the estimation/classification problem has been presented in [38]. Learning decision trees to perform object detection (by gradually reducing the uncertainty in a step-wise fashion, wherein each pruning step has probability of misdetection approximately zero while the probability of false alarm is reduced after each application of a decision rule) is discussed in [37] and [50]. Note that none of the works identifies how the sub-tasks (e.g. feature extraction steps) can be chosen automatically given an application context.

Our approach involves the following key steps:

**System Configuration choice:** The first step is to choose the modules for the system. This is done by use of context (in other words: application specific prior-distributions for object geometry, camera geometry, and error models, illumination models). Real-time constraints are satisfied by choosing pruning methods or indexing functions that restrict the search space for hypotheses. The choice of the pruning functions is derived from the application context and prior knowledge. The choice of the indexing function is not necessarily critical, except that the following criterion is met. The indexing function has to be of a form, which simplifies the computation of the probability of generating a false hypothesis or the probability of missing a true hypothesis as a function of the tuning constants. To satisfy the accuracy constraint we employ hypothesis verification and parameter estimation step. Bayesian estimation is used to evaluate candidate hypotheses and estimate object parameters by using a likelihood model, $P(measurements|hypothesis)$, that takes into account the effects of the pre-processing steps and tuning parameters. Note that this likelihood model is actually derived from the statistical characterization step that is described below. The indexing step provides computational efficiency, while the hypothesis verification and estimation step addresses accuracy.

**Statistical Model Derivations, Performance Characterization and Model Validation:** The second step involves the application of the methodology described above to derive statistical models for errors at various stages in the chosen vision-system configuration. That allows for quantifying the indexing step and tuning the parameters to achieve a given probability of miss-detection and false alarm rate. In addition, we perform a validation of theoretical models for correctness (through Monte-Carlo simulations) and closeness to reality (through real experiments). For our given system configuration choice, we perform statistical analyses to setup the tuning constants at the indexing steps, to derive likelihood models for feature measurements that are used in the hypothesis verification and in the estimation step, and to obtain the uncertainty of the estimate provided by the hypothesis verification step.

## 4.2 Dual-Camera Surveillance System – Performance Characterization

As a case study we have used the systematic engineering methodology to develop a real-time people detection and zooming system to meet given application requirements. We show in [55] that by judiciously choosing the modules for a dual-camera surveillance system and by performing a careful analysis of the influence of various tuning parameters on the system it is possible to: perform proper statistical inference, to automatically set control parameters and to quantify performance limits. The goal of the system is to continuously provide a high resolution zoomed-in image of a persons head at any location of the monitored area. An omni-directional camera video is processed to detect people and to precisely control a high-resolution foveal camera, which has pan, tilt and zoom capabilities. The pan and tilt parameters of the foveal camera and its uncertainties are shown to be functions of the underlying geometry, lighting conditions, background color/contrast, of the relative position of the person with respect to both cameras as well as of sensor noise and calibration errors. The uncertainty in the estimates is used to adaptively estimate the zoom parameter that guarantees with a user specified probability, $\alpha$, that the detected person's face is contained and zoomed within the image. Note, the higher the probability $\alpha$ the more conservative the zoom factor would be. We set $\alpha$ to 0.99 in our current system. Extensive amount of real and synthetic data experiments was used to validate the models derived. The system proved to be robust when tested under a variety of indoor situations (with varying backgrounds) without modification of system parameters. The system could adapt to variations in camera setup (for example: sub-optimal focusing/blurring) and automatically account for the varying uncertainty in the measurements (for example: with a sharper focused setup the zoom factors were consistently higher than with a blurred focus). In addition the statistical analyses enable us to optimize the system setup to obtain minimal variance in the control parameters over a large area. Figures 1.6 shows snapshots of the running system. The foveal camera control parameters as well as the zoom parameter are functions of the geometry, as well as of the current uncertainties in the estimates. The more certain the estimate the more we can afford to zoom in. The uncertainties are functions of the current scene, quality of segmentation, geometry, and calibration uncertainties. In these figures, the foot position estimate is displayed as a cross. Where the cross does not sit on the top of the toes, the camera does not zoom in too much. Precise estimation is characterized by stable positioning of the cross on the shoe. For a detailed description of the system and its performance characterization results please see [28].

## 5. Other Performance Characterization Research Efforts

We are actively pursuing research in performance characterization of other components needed for video surveillance. We summarize some of these efforts. These efforts involve the statistical characterization of background adaptation algorithms ( [52]) and the statistical modeling and calibration of the CCD sensor ( [94]). The first analysis enables the automatic setup of tuning constants for background adaption, while the second enables calibration of the CCD sensor and the derivation of uncertainty estimates for high-dynamic range images. The uncertainty estimates are used to automatically classify areas in the image that are primarily due to illumination change. Extensive simulations and tests on outdoor video suggests that the models derived are accurate and that long-term change detection (i.e. the detection of objects in the scene and the labeling of regions with illumination changes) can be achieved. Large scale tests to validate the system under a variety of real-world conditions is still pending.

## 6. Conclusion

In this paper we have described various real-time video analysis systems developed at the Real-time imaging group at Siemens Corporate Research. Our emphasis is on the use of rigorous statistical and systems engineering methodologies for the design and validation of video analysis systems. We have described real-world systems wherein statistical modeling plays a crucial role in the design of robust systems. In addition, we have described the use of statistical methodology for the characterization of performance of these systems. Ongoing research activities involve the research and development of systems for highway, tunnel monitoring and real-time video analysis for automotive settings.

## References

[1] Besag, J. (1986). On the statistical analysis of dirty images. *Journal of Royal Statistics Society*, 48:259–302.

[2] Chou, P. and Brown, C. (1990). The theory and practice of bayesian image labeling. *IJCV*, 4:185–210.

[3] Heitz, F., Perez, P., and Bouthemy, P. (1994). Multiscale minimization of global energy functions in some visual recovery problems.

*CVGIP: Image Understanding*, 59:125–134.

[4] Paragios, N. and Tziritas, G. (196). Detection and Location of Moving Objects, using deterministic relaxation algorithms. In *IARP ICPR*, volume 1, pages 201–205, Vienne, Austria.

[5] Paragios, N. and Tziritas, G. (1999). Adaptive Detection and Localization of Moving Objects in Image Sequences. *Signal Processing: Image Communication*, 14:277–296.

[6] R.T. Collins, A.J. Lipton, T. Kanade, "A System for Video Surveillance and Monitoring," *American Nuclear Society Eight Intern. Topical Meeting on Robotics and Remote Systems*, 1999.

[7] "Mean Shift Analysis and Applications," *IEEE Int'l Conf. Comp. Vis.*, Kerkyra, Greece, 1197–1203, 1999.

[8] "Real-Time Tracking of Non-Rigid Objects using Mean Shift," *IEEE Conf. on Comp. Vis. and Pat. Rec.*, Hilton Head Island, South Carolina, Vol. 2, 142–149, 2000.

[9] "Robust Detection and Tracking of Human Faces with An Active Camera," To appear, *IEEE Int'l Workshop on Visual Surveillance*, Dublin, Ireland, 2000.

[10] G. Sandini et al., "Image-based Personal Communication Using an Innovative Space-variant CMOS Sensor," *ROMAN'96*, Tsukuba, Japan, 1996.

[11] G. Sela, M.D. Levine, "Real-Time Attention for Robotic Vision," *Real-Time Imaging*, 3:223–194, 1997.

[12] C.W. Shin, S. Inokushi, "A New Retina-Like Visual Sensor Performing the Polar Transform," *IAPR Workshop on Machine Vision Applications*, Kawasaki, Japan, 52–56, 1994.

[13] H. Stark, "Sampling Theorems in Polar Coordinates," *J. Opt. Soc. Am.*, 69 (11): 1519–1525, 1979.

[14] H. Stark, J.W. Woods, I. Paul, R. Hingorani, "Direct Fourier Reconstruction in Computer Tomography," *IEEE Trans. Acoustics, Speech, Signal Processing*, ASSP-29(2):237–244, 1981.

[15] M. Woodward, F. Muir, "Hexagonal Sampling," *Stanford Exploration Project*, SEP-38:183–194, 1984.

[16] G. Wyszechi, W.S. Stiles, *Color Science: Concepts and Methods, Quantitative Data and Formulae*, Second Ed. New York: Wiley, 1982.

[17] N. Paragios and R. Visvanathan, *A MRF-based Real Time Approach for Subway Monitoring*, to appear in IEEE Computer Vision and Pattern Recognition, Dec. 01, Maui, Hawaii.

[18] J.K. Aggarwal, Q. Cai, "Human Motion Analysis: A Review," *Computer Vision and Image Understanding*, 73:428–440, 1999.

[19] J. Batista, P. Peixoto, H. Araujo, "Real-Time Active Visual Surveillance by Integrating Peripheral Motion Detection with Foveated Tracking," *IEEE Workshop on Visual Surveillance*, Bombay, India, 18–25, 1998.

[20] M. Bolduc, M.D. Levine, "A Real-Time Foveated Sensor with Overlapping Receptive Fields," *Real-Time Imaging*, 3:195–212, 1997.

[21] G. Bonmassar, E.L. Schwartz, "Space-Variant Fourier Analysis: The Exponential Chirp Transform," *IEEE Trans. Pattern Analysis Machine Intell.*, 19(10):1080-1089, 1997.

[22] D. Comaniciu, F. Berton, and V. Ramesh, "An Adaptive Visual Surveillance System," Siemens Corporate Research Technical Report, 2000.

[23] J.L. Crowley, F. Berard, "Multi-Modal Tracking of Faces for Video Communications," *IEEE Conf. on Comp. Vis. and Pat. Rec.*, Puerto Rico, 640–645, 1997.

[24] A. Eleftheriadis, A. Jacquin, "Automatic Face Location Detection and Tracking for Model-Assisted Coding of Video Teleconference Sequences at Low Bit Rates," *Signal Processing - Image Communication*, 7(3): 231–248, 1995.

[25] H.G. Feichtinger, K. Grochenig, T. Strohmer, "Efficient Numerical Methods in Non-Uniform Sampling Theory," *Numerische Mathematik*, 69:423–440, 1995.

[26] F. Ferrari, J. Nielsen, P. Questa, G. Sandini, "Space Variant Sensing for Personal Communication and Remote Monitoring," *EU-HCM Smart Workshop*, Lisbon, Portugal, 1995.

[27] W.S. Geisler, J.S. Perry, "A Real-Time Foveated Multiresolution System for Low-Bandwidth Video Communication," *Human Vision and Electronic Imaging*, B. Rogowitz and T. Pappas (Eds), SPIE 3299:294–305, 1998.

[28] Greiffenhagen et al, "Statistical Modeling and Performance Characterization of a Real-Time Dual Camera Surveillance System," *IEEE*

*Conf. on Comp. Vis. and Pat. Rec.*, Hilton Head Island, South Carolina, Vol. 2, 335–342, 2000.

[29] A.J. Jerri, "The Shannon Sampling Theorem - Its Various Extensions and Applications: A tutorial Review," *Proceedings of the IEEE*, 65(11):1565–1596, 1977.

[30] F. Jurie, "A New Log-Polar Mapping for Space Variant Imaging. Application to Face Detection and Tracking,", *Pattern Recognition*, 32:865-875, 1999.

[31] R.M. Lewitt, "Reconstruction Algorithms: Transform Methods," *Proceedings of the IEEE*, 71(3):390–408, 1983.

[32] S.J. McKenna, Y. Raja, S. Gong, "Tracking Colour Objects using Adaptive Mixture Models," *Image and Vision Computing*, 22:223–229, 1999.

[33] Moving Picture Experts Group, "Overview of the MPEG-4 Standard," *ISO/IEC JTC1/SC29/WG11*, 2000.

[34] D.W. Murray, K.J. Bradshaw, P.F. McLauchlan, I.D. Reid, P.M. Sharkley, "Driving Saccade to Pursuit using Image Motion," *Intern. J. Comp. Vis.*, 16(3):204–228, 1995.

[35] A.S. Rojer, E.L. Schwartz, "Design Considerations for a Space-Variant Visual Sensor with Complex-Logarithmic Geometry," *Int'l Conf. Pattern Recognition*, Atlantic City, New Jersey, 278–285, 1990.

[36] H.P. Rotstein, E. Rivlin, "Optimal Servoing for Active Foveated Vision," *IEEE Conf. on Comp. Vis. and Pat. Rec.*, San Francisco, 227–182, 1996.

[37] Y. Amit and D. Geman, "Shape quantization and recognition with randomized trees," Neural Computation, Vol. 9, pp. 1545-1588, 1997.

[38] Y. Amit and D. Geman, "A computational model for visual selection", Neural Computation, 1999.

[39] P.Allen and R. Bajcsy, "Two sensors are better than one: example of vision and touch", *Proceedings of 3rd International Symposium on Robotics Research*, pp. 48-55, Gouvieux, France, 1986.

[40] S. Baker and S. Nayar, "Global Measures of Coherence for Edge Detector Evaluation," Proceedings of the IEEE CVPR conference, Fort Collins, Vol. II, pp. 373-379, 1999.

[41] M. Boshra and B. Bhanu, "Predicting Performance of Object Recognition," IEEE Transactions on Pattern Analysis and Machine Intelligence, Vol. 22, No. 9, September 2000, pp. 956-969.

[42] Bowyer, K.W. and Phillips, P.J., editors, "Empirical Evaluation Techniques in Computer Vision", IEEE Press, 1998.

[43] F.J.Canny, "Finding edges and lines in images," Tech.Rep. 720, MIT AI Lab, June 1983.

[44] V. Chalana and Y. Kim, "A Methodology for Evaluation of Segmentation Algorithms on Medical Images," 1996, in Image Processing, Vol. 2710 of SPIE Medical Imaging pp. 178-189.

[45] K. Cho, P. Meer, J. Cabrera: "Performance assessment through bootstrap". IEEE Transactions in Pattern Anal. Machine Intelligence, 19, 1185-1198, 1997.

[46] P.Courtney, N.Thacker, A.Clark "Algorithmic Modeling for Performance Evaluation", Special Issue on Performance Characterization, Machine Vision & Applications Journal, Springer Verlag, 1998.

[47] B. A. Draper, "Learning Object Recognition Strategies," Ph.D. Dissertation, University of Massachusetts, 1993.

[48] M.-P. Dubuisson-Jolly, C.-C. Liang, and A. Gupta, "Optimal polyline tracking for artery motion compensation in coronary angiography", Proc. International Conference on Computer Vision, Bombay, India, pp. 414-419.

[49] B. Efron, R. Tibishirani "An Introduction to the Bootstrap", Chapman & Hall, New York, 1993.

[50] F. Fleuret and D. Geman, "Graded Learning for Object Detection," Proceedings of the IEEE Workshop on Statistical and Computational Theories in Vision, Fort Collins, CO, June 1999, (Published on the Web).

[51] Machine Vision & Applications, International Journal, Special Issue on Performance Evaluation; ed. by W. Forstner, Vol. 9, nos. 5/6, 1997; pp.229-239.

[52] X. Gao, T. Boult, F. Coetzee and V. Ramesh, "Error Analysis of Background Adaption," Proceedings of the IEEE CVPR Conference, Hilton Head, CA, 2000, pp. 503-510.

[53] C. Goad, "Special Purpose Automatic Programming for 3D Model-Based Vision," DARPA IU Workshop Proceedings, 1983, pp. 94-104.

[54] M. Greiffenhagen and V. Ramesh , "Performance Characterization of a People Detection and Tracking system," Special session on Video Surveillance, IAPR conference on Image analysis and Applications, Venice, September 1999.

[55] M. Greiffenhagen, V. Ramesh, D. Comaniciu, H. Niemannn, "Statistical Modeling and Performance Characterization of a Real-Time Dual Camera Surveillance System," in *Proceedings of the IEEE Conference on Computer Vision and Pattern Recognition (CVPR 2000)* , Hilton Head Island, South Carolina; June 13-15, 2000. IEEE Computer Society (publisher); Volume 2, pp. 335-342.

[56] W.E.L. Grimson and D. Huttenlocher, "On the Verification of Hypothesized Matches in Model-Based Recognition,"Lecture notes in computer science, 427, O.Faugeras (Ed.), Springer-Verlag, 1990.

[57] W.E.L.Grimson, D. Huttenlocher, "On the Sensitivity of Geometric Hashing," IEEE International Conference on Computer Vision, 1990, pp. 334 -338.

[58] R. M. Haralick, "Computer Vision Theory: The Lack Thereof," CVGIP 36, 1986, pp. 272-286.

[59] R.M. Haralick, "Performance assessment of near-perfect machines," Machine Vision and Applications 2, 1-16, 1989.

[60] R. Haralick, "Overview: Computer Vision Performance Characterization", *Proceedings of the DARPA Image Understand Workshop*, Vol.1, pp.663-665, 1994.

[61] "Dialogue: Performance Characterization in Computer Vision. With contributions from R.M.Haralick; L.Cinque, C. Guerra, S. Levialdi, J. Weng, T.S.Huang, P. Meer, Y. Shirai; B.A.Draper, J.R.Beveridge." CVGIP: Image Understanding 60, 245-265, 1994.

[62] R. Haralick, "Performance Characterization Protocol in Computer Vision," ARPA IUW94 Proceedings, Vol.1, pp. 667-674.

[63] R. M. Haralick, "Propagating Covariance In Computer Vision," IJPRAI, Vol. 10, pp. 561-572, 1996.

[64] Heath, M.D. et al, "A Robust Visual Method for Assessing the Relative Performance of Edge Detection Algorithms", PAMI (19), No. 12, December 1997, pp. 1338-1359.

[65] J. Hornegger and H. Niemann, "Probabilistic Modeling and Recognition of 3-D Objects," International Journal of Computer Vision, Vol. 39, No.3, 2000, pp. 229-251.

[66] Dialogue: Ignorance, Myopia, & Naivete in Computer Vision Systems, R. C. Jain and T. Binford, with contributions from: M.A.Snyder, Y.Aloimonos, A. Rosenfeld, T.S. Huang, K.W. Bowyer, and J.P. Jones. CVGIP, Image Understanding, Vol. 53, No. 1, Jan 1991.

[67] H. Joo, R. M. Haralick, and L. G. Shapiro, "Toward the Automating of Mathematical Morphology Procedures Using Predicate Logic," Proceedings of the ICCV, 1990, pp. 156-165.

[68] T. Kanade et al, "Advances in Cooperative Multi-Sensor Video Surveillance", *Proceedings of the DARPA Image Understanding Workshop*, Vol. 1, pp. 3-24, 1998.

[69] S. Konishi et al, "Fundamental Bounds on Edge Detection: An Information Theoretic Evaluation of Different Edge Cues", Proceedings of the IEEE CVPR Conference, Fort Collins, Vol. I, pp. 573-579, 1999.

[70] W. Mann and T. Binford, "Probabilities for Bayesian Networks in Vision,", Proceedings of the ARPA IU Workshop, 1994, Vol. 1, pp. 633-643.

[71] D.Mumford, Pattern theory: a unifying perspective, in "Perception as Bayesian Inference", edited by D.Knill and W.Richards, Cambridge Univ. Press, 1996.

[72] S. Nayar and T. Boult, "Omnidirectional Vision systems: 1998 PI Report", *Proceedings of the DARPA Image Understanding Workshop*, Vol. 1, pp. 93-100, 1998.

[73] S. Nayar, "Omnidirectional Video Camera", *Proceedings of the DARPA Image Understanding Workshop*, Vol. 1, pp. 235-242, 1997.

[74] J. Nielsen "Characterization of Vision Algorithms: An experimental Approach", *ECVnet Workshop on Benchmarking*, 1995.

[75] A. Papoulis, Probability, Random Variables, and Stochastic Processes, McGraw–Hill, 1986.

[76] L. Parra, V. Ramesh, S. H. Lai, "Recovering Alignment Errors via EM Algorithm," Manuscript submitted to Journal of Mathematical Imaging and Vision, 2000.

[77] D. Petkovic, "The Need for Accuracy Verification of Machine Vision Algorithms and Systems," Proceedings of the CVPR, 1989, pp. 430-440.

[78] K. E. Price, "Anything you can do, I can do better (No you can't)", CVGIP, Vol. 36, 1986, No. 2/3, 387-391.

[79] V. Ramesh and R. M. Haralick, "Random Perturbation Models and Performance Characterization in Computer Vision," Proceedings of the CVPR Conference, 1992, Champaign, IL, pp. 521-527.

[80] V. Ramesh and R. M. Haralick, "A Methodology for Automatic Selection of IU Algorithm Tuning Parameters," Proceedings of the ARPA IUW 1994, Vol. 1, pp. 675 - 687.

[81] V. Ramesh et al, "Automatic Selection of Tuning Parameters for Feature Extraction Sequences," Proceedings of the CVPR Conference 1994, Seattle, WA, pp. 672-677.

[82] V. Ramesh et al, "Computer Vision Performance Characterization," *RADIUS: Image Understanding for Imagery Intelligence*, edited by. O. Firschein and T. Strat, Morgan Kaufmann Publishers, San Francisco, 1997.

[83] V. Ramesh and R.M.Haralick, "Random Perturbation Models for Boundary Extraction Sequence", Special Issue on Performance Characterization, Machine Vision & Applications Journal, Springer Verlag, 1998.

[84] F. Sadjadi ed., "Performance Evaluation of Signal and Image Processing systems," SPIE 1993.

[85] C. Shekhar et al, "Knowledge-based Control of Vision Systems," Image and Vision Computing, Vol. 17, No. 9, July 1999, pp. 667-683.

[86] Shin, M.C. et al, "Comparison of Edge Detectors Using an Object Recognition task", Proceedings of the IEEE Computer Vision and Pattern Recognition Conference, Fort Collins, Colorado, Vol. 1, pp. 360-365, 1999.

[87] D.Stoyan, W.S.Kendall, J.Mecke, *Stochastic Geometry and its Applications*, John Wiley and Sons, 1987.

[88] T. M. Strat, "Natural Object Recognition," Springer, 1990.

[89] Z. Sun, V. Ramesh, and M. Tekalp, "Error Characterization of Factorization Technique," Proceedings of the ICCV Workshop on Geometic Algorithms: Validation and Practice, Korfu, September 1999, Springer Verlag.

[90] R. Vogt, "Automatic Generation of Simple Morphological Algorithms," Proceedings of the CVPR conference, 1988, pp. 760-765.

[91] S. Wang and T. Binford, "Local Step Edge Estimation – A New Algorithm, Statistical Model and Performance Evaluation," Proceedings of the ARPA IU Workshop, Wash DC, April 1993, pp.1063-70.

[92] S. Wang and T. Binford, "Generic, Model-Based Estimation and Detection of Discontinuities in Image Surfaces", Proceedings of the ARPA IU Workshop, 1994, Vol. 2, pp. 1443-1450.

[93] "Special Section on Empirical Evaluation of Computer Vision Algorithms," IEEE Transactions on Pattern Analysis and Machine Intelligence, Vol. 21, No. 4, April 1999, pp. 289-290.

[94] Y. Tsin, V. Ramesh, T. Kanade, "Statistical Calibration of the CCD Imaging Process," Proceedings of the IEEE International Conference on Computer Vision, Vancouver, BC, 2001, Vol. 1.

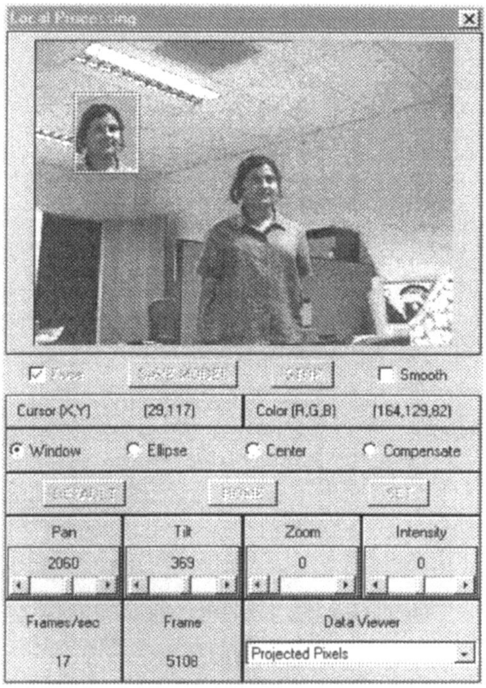


(a)

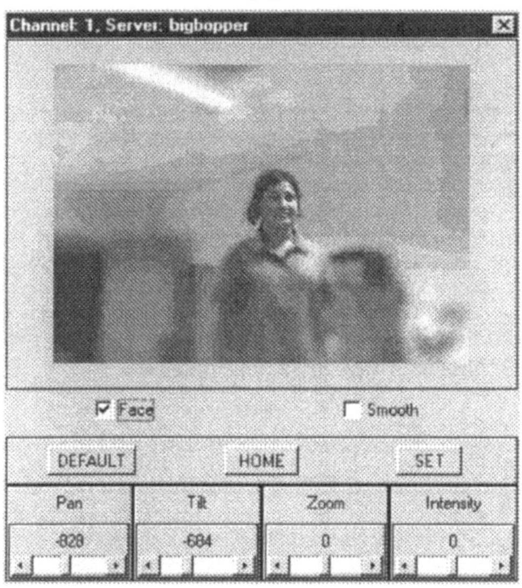


(b)

*Figure 1.3.* Intelligent Video Communication System: (a) User interface of the server. (b) User interface of the primary client.

(a)

(b)

*Figure 1.4.* (a) Original Input to Video Communication System. (b) Received.

*Figure 1.5. Crowd Density Estimation: Weighted Change Detection Results presented in raster scan order (duration: 5 min).*

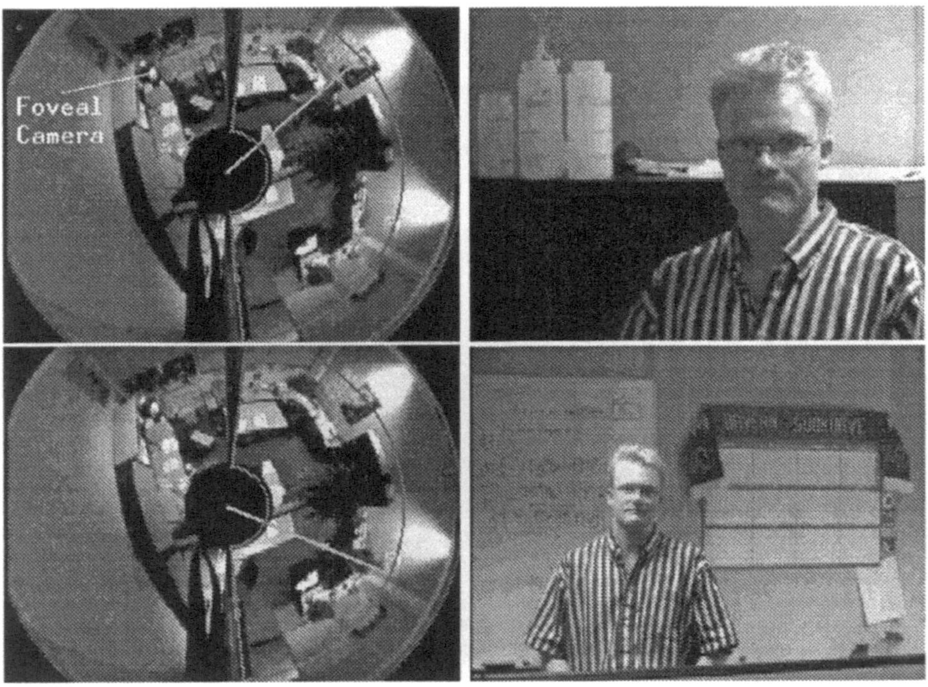


*Figure 1.6.* Dual Camera Surveillance System results: Left image corresponds to the Omni-image, the right corresponds to the foveal image illustrating the zoomed up face.

# Chapter 2

# Aerial Video Surveillance and Exploitation

Rakesh Kumar
*Sarnoff Corporation, Princeton, NJ, USA*

**Abstract**: There is growing interest in performing aerial surveillance using video cameras. Compared to traditional framing cameras, videos provide the capability to observe ongoing activity within a scene and to automatically control the camera to track the activity. However, the high data rates and relatively small field of view of videos present new technical challenges that must be overcome before videos can be widely used. In this talk, I will present a framework and details of the key components for real time, automatic exploitation of aerial video for surveillance applications. The framework involves separating an aerial video into the natural components corresponding to the scene. Three major components of the scene are the static background geometry, moving objects, and appearance of the static and dynamic components of the scene. In order to delineate videos into these scene components, we have developed real time, image-processing techniques for 2D/3D frame-to-frame alignment, change detection, camera control and tracking of independently moving objects in cluttered scenes. The geo-location of video and tracked objects is estimated by registration of the video to controlled reference imagery, elevation maps and site models. Finally static, dynamic and re-projected mosaics may be constructed for compression, enhanced visualisation and mapping applications.

**Key words:** Video Surveillance, Aerial Images, Image Processing, Computer Vision, Registration, Mosaicing, Change Detection, Geo-location, Tracking, Visualisation, and Camera Control.

## 1. Introduction

Aerial surveillance has a long history in the military for observing enemy activities and in the commercial world for monitoring resources such as forests and crops [1]. Until recently aerial surveillance has been performed primarily using film or electronic framing cameras. The objective has been to gather high resolution still images of an area under surveillance that could later be examined by human or machine analysts to derive information of interest. Currently there is growing interest in using video cameras for these tasks. Video captures dynamic events that cannot be understood from aerial still images. It enables feedback and triggering of actions based on dynamic events, and provides crucial and timely intelligence and understanding that is not otherwise available. Video observations can be used to detect and geo-locate moving objects in real time and to control the camera, for example to follow detected vehicles or constantly monitor a site.

However, video also brings new technical challenges. Video cameras have lower resolution than framing cameras. In order to get the resolution required to identify objects on the ground it is generally necessary to use a telephoto lens, with a narrow field of view. This leads to the most serious shortcoming of video in surveillance – it provides only a "soda straw" view of the scene. The camera must then be scanned to cover extended regions of interest. An observer watching this video must pay constant attention, as objects of interest move rapidly in and out of the camera field of view. The video also lacks a larger visual context – the observer has difficulty perceiving the relative locations of objects seen at one point in time to objects seen moments before. In addition geodetic co-ordinates for objects of interest seen in the video are not available

Further challenges of video relate to control and storage. A camera operator can have difficulty manually controlling the camera to scan a scene or to hold an object of interest in view because of the soda straw view of the world provided by video. Video contains much more data than traditional surveillance imagery, so it is expensive to store. Once stored in a database, surveillance video is difficult and tedious to search during subsequent analysis.

Before video can be generally employed in aerial surveillance, new video technologies must be developed that make it much easier for human operators to use and interpret video data. Technologies are needed to automatically control the camera and to detect and geo-locate objects of interest. New methods are needed to annotate and present video imagery to humans to provide an immediate, in depth, understanding of the observed scene. Technologies are also needed to compress and store surveillance video, and to give users easy access to archived video.

In order to serve a variety of different needs for surveillance applications, it is important to provide an underlying framework for spatio-temporal aerial video analysis. In the past decade or so, we have developed such a framework based on image alignment with progressively complex models of motion and scene structure. The framework involves delineation of video imagery into components that correspond to the static scene geometry, dynamic objects in the scene, and the appearance of both the static and dynamic parts of the scene. Progressive complexity in this framework provides us a handle on the model selection problem since always applying the most complex model (say, a 3D-alignment model) may lead to unstable and unpredictable results. In essence, such instabilities arise as the employed model results in over fitting the data at hand. For example, 3D estimation may produce unpredictable results when the scene may be largely flat. Furthermore, the alignment framework also includes situating the video components in a geo-referenced co-ordinate system within a reference imagery and model database. By aligning the static scene components in video frames to a reference database, static and dynamic objects, entities and locations in the video can be geo-referenced and annotated with respect to the information in the reference database.

The video analysis framework has been used to develop a number of key capabilities, some of which have been put together in a system [2] for aerial video surveillance. These key capabilities include:

- Frame-to-frame alignment and decomposition of video frames into motion (foreground/background) layers,
- Mosaicing static background layers to form panoramas as compact representations of the static scene,
- Detecting and tracking independently moving objects (foreground layers) in the presence of 2D/3D backgrounds, occlusion and clutter,
- Geo-locating the video and tracked objects by registering it to controlled reference imagery, digital terrain maps and 3D site models, and
- Enhanced visualisation of the video by re-projecting and merging it with reference imagery, terrain and/or maps to provide a larger context.

In the remainder part of the paper we very briefly describe these key capabilities. More details on the overall system can be found in [2] and details on each of the components can be found in the referred papers.

## 2. Alignment and CHANGE DETECTION

The displacement of pixels between subsequent frames in a video sequence may occur because of the motion of the video sensor, independent motion of objects in the scene, motion of the source of illumination and other factors. At the global motion estimation step, the displacement of pixels due to the motion of the sensor is computed.

At Sarnoff, we have constructed a family of techniques for alignment of video frames [4,5,6] using pyramid-processing [3]. These alignment techniques can lock into the motion of the background scene and can be computed in real time [13]. The motion parameters are used to warp the images into a common co-ordinate frame. Changes in the scene such as moving objects are detected by examining the set of aligned video frames and detecting pixels with poor correlation across the temporal domain [12].

An example of detecting change energy and moving objects from video is shown in Figure 1. Figure 1a shows one image of video sequence taken from a moving aerial camera. Figure 1b shows change energy without alignment of the background scene and Figure 1c shows change energy after alignment. Note in Figure 1b, many parts of the scene are highlighted as moving while in Figure 1c, only the pixels belonging to the truck are highlighted as moving.

*a* *b* *c*

Figure 1: Example change energy image. (a) Source frame, (b) change energy without alignment, and (c) change energy with alignment, moving truck appears as bright pixels in the change energy image.

## 3. MosAICING

Images are accumulated into the mosaic as the camera pans, effectively extending the field of view of the camera. Construction of a 2D mosaic requires computation of alignment parameters that relate all of the images in the collection to a common world (or mosaic) co-ordinate system [7,8,9,18]. The transformation parameters are used to warp the images into the mosaic co-ordinate system. These warped images are then combined to form a mosaic. In order to avoid seams between images in the mosaic, the warped video frames are merged in the Laplacian pyramid domain [10]. An example of mosaic construction from video is shown in Figure 3.

# 4. Tracking moving objects

Scene analysis includes operations that interpret the source video in terms of objects and activities in the scene. Moving objects are detected and tracked over the cluttered scene [11,21]. In [21] multi-object tracking is formulated as a 2D layer estimation and tracking problem where the state of each moving object is represented by its motion, appearance and shape. The state is updated at each instant of time using the Expectation-Maximization (EM) algorithm for MAP estimation. In Figure 3, we show key frames from an aerial video where three moving vehicles are successfully tracked. One of them eventually becomes stationary.

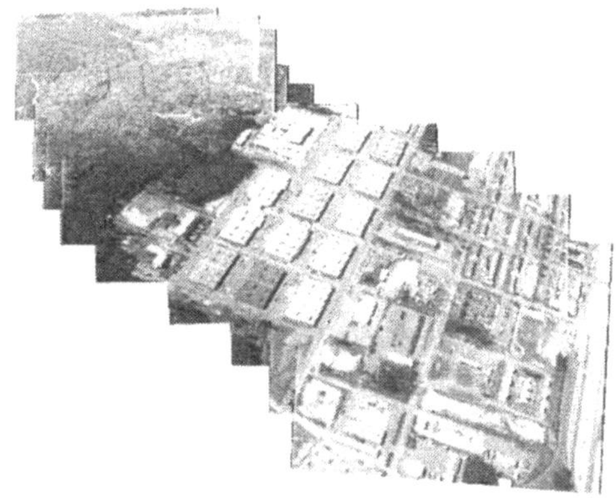

Figure 2: Mosaic constructed from video over Camp Lejeune, North Carolina.

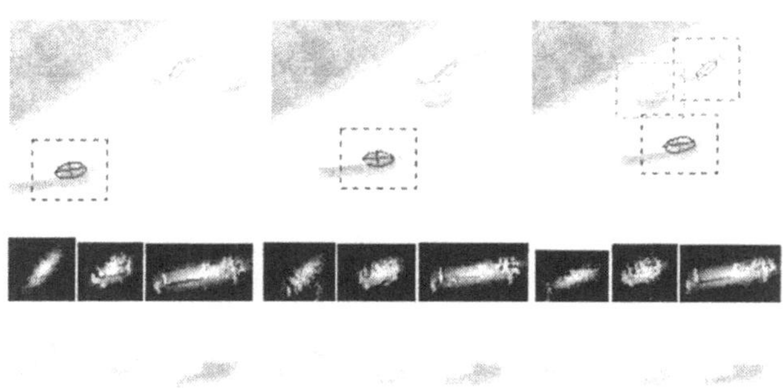

(a) (b) (c)

Figure 3: Vehicle passing and stationary vehicles. The first row shows the cut-outs of original video frames and the Gaussian shape priors. The second and the third rows show the segmentation and the appearance of the three vehicles.

# 5. Geo-Location

A video surveillance system must also determine the geodetic co-ordinates of objects within the camera's field of view. Rough geo-locations of points on the ground can be estimated from aircraft co-ordinates and camera orientation (ESD: engineering support data). However, as altitude increases, small errors in measured camera orientation result in larger geo-location errors. More precise geo-locations can be estimated by aligning video frames to calibrated reference images [14,15,16]. Further, image based geo-location allows video to be precisely localised in the presence of telemetry dropouts. Geo-locations are used in turn for precision targeting, target handoff between aircraft, updating imagery in geo-spatial databases and indexing video databases.

The process of aligning video sequences to reference imagery is described by the block diagram show in Figure 4. The ESD stream accompanying the video is used to index the reference imagery database and extract the correct portion of the imagery. The reference imagery is rendered from the point-of-view of the video frame as known in the ESD stream. Video frames are aligned to each other to create virtual video mosaics and provide larger context for matching. Both the video frames and reference imagery are pre-processed to highlight invariant structure. Next a process of coarse search is employed to locate the video frames within a few pixels in the reference imagery. Finally, a process of fine alignment is used to register the video data to the reference imagery to sub-pixel accuracy.

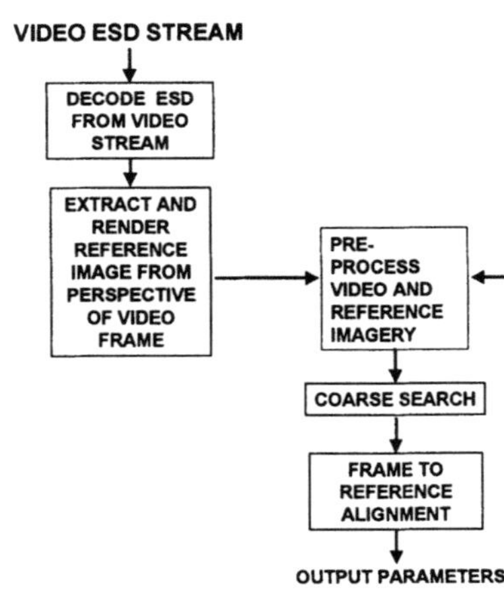


Figure 4: Geo-registration system

Figure 5 shows geo-registration results that were obtained from video captured from a X-drone UAV flying over Webster Field, Maryland. Key frames obtained at 3 Hz were used to geo-register the sequences. Figure 5a shows the overlay of the video mosaic over the reference imagery using ESD information alone. The video mosaic is not aligned and the geo-location error is about 1000 feet (100's of pixels). Figure 5b shows the geo-registration result after the coarse search step. The video mosaic is now aligned to within a few pixels from the reference image. Finally, Figure 5c shows the result after the fine alignment step [14] where it can be noted there is quite precise alignment.

(a) (b) (c)

Figure 5: Geo-registration of video, (a) Overlay of video mosaic over ortho-photo using telemetry information alone, (b) Overlay of video-mosaic over ortho-photo after coarse search step and (c) Overlay of video mosaic over ortho-photo after fine alignment step.

## 6. Enhanced VISUALIZATION

One of the most challenging aspects of aerial video surveillance is formatting video imagery for effective presentation to an operator. The soda straw nature of aerial video makes direct observation tedious and disorienting.

These shortcomings of video can be overcome, to a large extent, through the use of a "mosaic based display". Again, this makes use of the image processing functions and image representations that serve automated video analysis on board the aircraft, but now to support human interpretation on the ground. Elements of the mosaic display are shown in Figure 6. The mosaic display de-couples the observer's display from the camera [19,20]. An operator may scroll or zoom to examine one region of the mosaic even as the camera is updating another region of the mosaic.

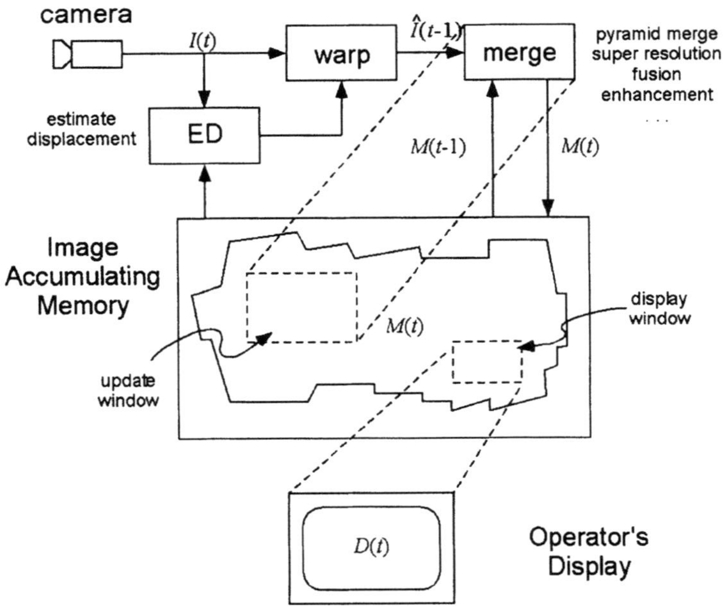


Figure 6: Elements of a mosaic display.

As a further step, the mosaic is re-projected to nadir view and aligned with map co-ordinates. This provides a link between current video and known geographic information about the scene, such as maps, prior imagery of the area, and names and classification data for specific cultural features in the scene such as roads and building, and natural features such as rivers and mountains. The current images can be overlaid on prior images or maps for direct comparison. The imagery can be annotated automatically with the names of landmarks of interest.

As a further generalisation, the mosaic can be overlaid on a terrain/ site model of the scene, then rendered from an arbitrary user selected viewing direction [17]. In this way the observer can "fly through" the scene independently of the aircraft's own motion. An example of this can be seen in Figure 7:

Figure 7: Video Flashlight: video frame is warped and overlaid over model; scene may be visualised from any viewpoint.

## References

[1] P.R. Wolf and B. A. Dewitt, "Elements of Photogrammetry with Applications in GIS", 3rd edition, *McGraw Hill*, ISBN 0-072-292454-3, 2000.
[2] Rakesh Kumar, Harpreet Sawhney, Supun Samarasekera, Steve Hsu, Hai Tao, Yanlin Guo, Keith Hanna, Art Pope, Rick Wildes, David Hirvonen, Mike Hansen and Peter Burt, "Aerial Video Surveillance and Exploitation," Proceedings of the IEEE, July 2001.
[3] P. Burt and E. Adelson. "The Laplacian pyramid as a compact image code," *IEEE Trans. on Communications*, April 1983.
[4] J.R. Bergen, P. Anandan, K. Hanna, R. Hingorani. "Hierarchical model-based motion estimation," *In Proceedings of the European Conference on Computer Vision*, 1992.
[5] Hanna, K.J., and Okamoto, N.E., "Combining Stereo and Motion Analysis for Direct Estimation of Scene Structure," *In Proceedings of the IEEE International Conference on Computer Vision,* pp. 357-365, Berlin 1993.
[6] R. Kumar, P. Anandan, and K. Hanna. "Direct recovery of shape from multiple views: a parallax based approach," *In Proceedings of the International Conference on Pattern Recognition*, Jerusalem, Israel, 1994.
[7] M. Hansen, P. Anandan, K. Dana, G. van der Wal, and P. Burt. "Real-time scene stabilization and mosaic construction," *In Proceedings of the IEEE Workshop on Applications of Computer Vision*, 1994.
[8] R. Kumar, P. Anandan, M. Irani, J. Bergen, and K. Hanna. "Representation of scenes from collections of images," *In Proceedings of the IEEE Workshop on Visual Representations*, Cambridge, MA, 1995.
[9] R. Szeliski. "Video mosaics for virtual environments," *IEEE Computer Graphics and Applications*, 16(2):22-30, March 1996
[10] P. Burt and T. Adelson. "A Multiresolution Spline with Application to Image Mosaics," *ACM Transactions on Graphics.* 1983

[11] P. Burt, J. Bergen, R. Hingorani, R. Kolczynski, W. Lee, A. Leung, J. Lubin, H. Shvaytser. "Object tracking with a moving camera: An application of dynamic motion analysis," *In Proceedings of IEEE Workshop on Motion*, 1989.

[12] H.S Sawhney, Y. Guo and R. Kumar, "Independent Motion Detection in 3D Scenes," *IEEE Transaction on Pattern Analysis and Machine Intelligence*, October 2000.

[13] G. van der Wal, M. Hansen, and M. Piacentino, "The Acadia Vision Processor," *In Proceedings of the IEEE International Workshop on Computer Architecture for Machine Perception,* Padua, Italy, Sept. 2000.

[14] R. Kumar, H.S. Sawhney, J. C. Asmuth, A. Pope and S. Hsu, "Registration of video to geo-referenced imagery," *In Proceedings of the International Conference on Pattern Recognition, ICPR'98,* Brisbane, Australia, Aug. 1998.

[15] R. Kumar, S. Samarasekera, S. Hsu and K. Hanna, "Registration of highly-oblique and zoomed in aerial video to reference imagery," *In Proceedings of the International Conference on Pattern Recognition*, Barcelona, Spain, 2000.

[16] R. Wildes, D. Hirvonen, S. Hsu, T. Klinedinst, R. Kumar, B. Lehman, B. Matei, W. Zhao, "Video georegistration: Algorithm and quantitative evaluation," *In Proceedings of IEEE International Conference on Computer Vision*, Vancouver, July 2001.

[17] S. Hsu, S. Samarasekera, R. Kumar, and H.S. Sawhney, "Pose Estimation, Model Refinement, and Enhanced Visualization using Video," *In IEEE Proceedings of Computer Vision and Pattern Recognition*, Hilton Head, South Carolina, pp. 488-495, 2000.

[18] H. Sawhney, S. Hsu, R. Kumar. "Robust video mosaicing through topology inference and local to global alignment," *In Proceedings of the European Conference on Computer Vision*, 1998.

[19] M. Irani, P. Anandan, J. Bergen, R. Kumar and S. Hsu, "Efficient represenations of video sequences and their applications," *In Signal Processing: Image Communication,* 8:327-351, 1996.

[20] A. Pope, R. Kumar, H. Sawhney, C. Wan. "Video Abstraction: Summarizing Video Content for Retrieval and Visualization," *In Proceedings of the Asilomar Conference on Signals, Systems, and Computers*, 1998.

[21] Hai Tao, Harpreet S. Sawhney, and Rakesh Kumar, "Dynamic Layer Representation with Applications to Tracking," *In Proceedings of the IEEE conference on Computer Vision and Patter Recognition*, Hilton Head, South Carolina, 2000.

Chapter 3

# TWO EXAMPLES OF INDOOR AND OUTDOOR SURVEILLANCE SYSTEMS: MOTIVATION, DESIGN, AND TESTING

Ioannis Pavlidis, Vassilios Morellas

*Honeywell Laboratories*

*3660 Technology Drive*

*Minneapolis, MN 55418*

*U.S.A. Partial funding provided by Honeywell Laboratories.*

{ioannis.pavlidis, vassilios.morellas }@honeywell.com

**Abstract** We examine the state of the security industry and market and underline the role that it plays in the R&D efforts. We also present a snapshot of the current state-of-the-art in indoor and outdoor surveillance systems for commercial applications. Then, we move on and describe in detail a prototype indoor surveillance system that we recently developed. The system is called Cooperative Camera Network (CCN) and reports the presence of a visually tagged individual throughout a building structure. Visual tagging is based on the color signature of a person. CCN is meant to be used for the monitoring of potential shoplifters in department stores. We also describe our prototype outdoor surveillance system, the DETER (Detection of Events for Threat Evaluation and Recognition). DETER can monitor large open spaces, like parking lots, and report unusual moving patterns by pedestrians or vehicles. To perform its function DETER fuses the field of views of multiple cameras into a super-view and performs tracking of moving objects across it. A threat assessment module with encoded suspicious behaviors performs the motion pattern identification. Both surveillance systems are good examples of technology transfer of state-of-the-art ideas from the research literature to the commercial domain. At the same time, they are good study cases for the extra engineering methodology and effort that is needed to adapt initial research concepts into a successful practical technology.

**Keywords:** Surveillance, security systems, threat assessment, color recognition, multinormal representation.

## 1. Introduction

The current security infrastructure could be summarized as follows: (a) Security systems act locally and they do not cooperate in an effective manner. (b) Very high value assets are protected inadequately by antiquated technology systems. (c) Reliance on intensive human concentration to detect and assess threats.

Today there is a chasm between what is commercially available and the security technology demonstrated in R&D labs. On one hand, the most sophisticated security product one can find is a camera with some rudimentary motion detection capability. On the other hand, complete multi-camera prototype systems that cooperate at several levels and feature automated threat assessment have proved themselves in realistic test environments. Why the security industry has been so slow in productizing concepts and designs fully developed and tested by academic and commercial labs over the last 10 years? An understanding of the industry's peculiarities and the forces that shape up its current profile is essential for anyone who is interested to perform technology transfer in the security domain. Below we enumerate what we consider the most important characteristics of the current security market and industry.

**Low Profit Margin** The security market is very cost sensitive. In an era, where quarterly profits make or brake corporate giants in areas with much higher profit margin, the security industry always struggles to "make the numbers." Its strategic horizon usually does not extend beyond six months.

**Resistance to Change** Like most traditional industries the security industry is not an advocate of innovation by nature. It is characteristic that most of the commercial security R&D was initiated and financed directly by corporate mandate and not by the individual business units.

**Low Tech Culture** The security industry is permeated by low tech culture. The management and the engineers of the security business units are trained and grown within a low tech environment and are ignorant and suspicious to state-of-the-art developments. Their users and customers are often underpaid and under-educated security guards and facility managers that also view high technology with skepticism.

**Hardware Mentality** The most advanced members of the security industry are probably the camera manufacturers. Even these, although they produce some advanced electronic products have difficulty outfitting them with the necessary software.

Despite the presence of many negative factors the future of the security industry can be viewed only in positive light. And, although the transformation of the industry and the market will take time to complete, it has already started happening in small steps. As a result of upcoming technology offerings the Freedonia group [1] is projecting significant growth of the security service market during the next several years. This growth will fuel further research and development and will hopefully bootstrap the process of incorporating the security industry to the new economy.

## 2. The State of the Art in Video-Based Security Technology

The computer vision community has performed extensive research in the area of video-based surveillance for the past 20 years. This research, although initially military in nature, it turned very quickly into civilian (security) with the end of the cold war. One can identify two rather distinct application categories: indoor surveillance and outdoor surveillance systems. From the technology point of view outdoor surveillance is much more challenging because of the greater variability in lighting conditions. It is ironic, however, that most of the R&D work has been directed towards outdoor rather than indoor surveillance. This trend may be partly due to the military heritage of the computer vision community and partly to the wider social acceptability of outdoor as opposed to indoor (e.g. home, workplace) surveillance and monitoring.

In indoor surveillance the state of the art [4, 2, 3, 11] features robust motion detection and tracking algorithms. Indoor environments are composed of many relatively small spaces that are separated with walls and communicate with each other through doors and corridors. In this situation it is important to associate the presence of a particular individual in different parts of the building structure. It is less important to track continuously the motion of an individual as this motion is bound to break quite often due to the building's topography.

In outdoor surveillance the state of the art [7, 10, 9] also features robust motion detection and tracking algorithms. The difference is that these algorithms are usually much more sophisticated than the corresponding indoor algorithms due to the complexity introduced by highly variable lighting. The topology of outdoor environments is also very different than that of indoor environments. The large open spaces invite for continuous object tracking. Moving objects are not only humans but vehicles as well, traveling at significantly higher speeds. Faster moving objects necessitate faster processing speeds, yet the algorithms are much more computationally intensive than those applied to indoor surveillance scenarios. These contradictory requirements simply add up to the technical challenges of an advanced outdoor security system.

## 3. CCN - A Prototype Indoor Surveillance System

We have developed a prototype indoor surveillance system that monitors human presence in and around our lab. We dubbed the system *Cooperative Camera Network or CCN* as it demonstrates cooperation between the different camera nodes through sharing of visual information. By camera node here we mean the combination of a camera with a PC. The hardware architecture of the system is depicted in Fig. 3.1. Each camera sends its live video feed to a networked PC. Depending on how powerful the processor is, the computational requirements of more than one live video feeds (cameras) can be accommodated by the same PC. The CCN architecture is highly modular allowing easy expansion at a low cost since it uses common off-the-shelf components. We consider that CCN type of systems should be preferably outfitted with USB or 1394 digital cameras. Both the USB and 1394 are ubiquitous serial interfaces and provide easy camera connectivity to the PC processors. Since the information pipeline is all digital no time is lost in Digital to Analog conversion and the quality of the video signal remains high. Also, full computer control of all the camera functions is possible, including the setting of brightness, contrast, aperture, and zoom values. In the incarnation of CCN in our lab we use the Sony 1394 DFW-VL digital camera model. The cameras sit atop Pan/Tilt devices that are controlled by the PCs through RS-232 connections. The Pan/Tilt devices facilitate easy camera repositioning and enable in-room tracking if desirable.

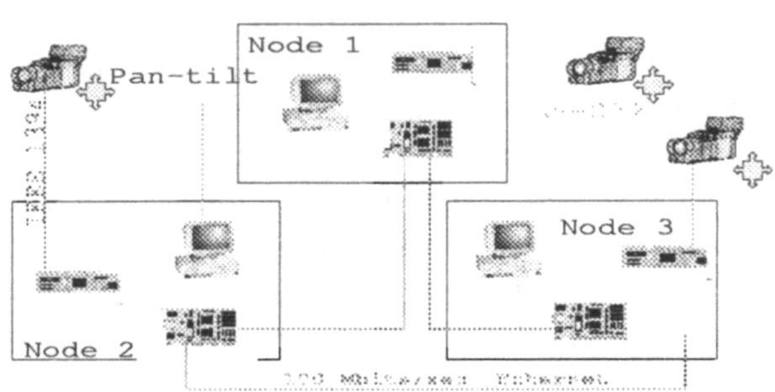


**Figure 3.1**: The CCN hardware architecture. More nodes can be added to the system to control larger portions of the building.

Currently, the CCN system consists of 3 camera nodes. One camera is physically located in our lab. The other two are located in two rooms consecutive to our lab (see Fig. 3.2). All the cameras are directed towards the rooms' doors. This is an exemplary setup for controlling human traffic in a subset of a building. We are interested in detecting and recording human presence in the rooms under surveillance. We are also interested in communicating detection information from one camera node to the other. This information could be as simple as the location and time of detection or as complicated as the color signature of the detected individual. Nominal location and time of incident information enables the system to reason as to where to expect the next move. For example, the information that an individual was detected in Room 1304 puts on alert the camera node in the adjacent Room 1303 (see Fig. 3.2). The color signature information enables the reporting of a visually tagged individual as he moves from room to room. Visual tagging of a particular individual is realized on demand by pointing and clicking on his silhouette in the corresponding live video stream at the central console of the system. CCN isolates the blob of the present human from the rest of the background thanks to its differencing and segmentation [8] algorithm. Then, CCN produces the color signature [5] for the individual and propagates it across the network of camera nodes. A copy of the signature is stored on each local model database under a single unique id. From that point on, every time a camera node detects human presence extracts the color signature of the detected individual and compares it against the color models stored in the database. If a match is found, the presence of the particular individual stamped by location and time is reported back to the central console of the system. A separate reporting window is maintained for each individual whose color signature has been stored in the database.

CCN is meant to enhance the effectiveness of a security guard that monitors human traffic in a commercial building. A very typical application is the monitoring of various parts of a department store for potential shoplifters. A security guard in such a case has to maintain awareness about video information displayed on dozens of screens. If at some point he notices a suspect it is very difficult to keep track of him as he moves out of the specific camera's field of view to another part of the store. It is typical in those cases for security the officer in charge to dispatch another guard, the so called 'shadow', to physically follow the suspect while he is present in the store. Since many suspect cases do not materialize this is an expensive use of security resources in a major department store. CCN allows the security officer to track not one but multiple suspects at one time semi-automatically, obviating the need of

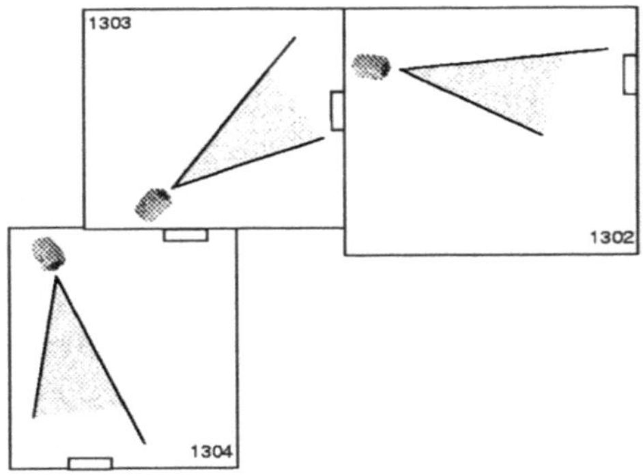


**Figure 3.2**: The topology of the CCN system in and around our lab (Room 1304).

'shadows'. Therefore, CCN does not only increase overall security in the department store scenario but also reduces its cost. Per our argumentation in Section 1 this would be a major selling point to conservative customers such as facility managers of department stores.

## 3.1 Testing of the CCN System

The CCN system has been tested extensively in a real building scenario. It performs flawlessly under the following two conditions:

1 The humans moving around the building are not dressed very similarly or the same.

2 The human traffic is sparse. Dense human traffic tends to occlude part of the lower body, thus reducing the discriminatory power of the color signature.

In a typical commercial building there is enough variability in people's apparel to render the first condition a non-issue. Of course, CCN would never be an ideal surveillance system for a boot-camp. Dense human traffic, however, is typical in department stores during the holiday season. Therefore, this is a condition that should be addressed before CCN becomes a viable product.

Fig. 3.3 shows some results from the segmentation and color recognition algorithm of CCN. It is worth noting that the CCN color recognition algorithm is sufficiently robust to orientation and size changes. While the subject's orientations differ $90^0$ the color recognition algorithm still performs successfully.

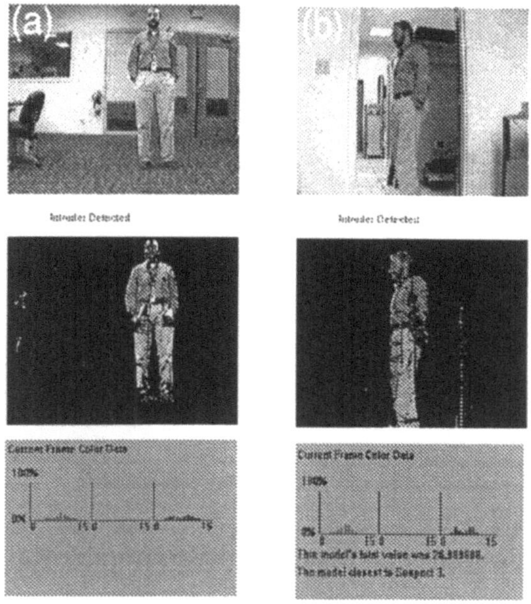


**Figure 3.3**: The color matching algorithm of CCN in action. (Column a) The top frame represents a snapshot from a live video feed. The mid frame shows the result of segmentation. The bottom frame shows the color signature of the segmented object as it was recorded in the CCN database. (Column b) Snapshot of the individual in a different room under surveillance. The mid frame shows the result of segmentation this time. The bottom frame shows the color signature of the segmented object this time and the successful matching result. At the time of this experiment a total of 3 color signatures were active in the model database.

## 4. DETER - A Prototype Outdoor Surveillance System

A comprehensive urban video surveillance system, such as DETER, depends primarily on two different technologies: computer vision and threat assessment. The computer vision part consists of the optical and system design, the moving object segmentation and tracking, and the multi-camera fusion stages. The threat assessment part consists of the feature assembly, the off-line training, and the threat classification stages (see Fig. 3.4). We will give a brief overview of each stage and compare our solutions to others proposed in the literature.

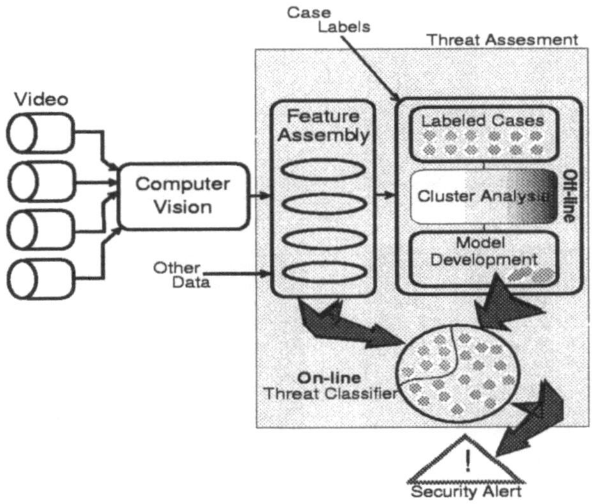


**Figure 3.4**: Architecture of the DETER system.

Our system is probably the only one that features a formal optical and system design stage. Most of the efforts reported in the literature had as their main objective to demonstrate the feasibility of a novel idea and they didn't pay any attention to the practical aspects of fielding a surveillance system. There is a number of requirements that a surveillance system needs to fulfill to function properly and be commercially viable. First, it should ensure full coverage of the open space or blind spots may pause the threat of a security breach. It is often argued in the technical literature that video sensors and computational power are getting cheaper and therefore can be employed in mass to provide coverage for any open space [6]. In reality things are not so rosy. Most of the cheap video sensors do not still have the required resolution to accommodate high quality object tracking. Both cheap and expensive cameras also need to become weather proof for employment outdoors, which increases their cost substantially. Then, it is the issue of installation cost that includes the provision of power and the transmission of video signals, sometimes at significant distances from the building. The installation cost for each camera is usually a figure many times its original value. Even if there were no cost considerations, cameras cannot be employed arbitrarily in public places. There are restrictions due to the topography of the area (e.g. streets, tree lines) and due to city and building ordinances (e.g. aesthetics). All these considerations severely curtail the allowable number and positions of cameras for an urban surveillance system.

In addition to optical considerations there are also system design considerations including the type of computational resources, the computer network bandwidth, and the display capabilities. Due to the cost sensitivity of the security market all these become critical issues and should be addressed in an optimal manner.

We achieve motion segmentation (see Fig. 3.5) through a multi-Normal representation at the pixel level. Our method resembles the method described in [9] with some interesting modifications. The method identifies foreground pixels in each new frame while updating the description of each pixel's mixture model. The labeled foreground pixels can then be assembled into objects using a connected components algorithm. Establishing correspondence of objects between frames (tracking) is accomplished using a linearly predictive multiple hypotheses tracking algorithm which incorporates both position and size.

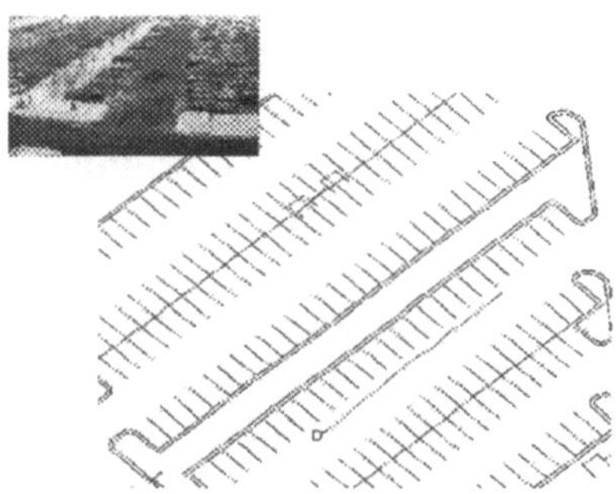

**Figure 3.5**: Live video snapshot of a car moving out of the parking lot and its itinerary (green line) as it is recorded by DETER at the CAD design level.

No single camera is able to cover large open spaces, like parking lots, in their entirety. Therefore, we need to fuse the Fields of View (FOV) of the various cameras into a coherent super picture to maintain global awareness. We fuse (calibrate) multiple cameras by computing the respective homography matrices. The computation is based on the identification of several landmark points in the common FOV between camera pairs.

The threat assessment portion of DETER consists of a feature assembly module followed by a threat classifier. Feature assembly extracts various security relevant statistics from object tracks and groups of tracks. The threat classifier decides in real time whether a particular point in feature space constitutes a threat. The classifier is assisted by an off-line threat modeling component (see Fig. 3.4).

DETER, like CCN, is a departure from the traditional proprietary security mentality. Both systems are implemented with off the shelf components. DETER can run on very moderate PC hardware (233 MHz) and its architecture provides for standard analog cameras that abound in commercial buildings. Therefore, PC surpluses and existing cameras can be used to build DETER, reducing dramatically its installation cost.

## 4.1 Testing of the DETER System

DETER is a fully operational prototype that performs perimeter surveillance in the Honeywell Laboratories building in Minneapolis. In addition to the qualitative testing performed by the actual users (security guards) we also performed quantitative testing for benchmarking purposes. Since August $11th$, 2000 we measured the tracking performance of DETER in the Honeywell Laboratories parking lot for 8 hrs. The testing was done in 1 hr increments spread over different days, times of day, and seasons. Table 3.1 shows the results of the DETER performance in the field tests. Parking lot activity included walking and running of a single individual, simultaneous walking of a number of individuals(following crossing or parallel paths), driving of a single and multiple cars, and finally a combination of cars and humans in motion. Some staged events included geometrically interesting walking patterns such as the ones we call *M-Patterns* (favorite by car thieves) and dangerous driving. These events were identified as suspicious by the Threat Assessment classifier.

| Perfect Tracks | Split Tracks | Joint Tracks | False Alarms | Missed Tracks |
|---|---|---|---|---|
| 554 | 77 | 16 | 5 | 3 |

**Table 3.1**: Experimental results for the 8-hour-long data set.

DETER detected and tracked perfectly 554 objects out of 666 - a remarkable performance. In 77 instances DETER has lost momentarily track of the object but regained it very quickly. The result was a split track. That was typically the case with pedestrians as they ventured momentarily under the tree lines (summer and early fall trials). In a few occasions (16) where pedestrians were moving next to each other (party of two) DETER correctly detected and tracked the motion but as a single object. This is a camera resolution problem. If we covered less area with each camera the resolution would have been better and the segmentation of closely spaced moving objects more accurate. DETER produced a small number of false alarms. Four of the five false alarms were produced in a snowy day as accumulated iced snow was hovering from the top cover of one of the cameras.

## 5. Summary

We have described two prototype surveillance systems, one for indoor use (CCN) and one for outdoor use (DETER). As a way of introduction we also discussed the state of the security industry and market and the role it played in our project decisions. CCN features a relatively simple visual motion detector based on image differencing and adaptive thresholding, which suffices for the locally stable illumination in indoor spaces. CCN can also record on demand the color signature of the segmented individual and propagate it across the camera network of the building. From that point on all the cameras are actively matching the color signatures of passing humans and report back to a central console if a match was found. We have adopted an illumination independent color matching method to accommodate for the different illumination intensities and spectral compositions across a typical commercial building. CCN is quite resilient in variations of object size and orientation. It has only been tested though in sparse human traffic scenarios. The challenge would be to extend the CCN functionality to dense crowd situations, typical in commercial buildings during peak times. This is an area our current work is focusing on.

DETER consists of a computer vision module and a threat assessment module. The two primary components of the computer vision module is the moving object segmenter and the associated tracker. The threat assessment module reports suspicious patterns detected in the annotated trajectory data at the CAD level. The threat assessor also uses the information produced by the computer vision module to perform some non-security functions, like monitoring the capacity of the parking lot. We are working towards the improvement of the threat assessment module with the inclusion of a clustering algorithm. The clustering algorithm will help in the partial automation of the off-line training, currently performed manually. DETER is scheduled for productization in 2002.

## References

[1] World security services to 2004. Technical Report 1348, The Freedonia Group, December 2000.

[2] Q. Cai and J.K. Aggarwal. Automatic tracking of human motion in indoor scenes across multiple synchonized video streams. In *Pro-*

*ceedings Sixth International Conference on Computer Vision*, pages 356–362, Bombay, India, January 4-7 1999.

[3] Q. Cai and J.K. Aggarwal. Tracking human motion in structured environments using a distributed camera system. *IEEE Transactions on Pattern Analysis and Machine Intelligence*, 21(11):1241–1247, 1999.

[4] Q. Cai, A. Mitiche, and J.K. Aggarwal. Tracking human motion in an indoor environment. In *Proceedings 1995 IEEE International Conference on Image Processing*, volume 1, pages 215–218, Washington D.C., October 23-26 1995.

[5] B.V. Funt and G.D. Finlayson. Color constant color indexing. *IEEE Transactions on Pattern Analysis and Machine Intelligence*, 17(5):522–529, 1995.

[6] W.E.L. Grimson, C. Stauffer, R. Romano, and L. Lee. Using adaptive tracking to classify and monitor activities in a site. In *Proceedings 1998 IEEE Conference on Computer Vision and Pattern Recognition*, pages 22–29, Santa Barbara, CA, June 23-25 1998.

[7] T. Kanade, R.T. Collins, A.J. Lipton, P. Burt, and L. Wixson. Advances in cooperative multi-sensor video surveillance. In *Proceedings DARPA Image Understanding Workshop*, pages 3–24, Monterey, CA, November 1998.

[8] N. Otsu. A threshold selection method from gray-level histograms. *IEEE Transactions on Systems, Man, and Cybernetics*, 9(1):62–66, 1979.

[9] C. Staufer and W.E.L. Grimson. Learning patterns of activity using real-time tracking. *IEEE Transactions on Pattern Analysis and Machine Intelligence*, 22(8):747–767, 2000.

[10] C. Stauffer and W.E.L. Grimson. Adaptive background mixture models for real-time tracking. In *Proceedings 1999 IEEE Conference on Computer Vision and Pattern Recognition*, volume 2, pages 246–252, Fort Collins, CO, June 23-25 1999.

[11] F. Ziliani and A. Cavallaro. Image analysis for video surveillance based on spatial regularization of a statistical model-based change detection. In *Proceedings 1999 International Conference on Image Analysis and Processing*, pages 1108–1111, Venice, Italy, September 27-29 1999.

# Chapter 4

# Visual Surveillance in Retail Stores and in the Home

Tomas Brodsky, Robert Cohen, Eric Cohen-Solal, Srinivas Gutta, Damian Lyons, Vasanth Philomin, Miroslav Trajkovic
*Philips Research USA, 345 Scarborough Rd, Briarcliff Manor, NY 10510*

**Abstract:** This paper presents an overview of video surveillance activities at Philips Research USA, which serves as a research arm of Philips Communications, Security and Imaging, a global leader in the design and manufacture of advanced automatic security systems. We concentrate on two application domains: professional security market, with emphasis on retail monitoring, and a low-cost, automated residential security. The main application for pan-tilt-zoom (PTZ) cameras is typically tracking of intruders throughout the facility. In addition, camera calibration is an important capability, because it allows geometric reasoning with respect to the floor plan as well as enhanced control of the camera. Another major component of our work concerns video content analysis, the detection of security related objects and events from video. The system typically processes video in real-time and can provide immediate alarms to alert the security operator. Relevant information is also stored in a database so that it can be efficiently retrieved later. The third and final topic discussed in the paper is a residential monitoring application, namely an intruder detection system. We describe detection of moving objects robust to changes in lighting and an object classification scheme based on radial-basis networks.

**Keywords:** video surveillance, retail monitoring, residential security

## 1. Introduction

Automated surveillance and monitoring holds great promise, but also presents challenging problems that need to be solved. We first look at several monitoring tasks and briefly illustrate the requirements for automated systems suitable for such tasks.

In the professional domain, security operator's tasks include active surveillance (e.g., tracking of suspicious individuals throughout the facility) as well as many time-consuming and mundane tesks (e.g., reviewing tapes from previous day to find out who stole a certain object). For active surveillance, the goal of the automated system is to simplify the operator's tasks, while leaving the person in control. For example, automatic tracking can be started when the operator chooses a certain individual on the screen and the system will follow that person without further intervention. However, the operator should have the ability to override the tracker and change the camera settings at any time.

In traditional commercial video surveillance systems, several hours or days worth of video is captured on videotape. Operators must then play or fast-forward through these tapes to locate incidents of interest. Manually searching through video in this way is so tedious that surveillance equipment operators frequently avoid searching tapes unless absolutely necessary. More current systems allow operators to set alarms. The time that each alarm occurs is stored so that an operator can skip through the video to view scenes at which an alarm was triggered. This alarm-based method is an improvement over the manual search method, but it is still limited in that alarms usually must be defined before the video is captured, and the search is limited to those predefined, often binary alarms. Searching through video can be cumbersome if these alarms are too simplistic, or if there are too many false alarms. To solve these problems, our system detects a whole range of events and a semantic indexing and retrieval process is used.

In the residential environment, low-cost is one of the most important factors. This may introduce extra constraints, such as grayscale cameras and a limited amount of computational power. Residential systems also run with no supervision and robustness to environmental changes is very important.

The rest of the paper is organized as follows: Sections 2 and 3 describe camera calibration and automated tracking using PTZ cameras. In Section 4 we discuss detection of events and objects and in Section 5 we describe our indexing and retrieval system. Section 6 describes a video motion detector suitable for the residential environment and finally, Section 7 talks about our classification technique based on radial-basis networks.

## 2. PTZ camera calibration

Pan-tilt-zoom (stationary, but rotating and zooming) cameras are often used in surveillance applications. The main advantage of a PTZ camera is that one camera can be used for the surveillance of a large area, yet it can also be used to closely look at the points of interest.

As the layout of the surveillance areas is prone to changes, the installer often wants to move a camera from one position to another. In this case, it is important to have a simple procedure to determine accurate camera position and orientation in reference to the surveillance area. The knowledge of camera position and orientation is crucial for geometric reasoning. This, in turn, enables some useful functionality, such as automatically pointing the camera to a certain location, by clicking on its position on the area map, or displaying the current field of view of the camera.

Knowledge of internal camera calibration parameters is also important for a variety of useful tasks, including tracking with a rotating camera, obtaining metric measurements, knowing how much to zoom to achieve a desired view, etc. Again, it is of utmost importance to develop a simple procedure that will enable the installer with little or no technical training to perform calibration of all the cameras covering the surveillance area, or even better, to develop a method that will perform calibration entirely automatically.

In following that goal, we developed a novel procedure for external and internal self-calibration of a PTZ camera. The algorithm assumes that the camera principal point and the center of rotation of the pan and tilt units coincide. The distance between these two points is usually small and therefore this assumption can be made. We also assume that the skew factor is zero, and that the principal point does not move while the camera is zooming. Finally, it is assumed that the maximum zoom-out factor $s$ of the camera is known.

To determine camera internal calibration, all the user needs to do is to point the camera to a texture rich area in the room. The camera then zooms in and out completely and we acquire two images $I_1$ and $I_2$. The principal point is then determined by scaling down for factor $s$ image $I_1$ and finding the best match for so obtained template in the image $I_2$. Once the camera principal point is determined, focal length at a particular zoom setting is estimated by taking two images at fixed pan and different tilt settings and finding the displacement $d$ of the principal point. Focal length is then calculated as:

$$f = -\frac{d}{\tan\alpha},$$

where $\alpha$ denotes a difference in the tilt angle. Once the camera focal length is known, we compute the mapping between zoom settings and focal length, by fitting the inverse of focal length (lens power) to the second order polynomial in zoom ticks. It can be shown that this fitting not only has desirable numerical properties (i.e., stability), but also yields linear solution.

The external calibration module assumes known camera height and allows the installer to determine camera position and orientation in the following manner:

1. The user points the camera at several points in the area and clicks on their respective position on the area map shown in the GUI in *Figure 1*.
2. Each time the user clicks on a point in the map, the system acquires current camera position and location on the map camera is pointing to.
3. The algorithm then computes camera position and orientation using data acquired at step 2.

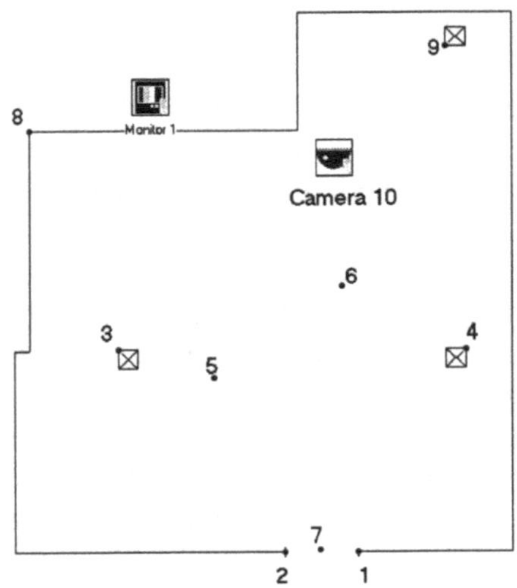


*Figure 1.* GUI used for CCTV surveillance. The operator can insert a camera at any position on the map and compute its position, orientation, and internal calibration using a simple user-friendly procedure

## 3. Intruder detection and tracking with a PTZ camera

To simplify the ease of use of Philips security equipment, we developed a software module that provides automated control of a PTZ camera to follow a person and keep his/her image centered in the camera view. Instead of manually following an intruder with the system joystick, the security operator simply needs to identify the subject and the system will track that person, leaving the operator free to closely observe the behavior of the subject.

The first step in the tracking process is target selection. In the current implementation, the operator selects the target manually, by placing a rectangle on torso (required) and the head and part of trousers (optional) of the person being tracked. The rectangle placed on the person is referred to as the Target Rectangle (TR), and we track the person by tracking its TR.

Once the TR is selected, the parameters of its color model are found by using its color histogram. In particular, we use hue and saturation model. However, since these parameters are unstable for colors close to gray, we model gray separately and represent the TR by its combined color/gray histogram.

We then use motion detection to find moving areas in the image, and use color matching to find the motion area that corresponds to the TR. Since histogram matching can be a costly procedure, we have developed a procedure for recursive and fast histogram matching.

Once the TR is identified, we issue velocity commands so that camera moves towards the TR and acquire the next image. In order to perform independent motion detection, we have to identify the global motion (due to camera movement) and then align images before and after camera motion. We have found that using camera feedback about its position usually does not result in reliable alignment. We have developed a procedure for feature based image alignment which does not require any information on camera calibration and camera motion.

## 4. Video Content Analysis

Automatic detection of surveillance related objects and events in video is an important technology in CCTV applications. Here we briefly describe our Video Content Analysis system. The system processes video in real-time, extracts relevant object and event information and indexes this information into a database. It can also issue alarms to alert the operator. Separate retrieval software can be used to search for specific data in the database and to quickly review associated video content.

We assume a stationary camera and use a background subtraction technique, where for each pixel the system maintains a non-parametric statistical model. We chose such a model instead of [1], because it is well suited for modeling of complicated scenes with cluttered (and possibly moving) background.

Each video frame is compared with the background model and pixels that do not fit the model are marked as foreground. Such pixels are further grouped into connected components and then tracked as they move through the video (see *Figure 2*). Multiple objects can be tracked simultaneously and the module handles object merging and splitting. The module also maintains a color model for each object. When object merge and later split, the correct tracking labels are re-assigned based on the color models.

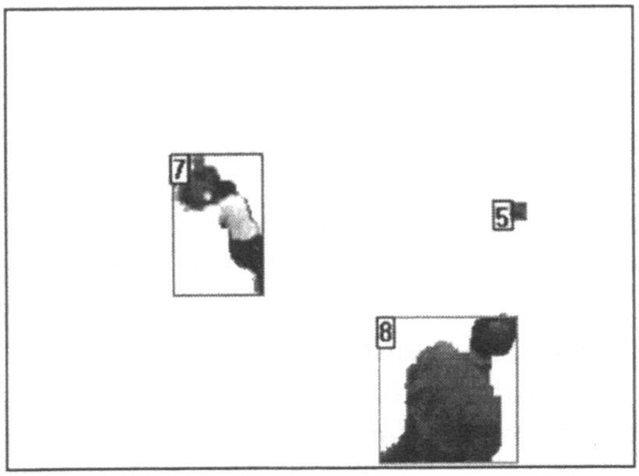


*Figure 2.* People and objects extracted from video.

For each object, we store information about its trajectory and a list of events the object was involved in. Simple events are based on appearance and disappearance of foreground regions, for example enter/leave and merge/split. A separate event-reasoning module generates more complicated events derived from the simple event stream, based on user provided rules. The rules can specify a sequence of events, an absence of an event, length of time intervals, etc. Hierarchies of events can be constructed, because the generated complex events are fed back into the event-reasoning module. The current system is similar to that in [2].

## 5. Indexing and retrieval

As described earlier, objects and events are identified while the video is being captured. Information related to these objects and events (e.g. time, location, size, color, etc.) are stored in a database. In addition, inter-object relations are indexed for events that involve multiple objects. This information can be retrieved at any time by performing a database query. The retrieval engine is linked with the video storage system, so video scenes can be easily and quickly viewed for any query.

By separating the indexing and retrieval process, an operator can perform queries for events that were not predefined when the video was captured. For example, the system can retrieve all instances in which a person took a specific object that was not identified earlier. Another system that uses semantic scene analysis to detect abandoned objects is described in [3]. Since we are also keeping track of the merging and splitting of moving objects, the operator can perform advanced queries such as showing all the people that an identified pickpocket interacted with. Another query can be to show scenes in which two or more employees are standing in a group without a customer, based upon their clothing color.

Since the total history of every object is stored in the database, we can perform a query to determine everything that a specific person did while in view of the camera. In addition to jumping to the video location in which that person entered the scene, a summary picture and list can be retrieved as shown in *Figure 3*. The retrieval results show the trajectory of the person through the scene, along with a list of every person or object that the person interacted with while in the scene.

| Obj 1 Type | Object 1 | Event Type | Obj 2 Type | Object 2 | Date/Time |
|---|---|---|---|---|---|
| Person | 2 | Enter | | | Apr 19, 2001 16:28:25.605 |
| Person | 1 | Merge | Person | 2 | Apr 19, 2001 16:28:32.235 |
| Person | 2 | Split | Person | 1 | Apr 19, 2001 16:28:40.907 |
| Person | 1 | People met | Person | 2 | Apr 19, 2001 16:28:42.319 |
| Person | 2 | Active | | | Apr 19, 2001 16:29:58.058 |
| Person | 2 | Deposit/Take | Thing | 4 | Apr 19, 2001 16:29:58.058 |
| Person | 2 | Merge | Person | 5 | Apr 19, 2001 16:31:12.755 |
| Person | 2 | People met | Person | 5 | Apr 19, 2001 16:31:22.910 |
| Person | 2 | Split | Person | 5 | Apr 19, 2001 16:32:48.363 |
| Person | 2 | Leave | | | Apr 19, 2001 16:33:03.495 |

*Figure 3.* Retrieval results showing a person's activities: The trajectory and the list of events that involved this person

Since objects and events are primarily defined in software, it is easy to add new items to index, even after the system has been deployed. For example, if we later develop an event indicating that a person hid an object in their clothing, the system could retrieve instances in which people take objects from shelves, hide the objects under their clothing, and leave the premises without placing those objects on a cashier's counter. This kind of system is limited only by the size of the database and the imagination of the person performing the queries.

## 6. Residential intruder detection

In the home, requirements for a surveillance system are quite different from the professional domain. One of the biggest challenges for such a system is elimination of false alarms caused by irrelevant changes in the environment, especially lighting changes. Changes in lighting and shadows most often change the intensity of the pixel, but do not change its color. Thus color components could be used (such as U, V components of the YUV space) in the background model, leading to a system insensitive to many lighting changes.

In this section, we describe our approach to handling lighting changes for low-cost, grayscale cameras, where the color components are not available. The detection of moving objects proceeds in two steps. First, a background subtraction technique is used to detect pixels that differ from the background model. Then we apply additional filters to classify such pixels into real objects or lighting changes. For example, it is possible to separate a person from his or her shadow (*Figure 4A*). Moreover, a flashing light directed, for example, at the sofa in the living room will produce a moving bright spot that is considered by most of the available surveillance system as a moving object. With the current system, this moving light is detected as a lighting change (*Figure 4B*).

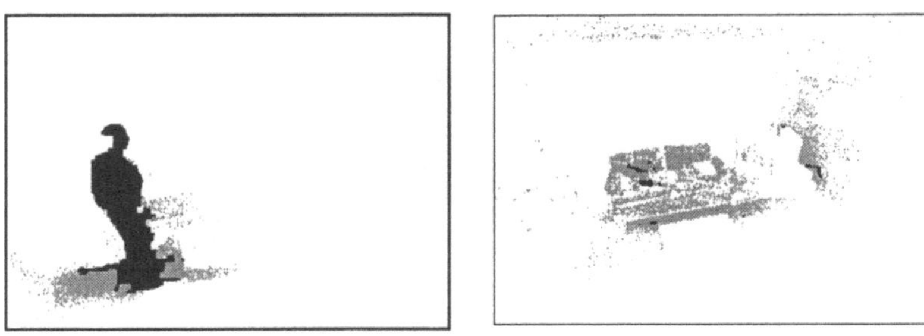

(A) (B)

*Figure 4.* (A) Detected person (black pixels) and his/her shadow (gray pixels). (B) Correctly classified light changes (gray) caused by a lamp shining on a sofa.

In addition to the non-parametric background model, the system maintains a reference image (for example, containing the mean or median value among the background model samples). For each detected pixel, we consider a small (3x3 or 5x5 pixel) neighborhood around that pixel and compare the gray-level structure with the corresponding region in the reference image. The comparison uses several filters, such as the normalized cross-correlation. When the gray-level structures of the input image and the reference image are similar, the pixel is marked as caused by lighting changes.

After all the filters are applied, the remaining foreground pixels can be grouped into objects. As described in the next section, the detected objects are further classified into people and animals, so that the system can suppress alarms caused by pets.

## 7. Object classification

In this section, we describe a classification module trained to distinguish between people, pets, and other objects. We extract horizontal and vertical gradient features from the input intensity image by first smoothing the input image to reduce the effects of noise and then applying a derivative operator over the entire image (as illustrated in *Figure 5*).

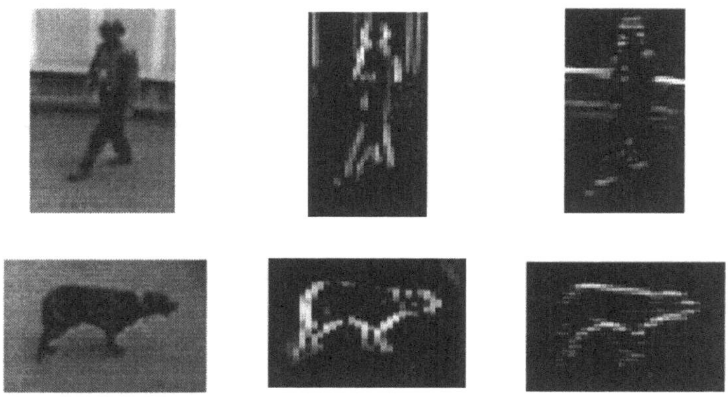

*Figure 5.* Left column: sample person and dog images. Middle column: horizontal gradients. Right column: vertical gradients

Since we would like to be able to classify objects under various poses and illumination conditions, it would be non-trivial if not impossible to model the entire space that the instances of a certain object class occupy given the fact that instances of the same class may look very different from each other (e.g. people wearing different clothes). Instead, we need to identify and model features that do not change much under these different scenarios. The gradient is one such feature since it reduces the dimension of the object space drastically by only capturing the shape information.

Next, the extracted gradients are used to train a Radial Basis Function (RBF) classifier. An RBF classifier has architecture very similar to that of a traditional three-layer back-propagation network [4]. Connections between the input and middle layers have unit weights and, as a result, do not have to be trained. Nodes in the middle layer, called BF nodes, produce a localized response to the input using Gaussian kernels. The basis functions (BF) used are Gaussians and each hidden unit can be viewed as a localized receptive field (RF). The hidden layer is trained using **k**-means clustering.

The RBF input consists of **n** normalized gradient images fed to the network as 1D vectors. The hidden layer yields the equivalent of a functional shape base, where each cluster node encodes some common characteristics across the shape space. The output (supervised) layer maps shape encodings ('expansions') along such a space to their corresponding class and finds the corresponding expansion ('weight') coefficients using pseudoinverse techniques. As an example, the output layer for the classification task would consist of nine nodes corresponding to nine classes. These classes include four corresponding to the different orientations of the person, one for groups of people, one for noise and the remaining three for different orientations of pets, in our case dogs.

Experiments were conducted by first training the RBF classifier on three sequences. Of the three sequences, one contained only the pet by itself, the second contained a single person walking in the first half and with other people in the second half of the sequence. Finally, the third sequence consisted of both the pet and human being present. A total of 750 objects corresponding to 9 classes were used for training. Preliminary experiments were conducted by testing on 7 sequences not used during training. The system showed an overall classification performance of 93.5 %.

## 8. Conclusions

The techniques presented in this overview provide new tools that efficiently assist the user in performing surveillance tasks and also augment the capabilities of current systems. Assistance is provided by simplifying or automating some of the tasks, such as camera calibration or PTZ tracking. Traditional video recorders can be significantly enhanced by indexing the video content and providing the means to efficiently retrieve relevant video segments. In addition to the professional security market, we are also developing tools that address the low-end market as shown for the case of residential security.

# References

[1] C. Stauffer, W. E. L. Grimson, "Adaptive Background Mixture Models for Real-time Tracking", *Proc. Computer Vision and Pattern Recognition*, pp. 246-252,1999.

[2] F. Bremond, M. Thonnat, "Object Tracking and Scenario Recognition for Video-Surveillance", *Proc. IJCAI*, 1997.

[3] E. Stringa and Carlo S. Regazzoni, "Real-Time Video-Shot Detection for Scene Surveillance Applications," *IEEE Trans. Image Processing*, vol. 9, no. 1, pp. 69-79, January 2000.

[4] R. P. Lippmann and K. A. Ng, "Comparative Study of the Practical Characteristic of Neural Networks and Pattern Classifiers", *MIT Technical Report 894*, Lincoln Labs, 1991.

II

# DETECTION AND TRACKING

Chapter 5

# DETECTING AND TRACKING PEOPLE IN COMPLEX SCENES

Yoshinori Kuno
*Saitama University*
*255 Shimo-okubo, Saitama, Saitama 338-8570, Japan*
kuno@ics.saitama-u.ac.jp

**Abstract** This paper presents the researches on visual surveillance by the author, and Prof. Shirai and his group at Osaka University. Our purpose is to develop visual surveillance systems that can detect and track humans and other moving objects in complex scenes. In this case, we cannot expect to extract a single type of feature reliably all the time. Thus we need to accumulate bits of fragmented information to realize robust detection methods. We adopt two approaches to this problem. One is to accumulate bits of evidence over time. We propose a predicted-path voting method, in which consistent optical flow data are accumulated in the space-time cube to detect humans. The second is to integrate multiple cues. We propose to combine motion (optical flow) and edge data, motion and depth (stereo) data, and motion and uniform brightness region data.

**Keywords:** Surveillance, evidence accumulation, multiple cue integration, optical flow

## 1. Introduction

Automatic visual surveillance systems are strongly demanded for various applications. We have several systems commercially available, most of which are based on the subtraction between consecutive frames or that between a current image and a stored background image. They can work as expected in simple scenes where environmental conditions do not change, such as indoors. However, they cannot work in complex scenes such as outdoors where there are various noises such as changes of lighting and movements of background objects. This paper describes the researches on detecting and tracking people to realize surveillance systems working in complex scenes by the author, and Prof. Shirai and his group at Osaka University.

In complex scenes, we cannot expect to extract a single type of feature reliably all the time. Thus we need to accumulate bits of fragmented information to realize robust detection methods. We adopt two approaches to this problem.

One is to accumulate bits of evidence over time. Simple subtraction based methods yield many false alarms for condition changes in complex scenes. To reduce such false alarms, Nakai [6] proposed a method using a statistical model of the background. The basic idea is that even if some changes are observed, they should not be reported if such changes often happen in the scene. The method reduces false alarm errors, but it also decreases the sensitivity of target detection.

We can solve this problem by accumulating bits of evidence over time. If a person is moving, the optical flow for the motion can be detected along his/her path if we observe this from a global viewpoint. This means that we can obtain bits of evidence for the motion along the path although in complex scenes it occurs at times that the optical flow for the object cannot be detected and that the flow for other background objects interrupts the target's flow. We have proposed a predicted-path voting method to view optical flow data globally [5].

The second is to integrate evidence from multiple cues. The rationale here is that even if any cues cannot be extracted properly at times, the others can cover the detection. There have been several methods using multiple cues or features. Etoh and Shirai [1] segmented an image into the regions from the motion and color distribution in the image. Thompson [11] combined motion and edges for extracting object regions. Still, what cues should be used and how they should be integrated are unsettled questions. We have proposed to combine motion (optical flow) and edge data [3], [4], motion and depth (stereo) data [7], [8], and motion and uniform brightness region data [12].

This paper presents our methods of detecting and tracking people based on the two approaches mentioned above and discusses the remaining problems.

## 2. Accumulating Evidence over Time

This section briefly shows a method of detecting people by accumulating bits of evidence over time.

Fig. 5.1(a) shows an example image. A human is passing in the scene where many trees are shaking in the background. Simple subtraction between consecutive frames yields Fig. 5.1(b). Since the trees are moving, we observe brightness changes almost everywhere in the image. Thus detecting the human separately seems almost impossible by this simple subtraction method. Our purpose here is to develop a method that can distinguish humans from background objects in such difficult conditions. We distinguish the former from the latter by making the following assumptions: humans move in constant directions at constant speeds whereas background objects move at random compared with humans.

We start with computing optical flow using the generalized gradient method [10]. Since we use it in all our researches described in this paper, we briefly show it here.

Two orientation-selective spatial Gaussian filters are applied to the original image $f(x, y, t)$: one is sensitive to vertical edges, the other to horizontal edges. Let $g(x, y)$ and $h(x, y)$ denote the two filters and $g \otimes f$ denotes the convolution of $f(x, y, t)$ and $g(x, y)$. From the two filtered images the following two constraint equations are derived at each pixel:

$$\begin{aligned} u \cdot (g \otimes f)_x + v \cdot (g \otimes f)_y + (g \otimes f)_t &= 0 \\ u \cdot (h \otimes f)_x + v \cdot (h \otimes f)_y + (h \otimes f)_t &= 0 \,, \end{aligned} \tag{5.1}$$

where $(u, v)$ denotes an optical flow vector and subscripts denote partial differentiation. By solving the equations, the flow vector is obtained.

Fig. 5.1(c) shows the optical flow for the example scene. We observe many flow vectors caused by the moving trees in addition to those by the human.

(a) Original image. (b) Subtraction result. (c) Optical flow.

*Figure 5.1.* Outdoor scene image.

We segment the flow image into uniform flow regions in a split-and-merge fashion. Then we use a predicted path voting process to detect target objects. We prepare a four-dimensional voting space $(x-y-t-\theta)$. For each uniform flow region, we predict a path of the region in a certain time interval of future. Fig. 5.2(a) shows the predicted path(only $x-y-t$ are shown). We assume that the region continues to move in the mean-flow direction ($\theta$) at its speed. We approximate each region by an ellipse whose center coincides with the region's centroid. Every point inside the ellipse is given weight, according to the two dimensional Gaussian as shown in Fig. 5.2(a). This weight is voted at the predicted position $(x, y)$ at the time $(t)$ in the direction $(\theta)$.

Since we assume that a target object is moving in a constant direction at a constant speed, the votes for the target will be accumulated as shown in Fig. 5.2(b) if observation continues for a certain length of time. On the other hand, since background objects move only around their original positions, the votes will not be accumulated as shown in Fig. 5.2(c). Note that Figs. 5.2(b), (c) and (d) show only the two-dimensional space $(x, t)$. The actual voting space has the dimensions for $y$ and $\theta$ in addition. $\theta$, which denotes the moving direction of the target, is required to avoid accidental accumulations of votes for regions with different directions. Fig. 5.2(d) shows an example of such cases. In the actual four-dimensional space, these paths do not intersect, thus cause no accumulation.

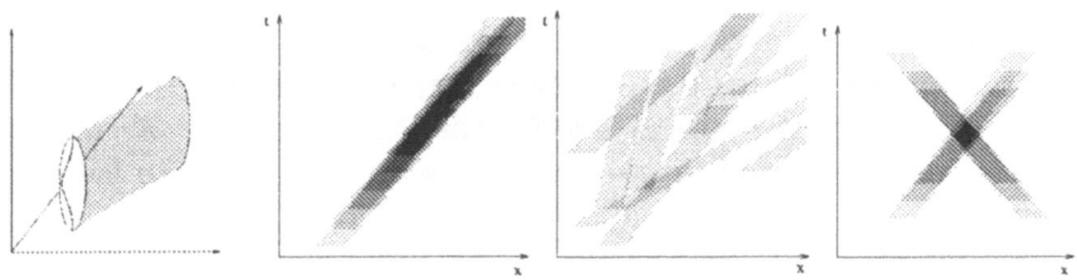

(a) (b) (c) (d)

*Figure 5.2.* Path voting.

Fig. 5.3 shows an experimental result for a scene where a man is walking from left to right in front of bamboo trees shaking heavily in the wind. In the detection result, darker regions indicate that more number of votes are accumulated. We can tell from this figure that the region corresponding to the man gets darker as time passes and that this method is able to detect him in this complex condition.

We have made other experiments on scenes with changing background, such as scenes with a fountain, a rotating signboard, and fluttering flags, obtaining successful results.

## 3. Integration of Multiple Cues

In this section, we describe three methods of detecting and tracking people based on the integration of multiple cues.

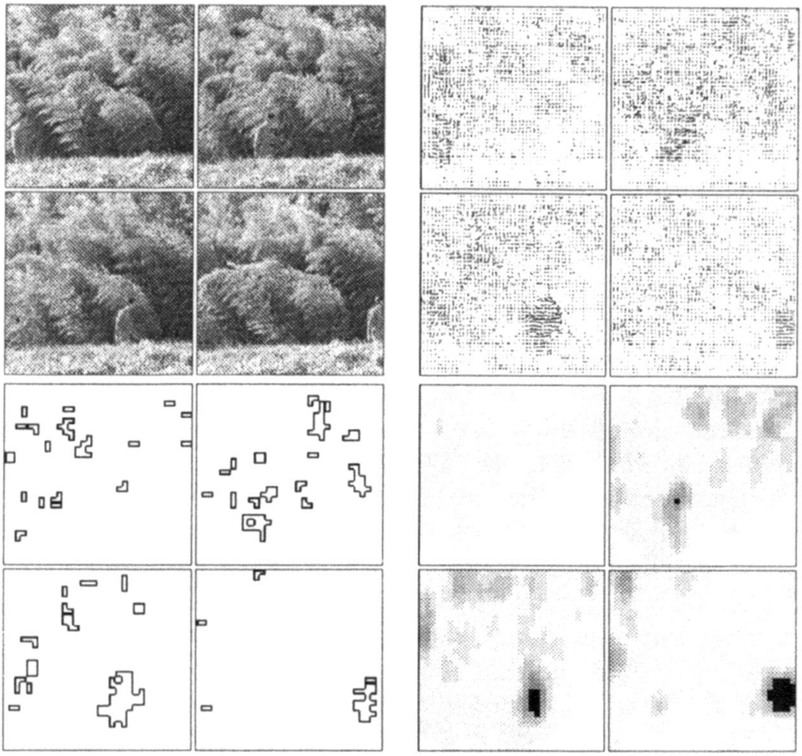

*Figure 5.3.* Experimental result for bamboo scene. Top left: original image, top right: optical flow, bottom left: segmentation result, bottom right: detection result.

## 3.1 Optical Flow and Edges

Mae et al. [3] have proposed a method of determining contours of moving objects in cluttered scenes by integrating optical flow and edges. Moving object detection based on optical flow such as described in Section 2 has a drawback that the detected motion boundary is not precise because the motion is not homogeneous near the motion boundary. In addition, the object contour cannot be determined when the motion of the object is similar to that of the neighborhood. Furthermore, if the background is cluttered, the contour cannot be determined only from edges because many edges may be extracted in the background and some edges on the contour may be missed. In the proposed method, the contour is determined by using optical flow and edges in a long sequence. The whole contour of a moving object is eventually obtained by accumulating edges near motion boundaries over an image sequence.

The method can also determine the occlusion relation of two overlapping objects by checking if edges exist on the predicted object contours.

The overview of the method is as follows. We consider the case that a target object is rigid and the object motion in the image can be approximated by translation in a short time. First of all, optical flow is extracted by the generalized gradient method [10]. Then neighboring pixels with similar flow are grouped to detect uniform flow regions.

The neighborhood of the motion boundary (the boundary of a uniform flow region) is searched for edges. Edges with similar direction (or opposite direction) to the direction of the motion boundary are extracted. The nearest edge to the motion boundary is considered as the object contour.

In general, enough edges are not necessarily extracted at one frame to compose the whole contour. Thus we accumulate the extracted edges over frames. If an edge is newly extracted which corresponds to a previously accumulated edge, the accumulated edge is replaced by the new edge. For the parts of the contour where no edges have been extracted, the current motion boundary is used as the contour. As the moving object is tracked over frames, new edges are extracted at various parts of the object contour, and the number of accumulated edges increases.

Fig. 5.4(a) shows an extracted motion boundary, which is represented by the white curve. There is a gap between the motion boundary and the true contour. Fig. 5.4(b) shows the edges extracted near the motion boundary in the current frame. Fig. 5.4(c) shows the previously accumulated edges. Fig. 5.4(d) shows the extracted contour consisting of the accumulated edges and parts of the motion boundary. The extracted contour is more precise in the parts where the edges are extracted.

If the flow of the object is similar to that of the neighborhood, the motion boundary cannot be detected. In this case, the contour is predicted from the previous contour and the current flow. If two objects overlap which have similar flow, we determine the occlusion relation by evaluating which predicted contour suits the observed edges in the image.

Although we assume 2-D translation here, the method has been extended to deal with 3-D motion cases [4].

## 3.2 Optical Flow and Depth

Okada et al. [7] have proposed to use two cues: optical flow and disparity, which corresponds to depth in the 3-D world. The method can work even if tracking with either velocity data or disparity data alone may fail.

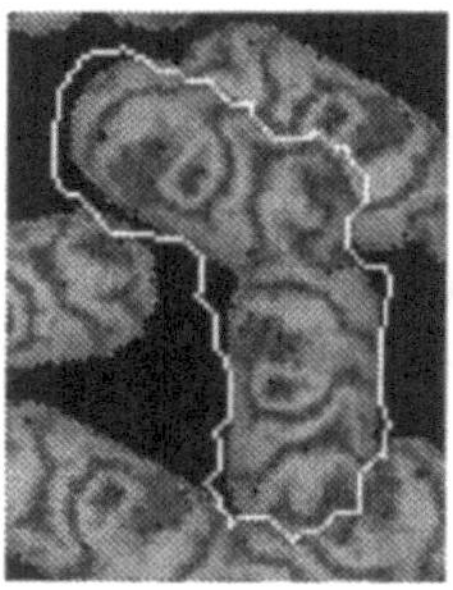

(a) Motion boundary.

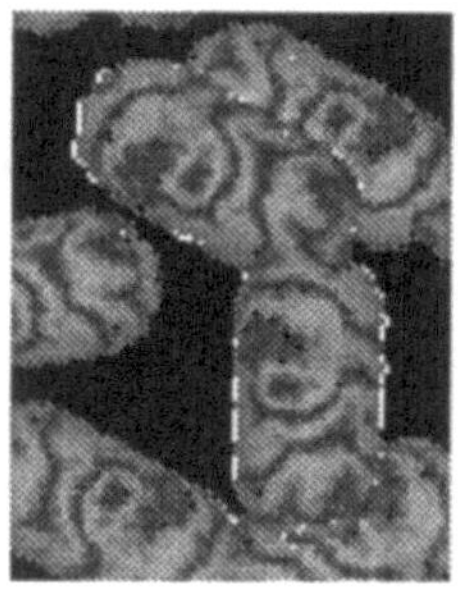

(b) Newly extracted edges.

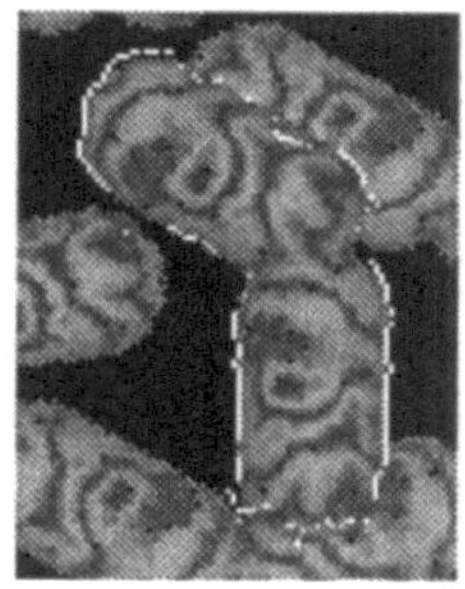

(c) Accumulated edges.

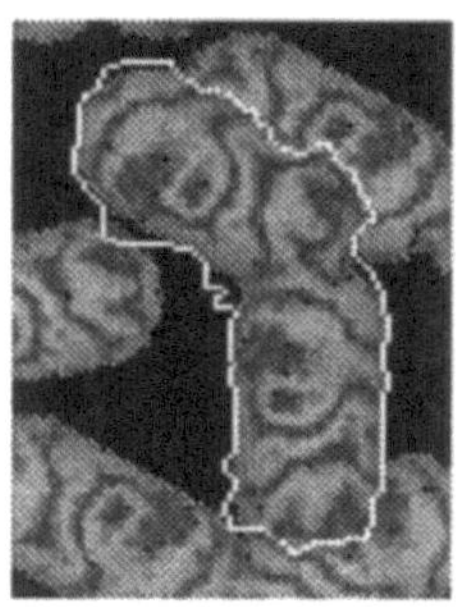

(d) Object contour.

*Figure 5.4.* Acquisition of object contour.

Also in this method, optical flow is calculated by the generalized gradient method[10] based on spatio-temporal filtering. The disparity can be computed from one-dimensional optical flow between a pair of stereo images because we can consider the left and right images as the images at two different times used in optical flow calculation. One-dimensional optical flow is calculated based on the gradient method using the constraint equation given by:

$$\frac{{f^R}_x(x,y) + {f^L}_x(x,y)}{2} d + f^L(x,y) - f^R(x,y) = 0 \tag{5.2}$$

where $f^L$ and $f^R$ denote the left and the right image, respectively, and $d$ denotes disparity.

Initially, a large uniform flow region is extracted. Then the rectangle circumscribing the region called a *target window* is determined.

In successive images, we track the region by the following way. First, a candidate target region is predicted from the target window of the previous frame. We compute the histograms of velocity $u$, $v$ and disparity $d$ in the target window to estimate the distribution of the target velocity and that of the target disparity. The probability of the pixel belonging to the target is determined from these histograms. Pixels with high probability are extracted in the candidate region as the points belonging to the target. These pixels are called *target pixels*. The rectangle circumscribing the target pixels is called a *temporal target window*.

The target pixels are not reliable because the extracted optical flow and disparity are noisy. The target window is divided into small fixed-sized rectangles at the initial frame. This rectangle is called a *block*. The block moves at the same velocity as that of the target window. If enough target pixels are extracted in a block in many frames, the block is determined to be a part of the target and called a *target block*. The rectangle circumscribing the target pixels and the target blocks becomes a new *target window*. In the next frame, the candidate target region is predicted by shifting the target window by the representative target velocity.

If the mean disparity in a target block becomes abruptly large, the region is determined to be occluded and its position is estimated by shifting the previous position by the representative target velocity. This block is called an *occluded block*. If the mean disparity in an occluded block becomes similar to the target disparity, the block is considered as a target block again.

If the target is completely occluded, the target window is estimated by assuming that the target moves with the constant velocity so that the target may be tracked again when it comes out of the occluding object.

Fig. 5.5 shows an experimental result. The white regions are the target pixels, and the black rectangles represent the target blocks and occluded blocks. The white rectangles indicate the target windows.

In this experiment, two persons walk from right to left in the scene. The target person is in the center of image No. 1. The person on the right side in image No. 1 is closer to the camera than the target person. As he walks a little faster than the target person, he overtakes the target person. In this situation, tracking using only the optical flow will fail because of similarity of the velocity. The target person is occluded from images No. 25 to No. 37. After the target object is overtaken, he passes by the standing person. Since the distances of the target person and the standing person are similar, the tracking using only depth data will fail.

By using the two cues, the target person was successfully tracked in our method. In the 28th frame, the system detects that the target person is completely occluded, and appears in the 38th frame.

Since the method assumes that the flow vectors in the target region are almost uniform, implying that the target translates almost parallel to the image plane, tracking may fail when the target moves with general 3-D motion. The method has been extended to treat such 3-D motion cases using a 3-D human model [8].

## 3.3 Optical Flow and Uniform Brightness Regions

Yamane et al. [12] have proposed a method of detecting and tracking a person based on the integration of optical flow and uniform brightness regions. Optical flow cannot be computed in textureless uniform brightness regions. Combining uniform brightness region data helps this situation.

Like in all other methods, optical flow is calculated by the generalized gradient method[10]. Fig. 5.6(a) shows a result of extracting optical flow. A pair of a white dot and a line denote an optical flow vector. Gray dots denote small contrast pixels where optical flow cannot be obtained. Then an area with large optical flow is found, considered as a part of the target person. The area is circumscribed by a rectangle called *motion window* (Fig. 5.7).

However, this method alone cannot discriminate the target from other objects with similar flow. In Fig. 5.6(a), two persons move to the left. Thus the motion window is incorrectly determined as the white rectangle circumscribing both persons as shown in Fig. fig:opticalflow(b). To solve this problem, the method searches the above and below of the motion window (the search region in Fig. 5.7) for uniform brightness regions,

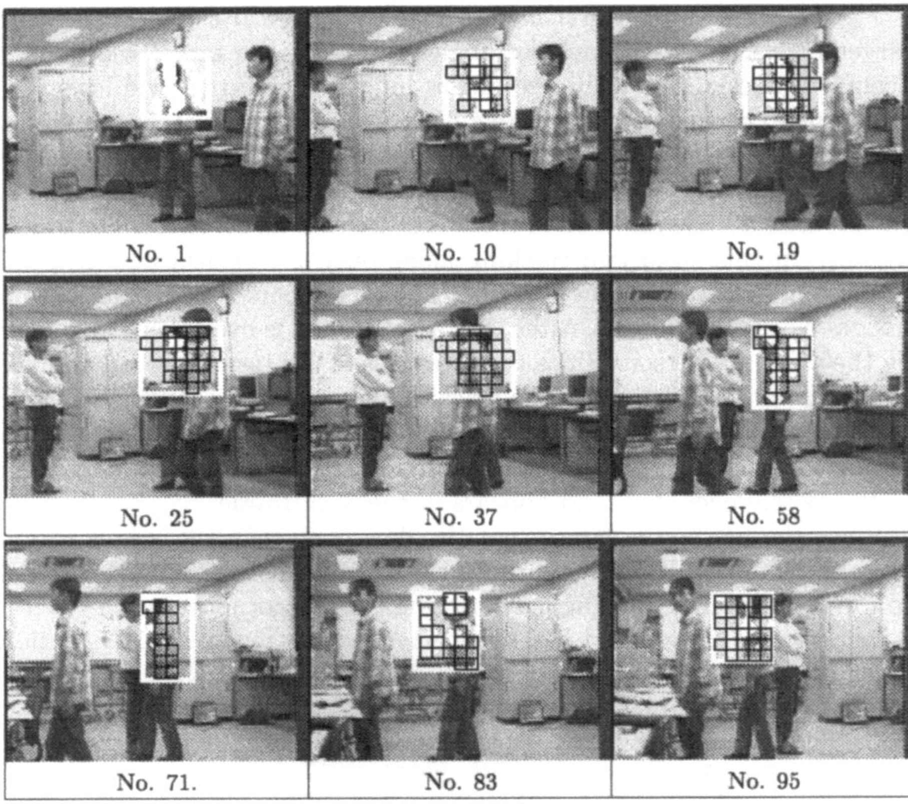


*Figure 5.5.* Experimental result.

which are defined as regions where optical flow cannot be obtained. The extracted regions are circumscribed by a rectangle called a *brightness window* (Fig. 5.7).

In tracking, the motion and the brightness windows are updated at each frame and the mean optical flow vector is calculated in the motion window. Then the motion and the brightness windows are predicted using the mean optical flow vector. The target motion region is extracted around the predicted position and the new motion window is set to circumscribe it. The target brightness regions are extracted around the predicted position and the new brightness window is set to circumscribe all of them.

Fig. 5.6(c) shows that the method can successfully track the target person among similarly moving persons.

(a) Optical flow. (b) Erroneous tracking result by flow data alone. (c) Tracking result using two cues.

*Figure 5.6.* Tracking using optical flow and uniform brightness regions.

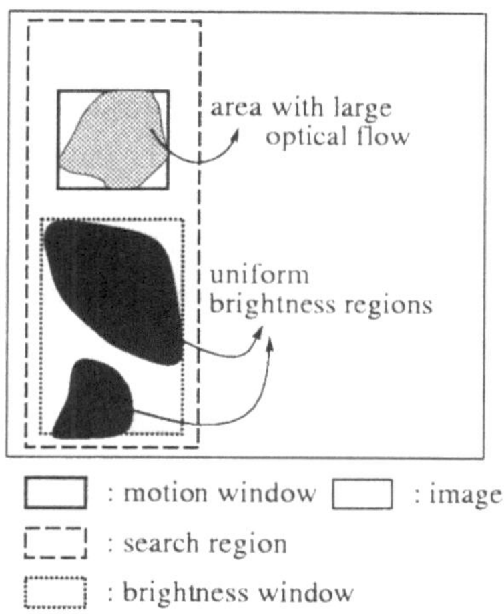


*Figure 5.7.* Windows used in tracking.

# 4. Discussion

All the methods except the one in Section 3.1 have been implemented on the DSP multiprocessor systems [9] and can work in real time.

Our next target is to integrate all the methods described in this paper. The method described in Section 2 accumulates bits of evidence over time, while the three methods in Section 3 integrate multiple cues in space domain. We need a new unified framework for integrating evidence in time and space domains. However, even in integrating cues in space domain, we have not adopted a unified approach.

In integrating flow and edges, we first examine the flow, then searching for brightness edges around the motion boundary. This is because there are many unrelated edges in images of complex scenes. Edges of target objects cannot be detected without the help of motion information. The edge information once detected, however, can help tracking the objects in occlusion cases.

In integrating flow and depth, both data are obtained at each pixel independently. Then target objects are extracted based on the probability of each pixel belonging to the target. Motion and depth are different properties in the real world and can be independently obtained from images. Thus they are treated equally. Objects with similar motion can be recognized by depth information, whereas objects existing at similar distances can be separated by motion information. However, image processing for both motion (optical flow) and depth (stereo matching) can work well for textured parts in images. Thus this combination cannot help the cases where target objects do not have texture.

In integrating flow and uniform brightness regions, moving regions and uniform brightness regions are obtained independently. (In the actual implementation, the uniform brightness regions are extracted from the regions where optical flow cannot be reliably computed around the moving regions.) Both are treated equally as in the case of flow and depth. However, the integration is done at the region (window) level rather than the pixel level. Optical flow can be computed well in textured regions, while uniform brightness regions can be obtained in textureless parts. Thus this combination can extend the class of applicable objects. However, it cannot improve anything if target objects do not have uniform brightness regions.

To sum up, we need to consider the followings to develop a unified integration framework.

1 Integration level: The integration needs multiple level assistance mechanisms.

   (a) Mutual assistance at the feature extraction level: Some features can be robustly extracted with the aid of other feature extraction results.

   (b) Mutual assistance at the final detection and tracking level: Appropriate integration level (pixel, region, or more global levels) may differ depending on the combination of features or cues.

2 Used cues: The integration mechanisms need to consider how to use cues from the following points.

   (a) Object dependence: Some objects may have the cue but others not.

   (b) Image feature dependence: Although different cues, they might be calculated from the same or similar image features.

## 5. Conclusion

This paper has presented visual surveillance methods that can detect and track humans and other moving objects in complex scenes. In these scenes, we cannot expect to extract a single type of feature reliably all the time. Thus we have proposed to accumulate bits of fragmented information to realize robust detection methods. We have adopted two approaches to this problem. One is accumulating bits of evidence over time. The second is to integrate evidence from multiple cues. We have proposed to combine motion (optical flow) and edge data, motion and depth (stereo) data, and motion and uniform brightness region data. Real time implementation of these methods has shown the effectiveness of our approaches. Integrating all these methods is left for our future work.

## References

[1] M. Etoh and Y. Shirai, "Segmentation and 2D motion estimate by region fragments," Proc. IEEE 4th International Conference on Computer Vision, pp. 192-199, 1993.

[2] A. Iketani, Y. Kuno, N. Shimada, and Y. Shirai, "Real-time surveillance system detecting persons in complex scenes," Proc. IAPR International Conference on Image Analysis and Processing, pp. 1112-1115, 1999.

[3] Y. Mae, Y. Shirai, J. Miura, and Y. Kuno, "Object tracking in cluttered background based on optical flow and edges," Proc. 13th International Conference on Pattern Recognition, Vol. 1, pp. 196-200, 1996.

[4] Y. Mae and Y. Shirai, "Tracking moving object in 3-D space based on optical flow and edges," Proc. 14th International Conference on Pattern Recognition, pp. 1439-1441, 1998.

[5] A. Nagai, Y. Kuno, and Y. Shirai, "Surveillance system based on spatio-temporal information," IEEE 1996 International Conference on Image Processing, Vol. 2, pp. 593-596, 1996.

[6] H. Nakai, "Non-parameterized Bayes decision method for moving object detection," Proc. 2nd Asian Conference on Computer Vision, pp. 447-451, 1995.

[7] R. Okada, Y. Shirai, and J. Miura, "Object tracking based on optical flow and disparity", Proc. IEEE/SICE/RSJ International Con-

ference on Multisensor Fusion and Integration for Intelligent Systems, pp. 565-571, 1996.

[8] R. Okada, Y. Shirai, and J. Miura, "Tracking a person by integrating optical flow and depth," Proc. IEEE International Conference on Automatic Face and Gesture Recognition, pp. 336-341, 2000.

[9] M. Shiohara, H. Egawa, S. Sasaki, M. Nagel, P. Sobey, and M.V. Srinivasan, "Real-time optical flow processor ISHTAR," in Proc. Asian Conference on Computer Vision, pp. 790-793, 1993.

[10] M.V. Srinivasan, "Generalized gradient schemes for the measurement of two-dimensional image motion," Biological Cybernetics, Vol. 63, pp. 421-431, 1990.

[11] W.B. Thompson, "Combining motion and contrast for segmentation," IEEE Trans. PAMI, Vol. 2, No. 6, pp. 543-549, 1980.

[12] T. Yamane, Y. Shirai, and J. Miura, "Person tracking by integrating optical flow and uniform brightness regions," Proc. IEEE International Conference on Robotics and Automation, pp. 3267-3272, 1998.

Chapter 6

# BAYESIAN MODALITY FUSION FOR TRACKING MULTIPLE PEOPLE WITH A MULTI-CAMERA SYSTEM

Ting-Hsun Chang, Shaogang Gong
*Department of Computer Science*
*Queen Mary, University of London, London E1 4NS, UK*
{ cth, sgg }@dcs.qmw.ac.uk

**Abstract** *We present a multi-camera system based on Bayesian modality fusion to track multiple people in an indoor environment. A Bayesian network is used to combine multiple modalities for matching subjects between multiple camera views. Unlike other occlusion reasoning methods, we use multiple cameras to obtain continuous visual information of people in either or both cameras so that they can be tracked through interactions. Results demonstrate that the system can maintain people's identities by using multiple cameras cooperatively.*

**Keywords:** Data Fusion, Tracking, Occlusions Reasoning.

## 1. Introduction

Human tracking in an indoor environment is of interest in a number of applications such as visual surveillance and human-computer interface. Occlusion is a significant problem which can not be ignored because identities of people can become ambiguous. An example of an occlusion scenario is shown in Figure 6.1. This paper attempts to solve the occlusion problem by using multiple uncalibrated static and widely-separated cameras.

Different solutions to the occlusion problem in human tracking have been proposed. Rosales and Sclaroff [1] used Kalman filters and Khan and Shah [2] used colour. However, neither of these methods work for all cases. Recently, Haritaoglu et al. [3] implemented a real-time

*Figure 6.1.* The task is to track people with identities even under occlusion using two widely separated cameras.

human-tracking system $W^4$ and suggested using a multi-camera system to analyse the occlusions. Using multiple cameras to solve the occlusion problem, the system needs to pass the subjects identities across cameras once the identities are lost in a certain view by matching subjects across camera views. To this end, Collins et al. [4] use the trajectory and normalised colour histogram of an object. Chang et al. [5] estimate the subjects' apparent height and apparent colour across cameras. This matching can also be done by geometric method, such as epipolar geometry [6], homography [9] and landmarks [5]. However, these feature-based matching methods can be unreliable due to the ambiguous positions of the extracted features resulting in inconsistencies. A framework is required to combine multiple visual modalities, or cues, to make the matching more reliable. Note that the method we present in this paper also can be used to track and follow multiple people as they move through the Field Of Views (FOVs) of different cameras. In order to track individuals continuously, the system assigns an identity to a new detected subject and keeps tracking it with this identity. If this subject has already appeared in the other cameras or loses the identity during Single Camera Tracking (SCT), the system then passes identity and assigns it to this subject by matching subjects across camera views, called Multiple Camera Cooperative Tracking (MCCT).

To track people with a single camera, the system first performs frame differencing. After thresholding and noise cleaning, connected component analysis is applied to the foreground pixels to find the moving blobs. To reliably maintain the identities of the detected people, the system integrates multiple modalities based on motion continuity and the apparent colour. A second-order Kalman filter is attached to each subject to estimate the motion of the blob centroid. Furthermore, the colour of the subject image is modelled as Gaussian mixture models in hue and saturation space. The conditional probability of a measured pixels,

$\boldsymbol{\lambda}$, being the subject, $\mathcal{S}$, modelled as a mixture with $m$ components is given by: $p(\boldsymbol{\lambda}|\mathcal{S}) = \sum_{i=1}^{m} p(\boldsymbol{\lambda}|i)P(i)$ where $P(i)$ is the prior probability of the component, the $i^{th}$ component is a Gaussian with mean $\boldsymbol{\mu}$ and covariance matrix $\boldsymbol{\Sigma}$, and:

$$p(\boldsymbol{\lambda}|i) = \frac{1}{2\pi|\boldsymbol{\Sigma}|^{1/2}} exp\{-\frac{1}{2}(\boldsymbol{\lambda} - \boldsymbol{\mu})^T \boldsymbol{\Sigma}^{-1}(\boldsymbol{\lambda} - \boldsymbol{\mu})\} \tag{6.1}$$

The method to define the closest match based on Kalman filter is by searching the minimum of $\mathcal{M}_m = \nu^T \ \mathbf{S}^{-1} \ \nu$ where the $\nu$ is the innovation and $\mathbf{S}$ represents the covariance of the innovation. The closest match based on colour can be found as the blob with the minimum $\mathcal{M}_c = \sum_{j=1}^{n} \sum_{i=1}^{m} [(\boldsymbol{\lambda} - \boldsymbol{\mu})^T \boldsymbol{\Sigma}^{-1}(\boldsymbol{\lambda} - \boldsymbol{\mu})]p(i)$ where $n$ is the number of pixels sampled from the blob. These $\mathcal{M}_m$ and $\mathcal{M}_c$ are the *Mahalanobis Distance* (MD) for each individual blob used to quantify the likelihood and decide the match. When the matching becomes ambiguous, determined by applying $\mathcal{X}^2$ test to the MD of each pair of the match [7], the system performs MCCT to pass identities between cameras.

## 2. Bayesian Networks for Building Correspondence

To define matching problem for MCCT, we firstly constrain the maximum number of subjects in each image to be $m$. To match subjects between 2 camera images, $I_i$ and $I_j$, we evaluate the matching globally, i.e. consider the matching for all subjects simultaneously. In each combination of assignment, every subject in $I_i$ is assigned a corresponding subject in $I_j$. After applying the uniqueness constraint, there could be $m!$ possible assignment combinations, $A_\alpha = \{A_1, \cdots, A_{m!}\}$. Given the visual evidence $\mathbf{e}$ from cameras, our goal is to find a most appropriate assignment combination which maximises the posterior $p(A_\alpha|\mathbf{e})$.

We employ Bayesian networks to probabilistically infer the correspondence of people in two images. The networks can capture the dependencies between the correspondence of the subjects between two images and multiple visual evidences in two images. A *Bayesian Belief Network* (BBN), also known as a *Bayesian Network*, is a graphical representation of a joint probability distribution over a set of random variables [8]. The defined matching problem can be probabilistically inferred by obtaining a probability distribution over the assignment combinations. In the discrete-variable BBN (Figure 6.2(a)), there are four different types of nodes: (1) Correspondence node which represents a multi-values variable and each value corresponds to a possible assignment combination $\{A_1, \cdots, A_{m!}\}$. (2) Comparison node. There are $m$ comparison nodes and each node compares one subject in $I_i$ against all $m$ subjects in $I_j$. (3)

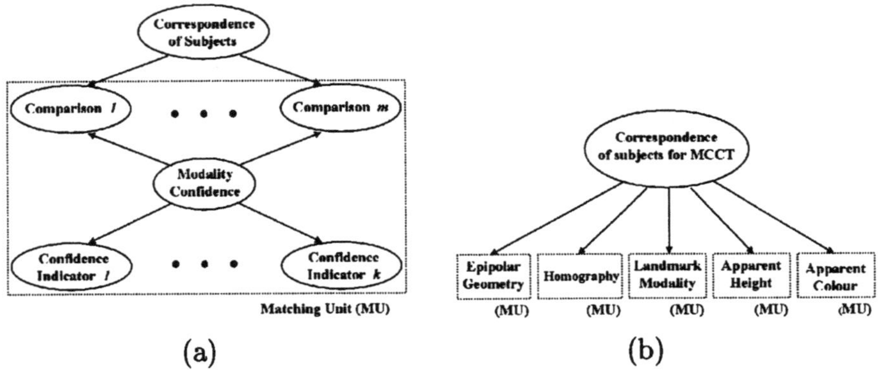


*Figure 6.2.* (a)The Bayesian Network for inferring the correspondence of subjects between two camera images based on a single modality. (b)The general representation of Bayesian Networks to fuse multiple modalities for matching subjects across camera views.

Modality confidence node which represents the confidence of the modality and constrains the influence of this modality on the correspondence. (4) Indicator node which indicates the modality confidence. In order to generalise the BBN for multiple modalities, we define a Matching Unit (MU) as the union of all comparison, modality confidence and confidence indicator nodes.

## 3. Multi-Camera Cooperative Tracking

In the following, we first introduce 3 geometry-based and 2 recognition-based modalities for MCCT and then describe how the system integrates multiple modalities.

***Epipolar geometry***: To apply epipolar geometry for matching, the topmost point of segmented blob in the first camera image $I_1$ is used to compute its associating epipolar line in the second camera image $I_2$. The distance between the epipolar line and topmost point of the subject in image $I_2$ is used as a match score. We assume that such a distance is a Gaussian variable with zero mean and a probability density function defined as:

$$f(x) = \frac{1}{\sqrt{2\pi\sigma^2}} exp\{-\frac{x^2}{2\sigma^2}\} \tag{6.2}$$

The likelihood of the subject in $I_2$ being the corresponding subject in $I_1$ is determined by the value of the density function for the measured distance. We define $\mathcal{M}_e = \frac{x^2}{\sigma^2}$ and use it to compare the candidate matches. The modality confidence indicator is defined by the mean distance between affine epipolar lines. Moreover, we also use the segmentation status of the topmost point to indicate the confidence.

***Homography***: We use the topmost point of a person's head and assume this point lies on the same virtual plane when he/she is moving. Once a person is matched in two views, the topmost point pairs are used to estimate the homography for this particular person. To match the subjects between two camera images $I_1$ and $I_2$, we first transform the feature point $(x, y)$ of a blob in $I_1$ to a point $(x', y')$ in $I_2$. This projected point is then used to compute $\mathbf{x}' = (x', y', \dot{x}', \dot{y}')$, called the *kinematic vector*, where $(\dot{x}', \dot{y}')$ is the spatial displacement between consecutive frames of $I_2$. The matching is based on the comparison of this kinematic vector. We again apply a Gaussian variable with zero mean to model the difference between the projected kinematic vector and the observed kinematic vector, $\mathbf{x}$, of its corresponding subject in $I_2$. We define $\mathcal{M}_h = [(\mathbf{x}-\mathbf{x}')^T\Sigma^{-1}(\mathbf{x}-\mathbf{x}')]$ and use it to compare the candidate matches. We also define the confidence indicator for homography modality in terms of the segmentation status of the topmost point again, and mean distance between subjects' topmost points in $I_2$.

***Landmark modality***: From the scene knowledge based on the vertical-line landmarks, the position of a subject with respect to the landmarks in an image , called Vertical Area (VA), can be used to constrain the positions of its corresponding subject in the other image. The modality confidence indicator is defined as the segmentation status of the topmost point used to determine the VA position of a subject. Moreover, the mean distance between the topmost point and the closest line landmarks is also used. This is because the VA position might not be reliable when a subject's topmost point is too close to the landmark due to wrong segmentation. However, geometric modalities alone do not provide enough constraints to match subject across cameras. In the next section, recognition-based modalities are described.

The recognition-based modalities are based on the the similarity. Since the appearance is view-variant, the system should estimate the appearance across camera views and use this "corrected" value for matching. We employ Support Vector Regression (SVR) for estimation. More detail can be found in [5].

***Apparent height***: The apparent height of a subject is defined as the longest distance in the vertical direction of a blob. This height is determined by a person's height and viewing geometry. Since our system is stationary, the correlation between the apparent height of a person in two views is fixed and can be used as a subject feature for matching. Again, we model this difference $h - h'$ between the estimated height and the observed height as a Gaussian variable with zero mean. We define $\mathcal{M}_{ht} = \frac{(h-h')^2}{\sigma^2}$ and use it to compare the candidate matches. We defined the confidence indicator in terms of the segmentation status of the feature points used for computing apparent height and the mean difference of subjects' heights in $I_2$.

***Apparent colour***: As mentioned in SCT, apparent colour of a subject clothes image is modelled as Gaussian mixture models. Similar to the apparent height, for a subject in $I_1$ the apparent colour of its corresponding subject in $I_2$ can be estimated from the learnt mapping. The estimation is done for each single Gaussian model of the apparent colour. The mapping between colours in two views is learnt for different colours and this mapping can generalise to an "unseen" colour. Then, the match likelihood, similar to SCT, can be obtained based on the estimated colour models for subjects in $I_2$. The MD and confidence indicator for this modality is defined as the same as those used for colour modality in SCT.

To fuse multiple modalities for matching with the BBN (Figure 6.2), our system use the accumulated evidence $\sum_{i=0}^{q} \alpha\mathcal{M}(k-i)$ to compare subjects where $k$ is the frame index and $\alpha$ is the weight to set more recent evidence with higher weights. To obtain consistency, the network is coupled indirectly over time through the specification of prior probability for correspondence node. As a consequence, the correspondence at each time instant is affected by the previous matching history. However, the matching might be incorrect when the visual information is not reliable. We apply $\mathcal{X}^2$ test, similar to SCT, to each pair of the assignment combination obtained from previous frame. If any pair fails the test, the system does not use the previous matching results in the correspondence node. Moreover, the number of frames of accumulated evidence used in comparison node is set as $q = 0$ for all modalities to prevent using wrong evidence. Once the system continues to infer the same assignment combination, it stops performing MCCT and assigns the identities to the matched subjects.

## 4. Results

A tracking example is used to demonstrate how the system match subjects across cameras in order to maintain the identity and solve the occlusion problem. To illustrate the modality fusion approach, we highlight a section of the sequence beginning from when person 1 is in both views and person 2 just enters the room already imaged by the right camera, $I_2$, and first appearing in the left $I_1$ (Figure 6.3.a). The system performs MCCT to obtain the identity for this new detected subject

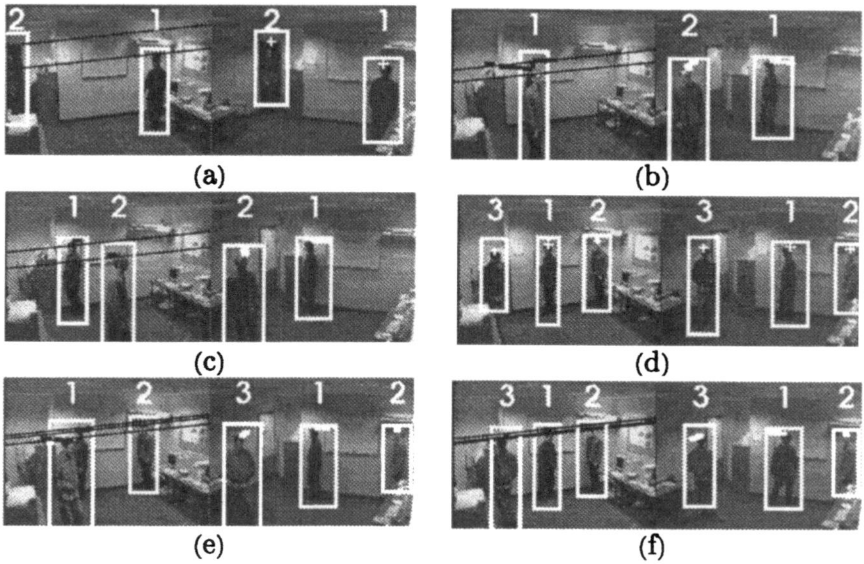


*Figure 6.3.* The system can track people with identities using two cameras cooperatively even occlusion is present.

from the right camera. From the topmost points of two subjects in $I_2$, the epipolar line (black) is used for searching subjects in $I_1$. The topmost point of person 1 is also transformed to $I_1$ (black dot on top of person 1) based on the on-line learnt homography. It also can be seen that the topmost point of person 2 $I_2$ in was incorrectly segmented, but the BBN can still effectively collect evidence and make a right match.

After entering, person 2 continues to walk towards the room centre and these two subjects meet in $I_1$. The system interprets that $I_1$ is ambiguous and relies on $I_2$ to disambiguate. The black dots in $I_1$ are the transformed points from the topmost points (white dots) of two subjects in $I_2$ based on its own stored estimated homography. From modality fusion, the merged blob in $I_1$ is matched to and interpreted as person 1 due to the top point of this blob corresponding to person 1. When the merged blob splits into two blobs, the system detects that the number of blobs changes and performs MCCT as shown in Figure 6.3.c and passes the identifier from $I_2$ to $I_1$. Then person 3 enters the room, occlusion happens in $I_1$ as shown in Figure 6.3.e, but two people change direction during occlusion. Note that tracking with a single camera can correctly resolve the ambiguity in the event of Figure 6.3.b, but can not maintain correct identities for the event Figure 6.3.d-f. Figure 6.4(a) illustrates the tracking failure with a single camera based on motion continuity for the latter event. The Kalman filters can follow people before occlusion, but fail to estimate correct positions of people after occlusion.

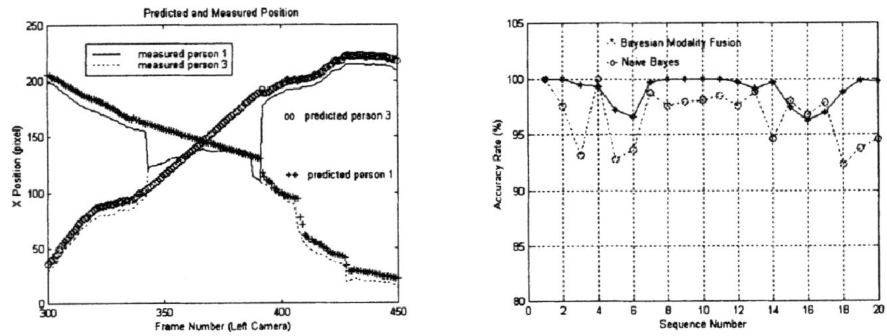


(a) (b)

*Figure 6.4.* (a) The measured (ground truth) and predicted (Kalman filter) blob centroids of person 1 and 3 during occlusion in the left view of the tracking example (Figure 6.3(d-f)). The Kalman filter fails due to change direction during occlusion. (b)The accuracy rate of matching subjects between two camera views based on two methods.

To highlight the strength of Bayesian modality fusion we compare it with a popular fusion method assuming all modalities are independent, often called the *naive Bayes.* Figure 6.4(b) illustrates the results of matching two people between two camera views. The accuracy rate of each sequence is the overall matching accuracy of all frames. The average accuracy of all 20 sequences is about 99.1% with deviation 1.2% for the Bayesian modality fusion and 96.5 % with deviation 2.4% for the naive Bayes method. We found that using BBN is more accurate in combining multiple visual evidences for matching subjects across cameras.

## 5. Discussion

We have demonstrated that our multi-camera tracking system can handle occlusion and maintain identities of multiple people. Handling occlusion using appearance and motion is in general hard because the image pattern of the subject appearance can experience severe variation during occlusion and the motion model might be violated during the estimating stage. From our experiments, we also found that wrong segmentation, such as shadow, causes the system to fail. This problem can be alleviated by using detection results of multiple cameras. This is the advantage of using a multi-camera system: more chances of obtaining unambiguous information.

## References

[1] R. Rosales and S. Sclaroff. (1998). *Improved Tracking of Multiple Humans with Trajectory Prediction and Occlusion Modeling.* IEEE International Conference on Computer Vision and Pattern Recognition.

[2] S. Khan and M. Shah. (2000). *Tracking People in Presence of Occlusion.* Asian Conference on Computer Vision.

[3] I. Haritaoglu, D. Harwood and L. Davis. (1998). *W4: Who? When? Where? What? A Real Time System for Detecting and Tracking People.* IEEE International Conference on Automatic Face and Gesture Recognition.

[4] R. Collins, A. Lipton, T. Kanadeand, H. Fujiyoshi, D. Duggins and Y. Tsin. (2000). *A System for Video Surveillance and Monitoring: VSAM Final Report.* Carnegie Mellon University. CMU-RI-TR-00-12.

[5] T. H. Chang, S. Gong and E. J. Ong. (2000). *Tracking Multiple People Under Occlusion Using Multiple Cameras.* British Machine Vision Conference.

[6] Q. Cai and J. K. Aggarwal. (1998). *Automatic Tracking of Human Motion in Indoor Scenes Across Multiple Synchronized Video Streams.* IEEE International Conference on Computer Vision.

[7] I. Cox. (1993). *A review of statistical data association techniques for motion correspondence.* International Journal of Computer Vision, 10(1):53-66.

[8] F. V. Jensen. (1996). *An Introduction to Bayesian Networks.* UCL Press.

[9] L. Lee, R. Romano and G. Stein. (2000). *Monitoring activities from multiple video streams: Establishing a common coordinate frame.* IEEE Transactions on Pattern Analysis and Machine Intelligence. Special Issue on Video Surveillance and Monitoring. 758-767.

Chapter 7

# TRACKING GROUPS OF PEOPLE FOR VIDEO SURVEILLANCE

Frederic Cupillard, Francois Bremond and Monique Thonnat
*INRIA Sophia Antipolis, ORION group*
*2004, route des Lucioles, BP93 - 06902 Sophia Antipolis Cedex, France*
{Frederic.Cupillard,Francois.Bremond,Monique Thonnat}@sophia.inria.fr

**Abstract** *We propose in this paper a method for tracking groups of people in a metro scene to recognise abnormal behaviours such as violence or vandalism. After presenting the overall system, we describe the tracking algorithm for groups of people. Finally we present results illustrating the algorithm.*

**Keywords:** Long Term Tracking, Human Group Model, Knowledge Based System

## 1. INTRODUCTION

In this paper, we present a method for tracking groups of people in a metro scene to recognise abnormal behaviours such as violence or vandalism [5]. Given a video sequence, our proposed algorithm is able to track real groups of people present in the scene. This algorithm is composed of three steps. First, from the current image of metro scene, a motion detector detects mobile objects in the scene, stores each of them in a *moving region* structure which is classified into different types, such as PERSON, GROUP or NOISE. Second, a tracking module detects and follows the real groups present in the scene by computing the trajectories of *moving regions* that can correspond to persons inside a real group. Third an interpretation module recognises the behaviour of the tracked groups. In this article, we focus on the second step, tracking groups of people. Contrary to traditional trackers [1, 2, 3, 6] that track each person individually, our algorithm tracks globally all the persons belonging to the same group and tracks these persons with a time delay to keep coherent the structure of the group during the entire sequence. We illustrate this algorithm with video sequences taken for the European project ADVISOR (http://www-sop.inria.fr/orion/ADVISOR/default.html).

# 2. MOTION DETECTOR

The goal of this step is to detect mobile objects in the scene and classify them into *moving regions* with a label corresponding to their type, such as PERSON. This task can be divided into three sub-tasks: detection of *moving regions*, extraction of features, classification of *moving regions.*

First for **detection of *moving regions***, we compute a difference image between a current image and a reference image [4]. Because the camera is fixed, the reference image is a still image representing the scene without mobile objects (also called background image). Second a thresholding of this difference image is realized, then a filtering and connected component analysis create all the *moving regions* (represented by their bounding boxes) that correspond to the mobile objects. The reference image is updated during the process to take into account illumination changes.

The **extraction of features** step consists in the computation for each *moving regions* of 8 parameters: centre of gravity, position, height and width all four defined both in 2D (in the image) and in 3D (in the scene).

The **classification of *moving regions*** step consists in labelling the *moving regions* into 8 semantic classes of objects: 4 classes for people (PERSON, OCCLUDED PERSON, GROUP, and CROWD) and 4 classes for other objects (METRO TRAIN, SCENE OBJECT, NOISE and UNKNOWN). An occluded person is for example a person partially occluded behind a pillar. A scene object is for instance a seat or a trash. The *moving regions* can be merged to improve the matching with a class. For instance, two *moving regions* corresponding to the head and the body of a person can be merged to form a new *moving region* belonging to the class PERSON.

# 3. GROUP TRACKING

## 3.1 Tracking Approach

A real group of people is defined as a set of persons who are close to each other. We detect real groups of people thanks to a structure that we call a *group.* This *group* structure is a set of *moving regions* characterised by four particularities:

- Size coherence: each *moving region* of a *group* has the dimensions of a person or bigger if several persons partially overlapp each other.

- Spatial coherence: all *moving regions* inside a *group* are close to each other.

- Temporal coherence: the speed of the *moving regions* inside a *group* cannot exceed the speed of a person.

- Structure coherence: The number and the size of *moving regions* inside a *group* should be stable. If a person goes away from the real group and come back several times, we would like that the *moving region* corresponding to this person stay inside the *group* structure in order to compute a stable behaviour analysis of the real group.

There are three main problems in tracking groups:

- Appearance/disappearance of real groups: a real group could appear and disappear anywhere in the observed scene.

- Dynamic of real groups: some persons or other groups could come and merge into an already existing group then leave it.

- Motion detector failure: persons are not always well detected by a *moving region* (head and body could be detected in two different *moving regions*) and well labelled (a person could be labelled as NOISE). A class GROUP for *moving regions* exists but characterises a real group only if persons partially overlapp each other. For example, two persons walking together will be characterised by two *moving regions* labelled PERSON and not GROUP.

To handle these problems, the tracking algorithm is divided in three tasks. The first task consists in tracking *moving regions* from frame to frame and in obtaining a graph (shown on figure 7.1) representing all possible trajectories of all *moving regions*. In this graph, each connected part (sub-graph) can correspond to a group of persons crossing each other. The second task consists in computing inside the sub-graph all possible *paths* (trajectories of a *moving region*) than can correspond to a person inside the real group. In the third step, we compute the *group* structure that gathers all these paths.

## 3.2 Frame to Frame Tracker

The goal of the frame to frame tracker is to link from frame to frame all *moving regions* computed by the motion detector. The output of the frame to frame tracker is a graph $\Gamma(t) = (X, U)$, where $X$ is a set of nodes containing the detected *moving regions* and $U$ is a set of links between two *moving regions* at successive times. This graph provides all the possible trajectories that a *moving region* may have. The link between a new *moving region* $M_{new}$ and an old one $M_{old}$ is computed depending of three criteria: their 2D (in the image) and 3D (in the real world) distance and the similitude between their bounding boxe size. If the criteria are met, we note: $link(M_{old}; M_{new}) = 1$ else $link(M_{old}; M_{new}) = 0$. As there could be many errors of detection, we use non-strict criteria to allow a large number of links. Moreover, an old *moving region* could be linked to several new ones (split case) or several old *moving regions* could be linked to the same new one (merge case). We define the set of old *moving regions* $\mathcal{O}$ and new ones $\mathcal{N}$ as following:

$$\mathcal{O} = \{M_i(t_c - q) \mid 1 \leq i \leq nb\_old, 1 \leq q \leq 3\}$$

$\mathcal{O}$ contains old *moving regions*, all those detected at $t_c - 1$ and also those that did not get linked (disappear, miss detection) at the previous q frames.

$$\mathcal{N} = \{M_j(t_c) \mid 1 \leq j \leq nb\_new\}$$

$\mathcal{N}$ contains new *moving regions* detected in current frame processed at time $t_c$.

We define a function $F$ (respectively $G$) to compute the links between $\mathcal{O}$ and $\mathcal{N}$ (respectively between $\mathcal{N}$ and $\mathcal{O}$).

$$F : \mathcal{O} \longrightarrow \mathcal{P}(\mathcal{N})$$

$$F(M_i(t_c - q)) = \{M_j(t_c),_{1 \leq j \leq nb\_new} \mid link(M_i(t_c - q); M_j(t_c)) = 1\}$$

$\mathcal{P}(\mathcal{N})$ is defined as the set containing all the possible sub-sets of $\mathcal{N}$

$$G : \mathcal{N} \longrightarrow \mathcal{P}(\mathcal{O})$$

$$G(M_j(t_c)) = \{M_i(t_c - q),_{\substack{1 \leq i \leq nb\_old \\ 1 \leq q \leq 3}} \mid link(M_i(t_c - q); M_j(t_c)) = 1\}$$

$\mathcal{P}(\mathcal{O})$ is defined as the set containing all the possible sub-sets of $\mathcal{O}$

## 3.3 Computing Paths

The purpose of this task is to select trajectories of *moving regions* (called *paths*) that can correspond to real persons inside a group during a temporal window. A *path* $P_k^{t_c}$ represents a temporal link in the graph $\Gamma(t)$ between *moving regions* during a temporal interval $[t_c - T; t_c]$, where $t_c$ is the time of the last (current) processed image and $T$ is the size of the temporal window used to analyse *groups*. $P_k^{t_c}$ is composed of a temporal sequence of *moving regions* $M_k(t), t \in [t_c - T; t_c]$.

A size coefficient $Size(P_k^{t_c})$ qualifies a path $P_k^{t_c}$:

$$Size(P_k^{t_c}) = \frac{\sum_{t \in [t_c - T; t_c]} Size_{3D}(M_k(t))}{T}$$

With $Size_{3D} = Height_{3D}.Width_{3D}$ of the bounding box of the *moving region.*

This coefficient characterises a *path*: if its size coefficient is bigger than the size of a person, then the *path* is likely to correspond to a real person inside a group. So this coefficient allows us to rank *paths.* This allows to distinguish the *paths* containing big *moving regions* (labelled as GROUP, PERSON) from the *paths* containing smaller ones (labelled as NOISE) and so less interesting.

### 3.3.1 Creation of Paths.

The most common case to create a *path* is when a *moving region* has been classified as a GROUP by the motion detector. In this case, this *moving region* is likely to represent several persons inside a real group. In order not to miss any person inside a real group, we also decide to create a *path* each time a trajectory (in the graph $\Gamma(t)$) is crossing an other one. The common *moving region* of both trajectories can represent several persons inside a real group if the *moving region* is at least as big as a person. More precisely, for each new processed image (at time $t_c$), *paths* are created if one of these three predicates is true:

- ***Moving region* labelled as a GROUP:**
  if a new *moving region* is classified as a GROUP a new *path* is created. More precisely:
  $\exists M_j(t_c)$, $Class(M_j(t_c)) = GROUP$,
  $\nexists P_{old}^{t_c - 1} \mid M_j(t_c) \in P_{old}^{t_c - 1}$,
  $\Rightarrow$ 1 *path* $P_{new}^{t_c}$is created:$P_{new}^{t_c} = (M_j(t_c))$
- **Splitted *moving region*:**
  if an old *moving region* is linked to $E$ new ones,
  $E$ new *paths* are created. More precisely:

$\exists M_i(t_c - q) \mid Size_{3D}(M_i(t_c - q)) \geq Size\_person,$
$E = card\, F(M_i(t_c - q)) \geq 2,$
$\nexists\, P_{old}^{t_c - 1} \mid M_i(t_c - q) \in P_{old}^{t_c - 1}$
$\Rightarrow E\; paths\; P_{new_l}^{t_c}, {}_{1 \leq l \leq E}$ are created:
$\forall\, l, {}_{1 \leq l \leq E}\, , P_{new_l}^{t_c} = (M_i(t_c - q), M_j(t_c)),$
$M_j(t_c) \in F(M_i(t_c - q))$

- **Merged *moving region*:**
  if a new *moving region* is linked to several old ones, a new *path* is created. More precisely:
  $\exists M_j(t_c) \mid Size_{3D}(M_j(t_c)) \geq Size\_person,$
  H$= card\, G(M_j(t_c)) \geq 2,$
  $\nexists\, P_{old}^{t_c - 1} \mid M_j(t_c) \in P_{old}^{t_c - 1}$
  $\Rightarrow 1\; path\; P_{new}^{t_c}$ is created:$P_{new}^{t_c} = (M_j(t_c))$

**3.3.2 Update of Paths.** For each new image processed at time $t_c$, each old *path* $P_k^{t_c - 1}$ is updated in $P_k^{t_c}$. If $M_{last}$ is the last *moving region* added in $P_k^{t_c - 1}$ and is linked to the *moving region* $M_{new}$ detected in the new frame, $P_k^{t_c - 1}$ is duplicated in $P_k^{t_c}$ and extended with $M_{new}$. If $M_{last}$ is not linked to any new *moving region*, the *path* $P_k^{t_c - 1}$ is only duplicated. As a result, the size coefficient of a such *path* decreases.

A special case appears when several old *paths* that have the same *moving region* $M_{last}$ could be extended with several new ones ($M_{last}$ is linked to several $M_{new_j}, {}_{j \geq 2}$). To limit the number of possible extensions that could increase significantly because of the large number of links given by the frame to frame tracker, we only keep extensions of the most interesting *paths* (*path* with the biggest size coefficient). The better old *path* is extended with all the new *moving regions* $M_{new_j}, {}_{j \geq 2}$ whereas the other old ones are only extended with the new *moving region* that has the biggest $Size_{3D}$.

**3.3.3 Removing Paths.** A *path* $P_i$ is removed if one of these two cases is verified:

- $P_i$ is totally overlapping another *path* $P_j$ and the size of $P_j$ is bigger than the size of $P_i$.

- $P_i$ does not belong anymore to a *group* (see below)

## 3.4 Group Structure

The *group* structure is a set of *paths* belonging to a connected subgraph of $\Gamma(t)$ that corresponds to a real group of people. The problem is that the number of *paths* (trajectories that can correspond to persons inside groups) can be important and that some *paths* can correspond to detection errors. So the goal of the *groups* computing step is to select the *paths* of the connected sub-graph that best match with the trajectories of real persons. A *group* $G_m$ is represented by its $N$ *paths* $P_{m,k}, {}_{1 \leq k \leq N}$.

If $t_c$ is the time of the last (current) processed image, *groups* are computed at time $t_c - T$ with a delay $T$. This delay is used to obtain a stable *group* structure. The **delay** $T$ constitutes a temporal window $[t_c - T; t_c]$. In this window, we first compute all possible future trajectories of *moving regions* detected at time $t_c - T$. This analysis allows us to select at time $t_c - T$ the *moving regions* that best match a real group that would be observed from time $t_c - T$ to $t_c$.

We characterise a *group* $G_m$ by a quality coefficient $Q_{G_m}$:

$$Q_{G_m}(t) = \alpha Q_{G_m}(t-1) + (1-\alpha) Q'_{G_m}(t) \quad t > 0$$

$$Q_{G_m}(0) = Q'_{G_m}(0)$$

$$Q'_{G_m}(t) = \sum_{k=1}^{N_{paths}} Prox(P_{m,best}; P_{m,k}).Size(P_k^{t+T})$$

$Q'_{G_m}(t)$ characterises the new set of *paths* selected at time $t$ and is called instantaneous quality coefficient.

$\alpha$ is an update coefficient $\in [0;1]$ to balance the previous quality coefficient against the instantaneous quality coefficient. $\alpha$ allows to obtain a stable quality coefficient over time.

$P_{m,best}$ is the *path* with the biggest size coefficient.

$N_{paths}$ is the number of *paths* selected in the *group*,

We use the best *path* $P_{m,best}$ trajectory to give the notion of direction to the *group* $G_m$. Usually a *group* is formed by a dense part and a less compact part around. A dense part can be for example several persons overlapping each other and detected as only one *moving region* bigger than the other *moving regions* around. As we work in a temporal window, we approximate the direction of the *group* $G_m$ by the direction of its biggest *path* $P_{m,best}$. $Prox(P_{m,best}; P_{m,k})$ computes the proximity between $P_{m,best}$ and $P_{m,k}$ and is defined as following:

$$Prox(P_{m,best}; P_{m,k}) = e^{-dist(P_{m,best}; P_{m,k})}$$

This proximity is equal to 1 when both *paths* are connected (own a common *moving regions*) and decreases exponentially to zero when the distance $dist(P_{m,best}; P_{m,k})$ increases. The distance $dist(P_{m,best}; P_{m,k})$ is defined as the minimum $3D$ distance between their respective *moving regions* detected at the same time:

$$dist(P_{m,best}; P_{m,k}) = \min_{t \in [t_c - T; t_c]} (3D\_distance(M_{P_k}(t); M_{P_{best}}(t)))$$

This proximity allows us to compute $Q'_{G_m}(t)$ by giving more weight to the paths which are close to the best path. By this way, $Q_{G_m}(t)$ characterises the density of the group over the time.

**3.4.1 Creation of *Groups*.** As we mentioned above, we wait a delay $T$ between creating a *group* at time $t_c - T$ and detecting the corresponding *moving regions* at time $t_c$. The creation of a *group* $G_m$ consists in selecting the *paths* (that do not yet belong to a *group*) with the biggest size coefficient among a tree of the graph $\Gamma(t)$. In this tree, the root is a *moving region* detected at time $t_c - T$ and corresponds to the appearence of the group in the scene. To limit memory space and processing time, we just keep the $N_{max}$ *paths* with biggest size coefficient. To avoid the creation of *groups* that do not correspond to real groups of people in the scene, we check before the creation of the *group* if the quality coefficient is higher than a threshold. This threshold corresponds to the smallest size that a real group could have.

**3.4.2 Update of *Groups*.** During the update step of a *group* $G_m$ at time $t_c - T$, the main task consists in adding, extending or removing the *paths* composing the *group*. The goal is to obtain the most dense (with the biggest $Q'_{G_m}(t)$) set of *paths* contained in the same connected sub-graph (connected before the time $t_c - T$) of the graph $\Gamma(t)$. We propose to approximate this goal by first removing all the *paths* $P_{m,i}$ too far from the best *path* $P_{m,best}$ ($dist(P_{m,best}; P_{m,i}) > threshold$) and second to select the remaining *paths* with the best size coefficients (e.g. on figure 7.1, *path 1* belongs to $G_1$ at time $t_c - T - 1$ but is not selected in $G_1$ at time $t_c - T$). Finally, the new best path $P_{m,best}$ is reselected and the quality coefficient recomputed.

There is a special case when several *groups* **merge**. Two *groups* $G_r$ and $G_s$ merge together at time $t_c - T$ if:
$\exists$ path $P^{t_c}_{r,i}, \in G_r$,
$\exists$ path $P^{t_c}_{s,j} \in G_s \mid M_{P_{r,i}}(t_c - T) = M_{P_{s,j}}(t_c - T)$

In this case, we first collect the *paths* from all the merged *groups* and select the best *path* with the biggest size coefficient (among the different best *paths*). And then we continue to update the merged *groups* as in the general case.

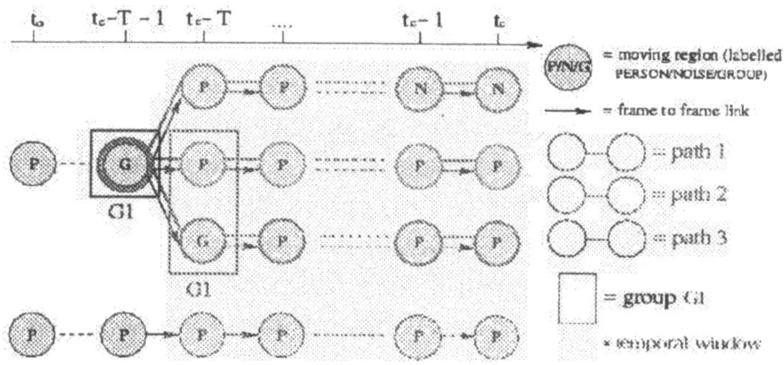


*Figure 7.1.* The 3 *paths*: *path*1, *path*2, *path*3 are updated at time $t_c$ whereas *group* $G_1$ is updated with a delay $T$

**3.4.3 Removing *groups*.** A *group* is removed if the quality coefficient (corresponding to the *group* density) is lower than a threshold (the same one used in the creation step). For instance, when persons inside a group split far away from each other, the quality coefficient of the *group* becomes too small. When a *group* is removed, all its *paths* are also removed.

# 4. RESULTS

This tracking module has been tested on several metro sequences. The longest sequence lasts 5 minutes (more than 6500 frames). In this section we are showing different image samples of the processed videos. In these images, a red box corresponds to a *moving region* classified as a PERSON, a green box corresponds to a *moving region* classified as a GROUP and a blue box corresponds to *moving regions* tracked globally as a group. The three images on figure 7.2 focus on the creation of a *group* in two situations: the first occurs when a *moving region* is labelled as a GROUP (left images) and the second situation occurs when an old *moving region* splits into two new *moving regions* (both right images). The three images on figure 7.3 illustrate the use of the temporal window to help the tracker to keep coherent the structure of a *group* during the tracking: a person first goes away from another one then comes back close to the second person. The next two images of figure 7.4 demonstrate that the algorithm is able to track a *group* during a long period of time. In this sequence, the group is tracked during more than 800 frames. Then, figure 7.5 shows an example of a *group* being removed when the persons inside the *group* split far away from each other. Currently, the main limitation of the system is an imperfect estimation of real group size due to errors in the motion detector. The system over estimates a group of persons when there are shadows or reflections strongly contrasted. The system under estimates a group of persons when the persons are occluded, overlapping each others or in case of miss detection (person has the same colour than the background). However in most cases, these imperfect estimations do not induce errors in the tracker thanks to the temporal delay: by default the tracker creates *paths* than can correspond to person trajectories in a group over a temporal window and only create the *group* (containing a selection of these *paths*) with a quality coefficient good enough.

(a) (b1) (b2)

*Figure 7.2.* Creation of a *group*: one *moving region* is detected and labelled as a GROUP in the top right on the left image (a), thus after checking the $T$ following frames, a *group* is created. On the image (b1), two persons in close contact are detected as one *moving region* (miss labelled as a PERSON), then on the image (b2), these two persons are detected as two *moving regions*. Because the old *moving region* is splitting into two new ones and after checking the $T$ following frames, a *group* is created even if no *moving region* is classified as a GROUP.

(a1) (a2) (a3)

*Figure 7.3.* The three images show two persons inside a group (a1) where a first person goes away from the other person (a2) then comes back close to the second (a3). Even when they are far away from each other, the group containing both persons is not removed (a2). This illustrates the use of the temporal window to keep coherent the structure of the *group*.

(a1) A *group* is tracked at frame 661

(a2) The *group* continues to be tracked at frame 1303

*Figure 7.4.* The algorithm of *group* tracking is able to track real groups of persons during a long period of time. In this case the group is tracked during more than 800 frames (32s).

(a1) Several *moving regions* compose a *group*

(a2) These *moving regions* are not in a *group* anymore

*Figure 7.5.* On the image (a1), several *moving regions* (close to each others) are composing a *group*. Then on the image (a2), as the persons split far away from each others, the density of the *group* becomes too low and the *group* is removed.

# 5. CONCLUSIONS

We proposed in this paper an algorithm to track real groups of people in a metro scene. The originality of the apporach consists in tracking globally all the persons belonging to the same group and in using a delay to keep coherent the structure of this group. The algorithm allows to track efficiently groups of people in several video sequences, compensating detection errors. In metro scene, the persons inside a group cannot be tracked individually because they cannot be segmented when they cross each other. However our algorithm can track correctly groups of people from beginning to end. This is essential for the interpretation module to be able to recognise the behaviours of groups. Currently, we are planning to extend the model of group by computing relevant informations in order to keep track of groups of people in special cases. For example, if two persons in a group overlap each other for a long period of time, the dimension of the group becomes lower than the size that a real group should have and this group will not be tracked anymore. In this case, the knowledge of the number of persons inside the group could help the system to not lose this group. Future developments include computation of group trajectory, speed and events inside the group (for example: two persons fighting) in order to recognise abnormal behaviours such as violence or vandalism in a metro scene

## References

[1] Brémond, F. and Thonnat, M. (1997). Tracking multiple non-rigid objects in a cluttered scene. In *proc. of the 10th Scandinavian Conference on Image Analysis (SCIA)*, Lappeenranta (Finland).

[2] Cox, I. and Hingorani, S. (1996). An efficient implementation of reid's Multiple Hypothesis Tracking algorithm and its evaluation for the purpose of visual tracking. In *IEEE Transactions on pattern analysis and machine intelligence*, volume 18.

[3] Meyer, F. and Bouthemy, P. (1992). Region-based tracking in an image sequence. In *Proc. of European Conference on Computer Vision (ECCV)*, pages 476–484.

[4] Rota, N., Stahr, R., and Thonnat, M. (2000). Tracking for visual surveillance in VSIS. In *First IEEE International Workshop on Performance Evaluation of Tracking and Surveillance PETS2000*, Grenoble, France.

[5] Thonnat, M. and Rota, N. (1999). Image understanding for visual surveillance application. In *Third international workshop on cooperative distributed vision CDV-WS'99*, pages 51–82, Kyoto, Japan.

[6] Zhang, Z. (1994). Token tracking in a cluttered scene. In *Int'l J. of Image and Vision Computing*, volume 12.

Chapter 8

# Colour-Invariant Motion Detection under Fast Illumination Changes[1]

Ming Xu and Tim Ellis
*Department of Electrical, Electronic & Information Engineering, City University, London EC1V 0HB, UK*

Abstract: This paper tackles the problem of robust change detection in image sequences from static cameras. Motion cues are detected using frame differencing with an adaptive background estimation modelled by a mixture of Gaussians. Illumination invariance and elimination or detection of shadows is achieved by using a colour chromaticity representation of the image data.

Key words: Colour, motion detection, segmentation.

## 1. Introduction

Motion detection algorithms are often based on the differencing operation of image intensities between each frame and a background image. The background image reflects the static elements in a scene. It needs to be continuously updated because of the lack of a target-free training period, gradual illumination variations and background objects (e.g. a parked car) which then move. The updating schemes usually involve linear interpolation between the previous background value and the new observation if their difference is within some tolerance; otherwise a foreground pixel is declared. The Gaussian mixture model [6] is one such background updating scheme. Although this method, which is based on grey-level or RGB colour intensities, can account for a large proportion of changes, it cannot readily "follow" fast illumination changes caused by moving clouds, long shadows and switching of artificial lighting. These fast illumination changes can make

[1] This work was supported by the EPSRC under grant number GR/M58030.

a large region in a frame be incorrectly identified as a "moving" object, within which the genuine moving targets are overwhelmed and lost.

Robustly identifying a particular object surface under varying illumination has received considerable attention in colour invariance research [1][3][5]. For example, Marchant and Onyango [3] proposed a physics-based method for shadow compensation in scenes illuminated by daylight. They represented the daylight as a black body and assumed the colour RGB camera filters to be of infinitely narrow bandwidth. They found that the ratio, $(R/B)/(G/B)^A$ , only depends on surface reflection as the illumination changes ($A$ can be pre-calculated from the daylight model and camera). Under the same assumptions, Finlayson *et al.* [1] found that the log-chromaticity differences (LCDs), $\ln(R/G)$ and $\ln(B/G)$, are independent of light intensity and there even exists a weighted combination of LCDs which is independent of both light intensity and light colour.

There exist some adaptive schemes in colour-invariant detection of motion under varying illumination. Wren *et al.* [7] used the normalised components, U/Y and V/Y, of a YUV colour space to remove shadows in a relatively static indoor scene. A single, adaptive Gaussian was used to represent the probability density of each pixel belonging to the background. Therefore, the scene without any person has to be learned before this system locates people. Raja *et al.* [4] used the hue (H) and saturation (S) of an HSI colour space to decouple the influence of illumination changes in an indoor scene. A Gaussian mixture model was used to estimate the probability densities of each pixel belonging to a multi-coloured foreground object. Each Gaussian models one colour in the foreground object and was learned in a training stage.

In this paper, the motion detection in outdoor environments illuminated by daylight is emphasised. A reflection model influenced by ambient objects has been used, because large-scale illumination changes mainly arise from varying cloud cover and the dominant illumination comes from either direct sunlight or reflection from clouds. In this paper the normalised rgb colour space is used to eliminate the influence of varying illumination. A Gaussian mixture model has been used to model each pixel of the background and thus provides the bootstrapping and multi-background modelling capabilities necessary for complex outdoor scenes.

## 2. Colour Fundamentals

An image taken with a colour camera is composed of sensor responses as:

$$f_K = \int I(\lambda)\rho(\lambda)S_K(\lambda)d\lambda \qquad (K = R, G, B) \tag{1}$$

where $\lambda$ is wavelength, $I$ is the illumination, $\rho$ is the reflectance of an object surface, and $S_K$ is the camera sensitivity. Given a particular colour camera, the image intensity depends only on the reflected light from the object surface:

$$I_{reflected}(\lambda) = I(\lambda)\rho(\lambda) \tag{2}$$

Therefore, the appearance of objects in an image is a result of interaction between illumination and reflectance. Either the emergence of an object or illumination variation can cause the image intensity to vary. To be able to identify and track the same object surface (e.g. a background pixel) under varying illumination, it is desirable to separate the variation of the illumination from that of the surface reflection.

In an outdoor environment, fast illumination changes tend to occur at the regions where shadows emerge or disappear. These shadows may be either large-scale (e.g. those arising from moving cloud) or small-scale (e.g. those arising from objects themselves). Here a shadow model derived from that in [2] has been used and is shown in Fig. 1. There is only one illuminant in the scene. Some of the light does not reach the object because of a blocking object, thus creating a shadow region and a directly lit region on the observed object. The shadow region is not totally dark but illuminated by the reflection from each ambient object $j$:

$$I_{ambient,j}(\lambda) = I_{incident}(\lambda)\rho_{ambient,j}(\lambda) \tag{3}$$

For the directly lit region the reflected light from the object surface is:

$$\begin{aligned} I_{reflected}(\lambda) &= \left[ I_{incident}(\lambda) + \sum_j I_{ambient,j}(\lambda) \right] \rho(\lambda) \\ &= I_{incident}(\lambda)\rho(\lambda) \left[ 1 + \sum_j \rho_{ambient,j}(\lambda) \right] \end{aligned} \tag{4}$$

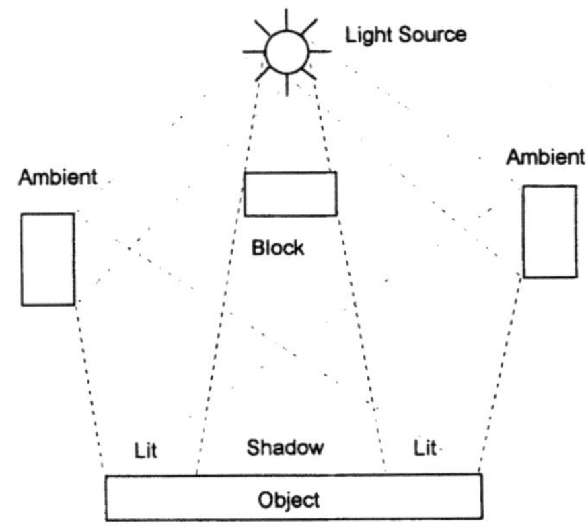


*Figure 1.* A shadow model.

For the shadow region this becomes:

$$I_{reflected}(\lambda) = I_{incident}(\lambda)\rho(\lambda)\sum_j \rho_{ambient,j}(\lambda) \tag{5}$$

To ensure the reflected lights from the directly lit and shadow regions have the same spectral distribution, we assume that the chromatic average of the ambient objects in a scene is nearly grey, i.e. it is relatively balanced in all visible wavelengths and:

$$\sum_j \rho_{ambient,j}(\lambda) = c \tag{6}$$

where $c$ is independent of $\lambda$ and may varies over space.

This assumption is realistic for the fast-moving cloud case, in which the only illuminant is the sunlight and both the blocking and ambient objects are grey (or white) clouds. Under such an assumption, the reflected light, $I_{reflected}(\lambda)$, from directly lit and shadow regions will stay in proportion for a given object surface. This leads to the image intensities, $f_K$, at all colour channels being in proportion, no matter whether the object surface is lit or shadowed. The proportionality between RGB colour channels can be better represented using the normalised colour components:

$$f_k = f_K \bigg/ \left( \sum_{i=RGB} f_i^2 \right)^{1/2} \tag{7}$$

where each component of $f_k$ will keep constant for a given object surface under varying illumination.

## 3. Colour-Invariant Motion Detection

For one channel of the RGB components resulting from a particular surface under particular lighting, a single Gaussian is sufficient to model the pixel value and account for acquisition noise. If lighting changes gradually over time, a single, adaptive Gaussian would be sufficient to model each RGB channel, in which the estimated background value is interpolated between the previous estimation and the new observation. However, an adaptive Gaussian cannot readily follow an RGB component under fast lighting changes. In contrast, a normalised colour component (rgb) for a given object surface tends to be constant under lighting changes and is appropriate to model using an adaptive Gaussian. In practice, multiple object surfaces may appear as the backgrounds at a particular pixel, e.g. swaying trees. Therefore, multiple, adaptive Gaussians (a mixture of Gaussians [6]) are necessary to model such a pixel.

Let the pixel value at time $t$ be $X_t = (f_r, f_g, f_b)^T$ and modelled by a mixture of $N$ Gaussian distributions. The probability of observing the pixel value is:

$$P(X_t) = \sum_{i=1}^{N} \omega_{i,t} G(X_t, \mu_{i,t}, \sigma_{i,t}) \tag{8}$$

where $G$ is the Gaussian probability density function of the $i$-th background $B_i$, i.e. $P(X_t \mid B_i)$, $\mu_{i,t}$ and $\sigma_{i,t}$ are the mean value and standard deviation of the i-th distribution, respectively. $\omega_{i,t}$ is the prior probability for the distribution, i.e. $P(B_i)$, reflecting the likelihood that the distribution accounts for the observed data.

Every new observation, $X_t$, is checked against the $N$ Gaussian distributions. A match is defined as an observation within about 3 standard deviations of a distribution. If none of the $N$ distributions match the current pixel value, the least probable distribution is replaced by the new observation. For the matched distribution, *i*, the parameters are updated as:

$$\begin{aligned}\mu_{i,t} &= (1-\varphi)\mu_{i,t-1} + \varphi X_t \\ \sigma_{i,t}^2 &= (1-\varphi)\sigma_{i,t-1}^2 + \varphi(X_t - \mu_{i,t})^T (X_t - \mu_{i,t}) \\ \omega_{i,t} &= (1-\phi)\omega_{i,t-1} + \phi\end{aligned} \tag{9}$$

where $\phi$ and $\varphi$ control the updating rate. For the unmatched distributions, *j*, $\mu_{j,t}$ and $\sigma_{j,t}$ remain the same, and:

$$\omega_{j,t} = (1-\phi)\omega_{j,t-1} \tag{10}$$

The distribution(s) with greatest weight, $\omega_{i,t}$, is (are) considered as the background model.

## 4. Experimental Results

To assess the significance of the colour-invariant motion detection, we evaluated it at both pixel and frame levels using a set of image sequences. The image sequence shown here was captured at a frame rate of 2Hz. Each frame was lossily compressed in JPEG format and has a frame size of 384×288 pixels. This sequence well represents the abundant contexts of a day lit outdoor environment, such as fast illumination changes, waving trees, shading of tree canopies, highlights of specular reflection, as well as pedestrians.

### 4.1 Colour invariance at pixel level

Fig. 2 shows the absolute (RGB) and normalised (rgb) colour components at selected pixels through time. To incorporate the two sets of curves into a single graph, the absolute colour components, RGB, were linearly transformed between [0, 1]. Each arrow at the bottom of the graph indicates a "ground truth" foreground object present at the relevant frame.

The absolute colour components (RGB) change greatly with the illumination, even when no foreground object is present (Fig. 2(a)). The gradual slope in the RGB curves, e.g. at frame 20 in Fig. 2(b), is difficult to model with a Gaussian distribution. The sharp RGB edge, e.g. at frame 75 in Fig. 2(b), is due to the illumination change but resembles that for an emerging or disappearing foreground object. If one uses a fast parameter updating rate to differentiate these two types of sharp edges, the background estimation may follow the observation so closely that many foreground pixels are classified as background pixels. On the other hand, the normalised colour components (rgb) for a background pixel have flat profiles under varying illumination. For each foreground pixel, at least one rgb component appears as an apparent spike. Therefore, an adaptive Gaussian is more appropriate for modelling a normalised colour component.

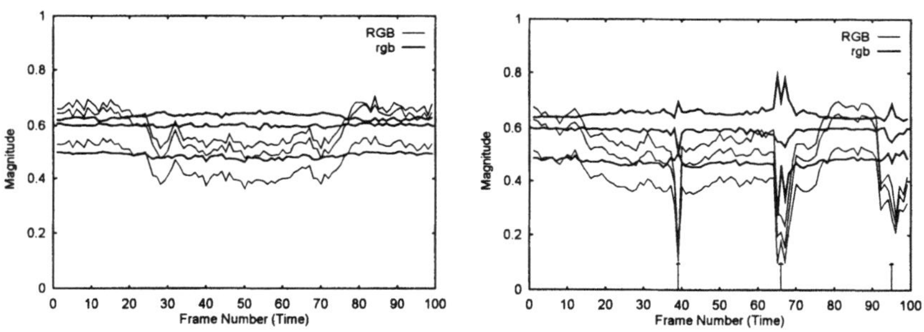


(a) (b)

*Figure 2.* The absolute (RGB) and normalised (rgb) colour components at selected pixels through an image sequence, when (a) no foreground object is present and (b) foreground objects are present.

## 4.2 Foreground detection at pixel level

Fig. 3 shows the parameter updating procedure of the Gaussian background model for one rgb component at selected pixels. $\varphi$ and $\phi$ were set to 0.2. The middle and upper (lower) profiles in thin lines represent the estimates, $\mu$ and $\mu\pm3\sigma$, respectively.

Fig. 3(a) corresponds to a lit region with foreground objects. Each "ground truth" target has an observation going beyond $\mu\pm3\sigma$ of the estimated background model and thus is correctly identified. The first target in Fig. 3(a) is nearly missing, because the colour component shown here is not the one with the greatest variation, see Fig. 2(b). In fact, the variations of all three colour components are considered in our algorithm and those pixels with a major change in only one component are reliably detected.

Suppose the noise in the RGB components arises from acquisition noise and has a zero-mean Gaussian distribution with a uniform variation over space and time. Because the rgb components are the RGB components normalised by the intensity, the absolute noise level in the rgb components appears "amplified" for a dark region, compared with that for a bright region (the signal-to-noise ratio is consistent in the rgb and RGB components). This is shown in Fig. 3(b) taken from a shaded region, under the tree canopy, without any foreground object passing through. However, no background pixel is falsely detected as the foreground, even when the variation of the noise in Fig. 3(b) is stronger than the signal (the spikes for the foreground targets) in Fig. 3(a). This indicates that the Gaussian model can be adapted to the local noise level from an initial deviation value larger than the noise variation. Currently, this initial deviation is manually selected and spatially uniform (see the Discussion).

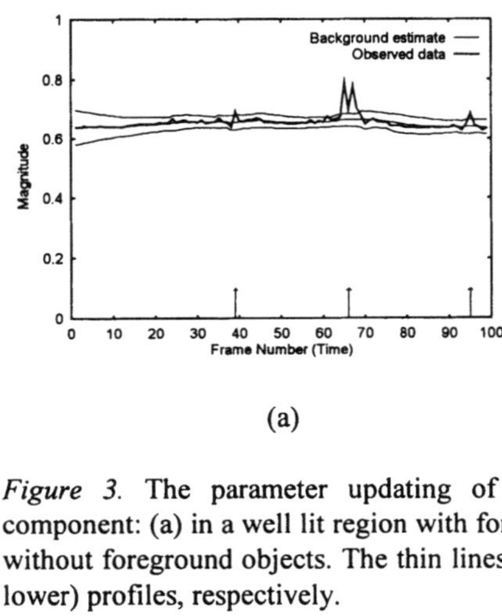


(a)

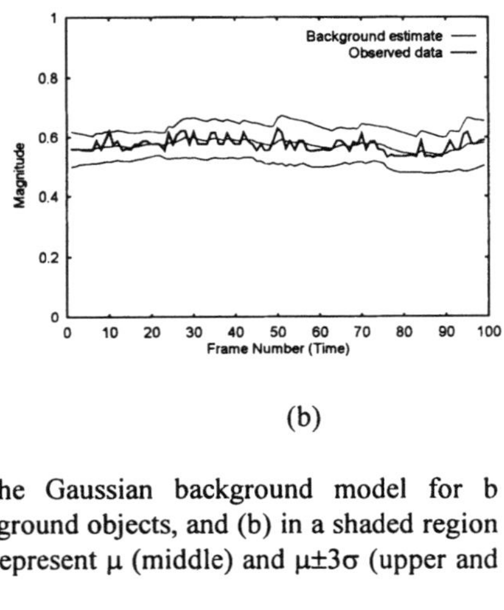


(b)

*Figure 3.* The parameter updating of the Gaussian background model for b component: (a) in a well lit region with foreground objects, and (b) in a shaded region without foreground objects. The thin lines represent $\mu$ (middle) and $\mu \pm 3\sigma$ (upper and lower) profiles, respectively.

## 4.3 Foreground detection at frame level

Figs. 4 and 5 show the results of the motion detection at two frames of the image sequence. To adapt the variations across the image, a higher threshold was used to select "foregrounds". The foreground pixels in the rgb results are those that go beyond $[\mu-3.5\sigma, \mu+3.5\sigma]$ of the most probable Gaussians. The foreground pixels in the RGB results arise from a global threshold on the difference between the observation and the mean of the most probable Gaussian. The thresholding level is selected so as to produce "blobs" of similar sizes to those in the corresponding rgb results. A 1×3 closing operation has been applied to the binary image of detected foreground "blobs". The grey-level intensity images shown here were obtained using $I = \sqrt{\left(f_R^2 + f_G^2 + f_B^2\right)/3}$ .

Fig. 4 (frame 47) is an example comparing the RGB and rgb results under little illumination change. The foreground "blobs" extracted using rgb space are as coherent as those using RGB space. Because of the different emphasis of image contexts for both the colour spaces, the corresponding blobs in Figs. 4(a) and (c) may appear as different shapes.

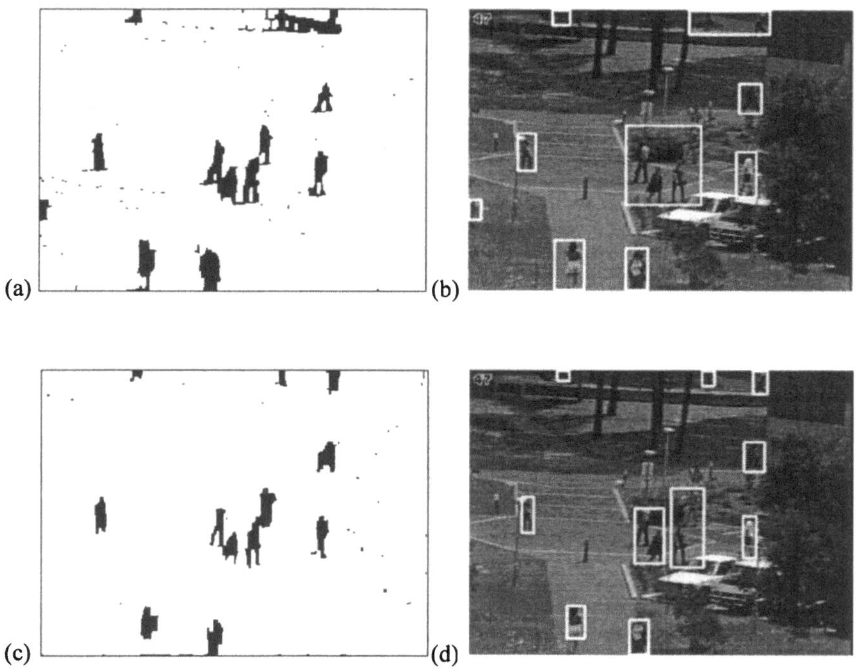


*Figure 4.* Motion detection at frame 47 with little illumination change: the detected blobs (left) and corresponding bounding boxes overlaid on the frame (right) using the RGB (top) and rgb (bottom) spaces.

Fig. 5 (frame 78) shows the RGB and rgb results under a major illumination change (refer to the RGB curves in Fig. 2(a)). In the RGB result, Fig. 5(a), a large area of the background is detected as a huge foreground object, in which the "ground truth" targets (pedestrians) are submerged and lost. On the other hand, in the rgb result, Fig. 5(c), fast illumination changes give no additional "foreground" blob and the "ground truth" targets are clearly visible. Note the poor detection of some foreground blobs on the left of the frame is caused by the stationary pedestrians that are being absorbed into the estimated "background" by the adaptive model.

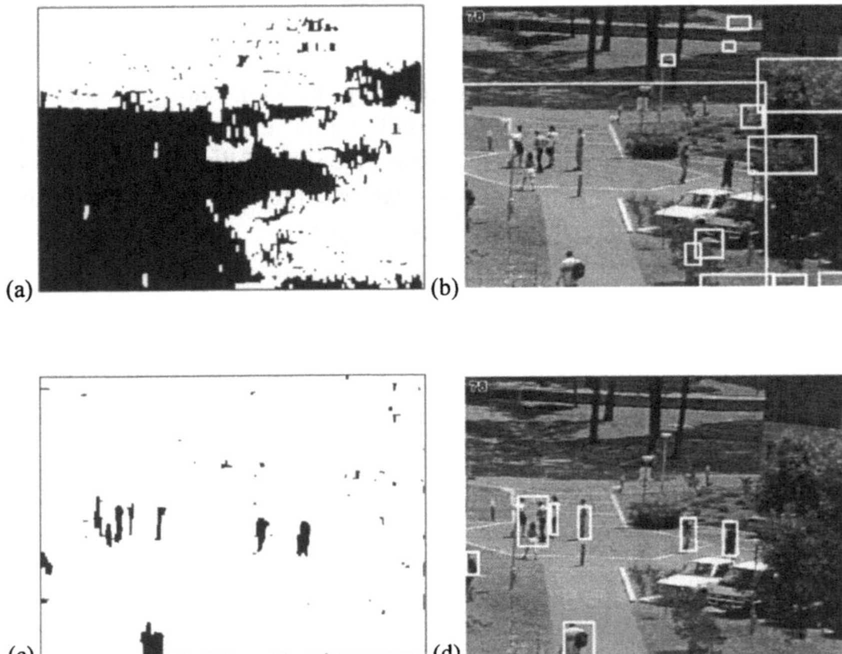


*Figure 5.* Motion detection at frame 78 with a major illumination change: the detected blobs (left) and corresponding bounding boxes overlaid on the frame (right) using the RGB (top) and rgb (bottom) spaces.

## 5. Discussion

The convergence rate of the colour mixture model depends on an appropriate selection of the initial deviation for each Gaussian distribution. An underestimate of the initial deviation prohibits many "ground truth" background pixels from being adapted into background models and produces a noisy result for change detection. On the other hand, an overestimate of the initial deviation needs a longer learning period at the start of an image sequence. Currently, the initial deviation is manually selected and globally uniform according to the noise level in shaded regions where the absolute noise level in rgb components is high. In future, the initial deviation may be automatically selected according to the local spatial variation in rgb components at the start frame.

We have also combined the intensity, *I*, with the rgb colour space. Such an rgbI colour space is an invertible transformation from RGB space and avoids the loss of intensity information, which may bring about some promising applications. One example is for shadow detection that can guide the positioning and orientating of light sources in a scene. In an environment without large-scale illumination changes, a region can be determined as being shadowed if the rgb components are stable but the I component becomes significantly lower. In practice, the RGB components may be saturated under illumination changes, which can make the corresponding rgb components unconstrained even when no foreground object is present. On the other hand, the rgb components in over-dark regions are very noisy. Therefore, pixels where the intensity *I* were zero or saturated may be excluded from consideration. Using cameras with auto iris control or Gamma correction may alleviate this problem.

## 6. Conclusions

A Gaussian mixture model based on the rgb colour space has been presented for maintaining a background image for motion detection. This scheme is especially successful when applied to outdoor scenes illuminated by daylight and is robust to fast illumination changes arising from moving cloud and self-shadows. The success results from a realistic reflection model in which shadows are present.

## REFERENCES

[1] G. D. Finlayson and S. D. Hordley, "Colour invariance at a pixel", *Proc. British Machine Vision Conf.*, pp. 13-22, 2000.

[2] R. Gershon, A. D. Jepson and J. K. Tsotsos, "Ambient illumination and the determination of material changes", *J. Optical Society of America*, 3(10):1700-1707, 1986.

[3] J. A. Marchant and C. M. Onyango, "Shadow invariant classification for scenes illuminated by daylight", to appear in *J. Optical Society of America*, 2000.

[4] Y. Raja, S. J. McKenna and S. Gong, "Segmentation and tracking using colour mixture models", *Proc. Asian Conf. on Computer Vision*, 1998.

[5] J. M. Rubin and W. A. Richards, "Color vision: representing material changes", AI Memo 764, MIT Artificial Intelligence Lab., 1984.

[6] C. Stauffer and W. E. L. Grimson, "Adaptive background mixture models for real-time tracking", *Proc. IEEE Conf. on Computer Vision and Pattern Recognition*, 1999.

[7] C. Wren, A. Azarbayejani, T. Darrell and A. Pentland, "Pfinder: real-time tracking of the human body", *IEEE trans. on Pattern Analysis and Machine Intelligence*, 19(7):780-785, 1997.

Chapter 9

# FACE AND FACIAL FEATURE TRACKING

*Using the Active Appearance Algorithm*

Jörgen Ahlberg
*Image Coding Group, Dept. of Electrical Engineering, Linköping University*

Abstract: This paper describes an approach to tracking a face and its facial features in an input video sequence using the Active Appearance Algorithm. The algorithm adapts a wireframe model to the face in each frame, and the adaptation parameters are converted to MPEG-4 Facial Animation Parameters. The results are promising, and it is our conclusion that a real-time model-based coder running on consumer hardware is possible to implement this way.

Key words: Face tracking, facial feature extraction, MPEG-4 Face Animation, model-based-coding, active appearance models.

## 1. Introduction

Our goal is to find and track a human face and its features in a video sequence. We want to do this by adapting, in each frame, a wireframe model to the face in the image. This should be done accurately enough to allow a realistic-looking facial animation to be created from the extracted face model parameters, and we also want the algorithm to have real-time performance. This paper describes how we use the Active Appearance Algorithm to adapt the model to the image, and then track it through an image sequence.

The concept of Active Appearance Models (AAMs) was introduced by Cootes *et al* [4] a few years ago, and has been the subject of several reports and investigations, especially by the original inventors [3]. Together with the AAMs came a search algorithm for adapting the model to an image, here referred to as the Active Appearance Algorithm (AAA). The AAA can be used on a complete AAM, or, as here, on a simpler model being parameterized in geometry and texture separately.

The AAA is a greedy algorithm, finding locally optimal face model parameters. Thus, a quite accurate à priori estimate of the face model parameters (at least the size and position of the face) must be available before starting the AAA. In a video sequence, we can use the parameters extracted from the previous frame as the à priori estimate. For the first frame of the sequence, we need an additional algorithm, for example a colour and/or motion based face candidate finder. Such algorithms are not treated in this paper. Section 2 treats the parameterization of the face model and how to find the optimum parameters to adapt the model to the image sequence. In the following section, the training process as performed in this experiment is described, followed by details on the implementation and examples of the results. In Section 5, a few directions of our continuing work are mentioned.

## 2. MODEL PARAMETERIZATION AND THE ACTIVE APPEARANCE ALGORITHM

The face model is a wireframe model with a texture mapped on its surfaces. The texture is represented as a standard-shaped image, being a linear combination of a set of *texture modes* consisting of *geometrically normalized eigenfaces* [6]. We formulate this as

$$x = \bar{x} + X\tau$$

where $\bar{x}$ is the mean texture, the columns of $X$ are the texture modes and $\tau$ is the vector of texture parameters. The synthesized texture $x$ is mapped on the wireframe model. The geometry of the wireframe model is parameterized according to

$$g(\sigma, \alpha) = \bar{g} + S\sigma + A\alpha$$

where the resulting vector g contains the $(x, y, z)$ coordinates of the vertices of the model. $\bar{g}$ is the standard shape of the model, and the columns of $S$ and $A$ are the Shape and Animation Units respectively, and thus $\sigma$ and $\alpha$ contain the shape and animation parameters. Since we also want to perform global motion, we need six more parameters for rotation, scaling, and translation. Thus, we replace the last equation with

$$g = sR(\bar{g} + S\sigma + A\alpha) + t$$

where $R = R(r_x, r_y, r_z)$ is a rotation matrix, $s$ is the scale, and $t = t(t_x, t_y)$ is the 2D translation vector (quaternions is a better representation of global rotations, but Euler angles works well at the small angles we have been working with so far).

The geometry of our model is thus parameterized by the parameter vector

$$p = \lfloor r_x, r_y, r_z, s, t_x, t_y, \sigma, \alpha \rfloor.$$

Note that this differs from the original AAM formulation, where out-of-plane rotation is rather built-in in the Shape Units than a separate parameter. Another difference is that in the original formulation, there is no distinction between Shape Units and Animation Units.

When adapting a model to a video sequence, the shape parameters $\sigma$ should only be changed in the first frame(s); the head shape does not vary during a conversation. The shape parameters can be converted to MPEG-4 Facial Definition Parameters (FDPs) and the animation parameters to MPEG-4 Facial Animation Parameters (FAPs).

## 2.1 Matching the Model and the Image

Our goal is to find the optimal adaptation of the model to the input image, that is to find the p that minimizes the distance between the model and the image. As distance measure, we choose the summed squared error (SSE) between the remapped input image and the synthesized texture. We compute this by, for a given *p*, reshaping the model according to $g(p)$, and mapping the input image i onto the model. We then reshape the model to the standard shape, $\bar{g}$, and get the resulting image as a vector $j(p) = j(i, g(p))$. This image can be approximated by the texture modes by computing the texture parameters according to

$$\tau(p) = X^T (j(p) - \bar{x}).$$

Inserting this in the expression for synthesizing the texture above, we get

$$x(p) = \bar{x} + XX^T (j(p) - \bar{x})$$

and we compute the residual image

$$r(p) = j(p) - x,$$

and thus the SSE as

$$e(p) = \|r(p)\|^2$$

The entire process is illustrated in Figure 1.

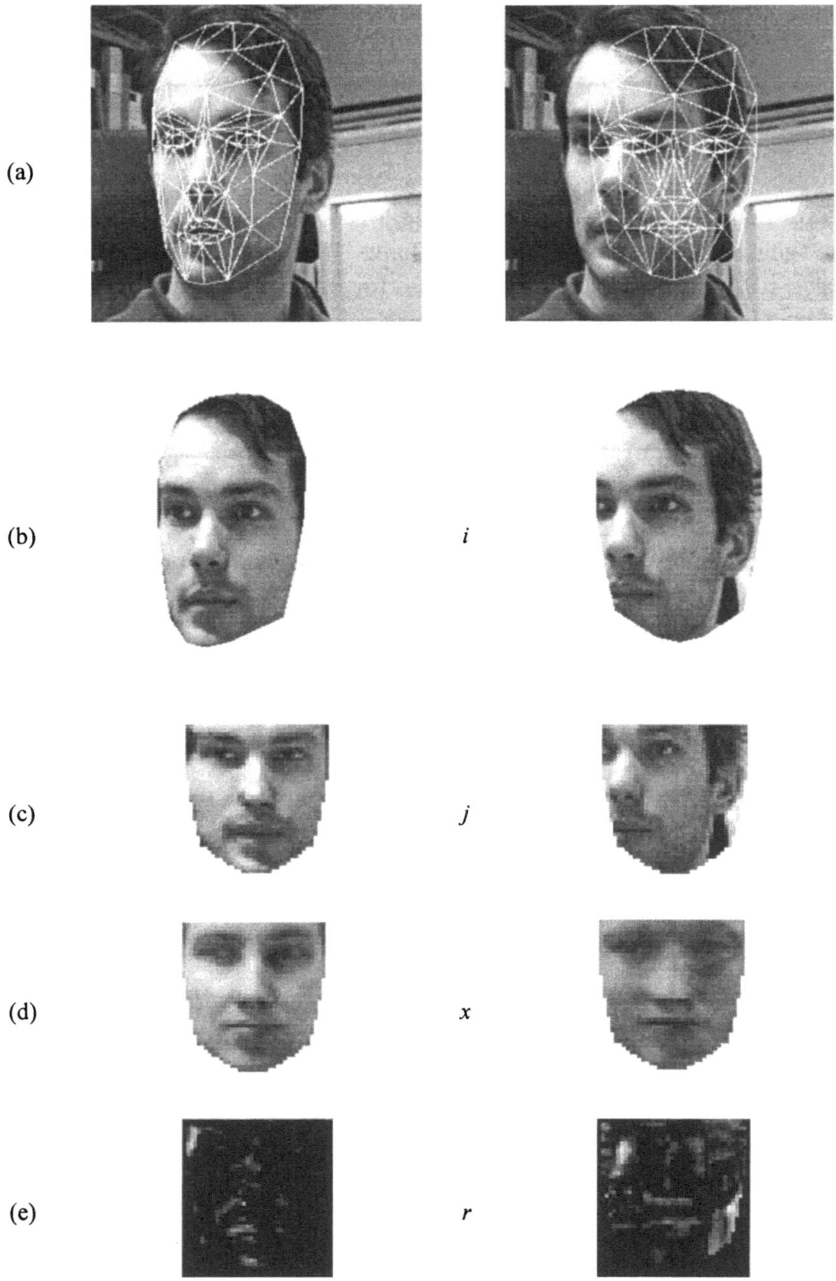


*Figure 1.* The model matching and texture approximation process. A good and a bad (left and right column respectively) model adaptation is shown (a); The image mapped onto the model (b); The model is reshaped to the standard shape, producing the image $j$ (c); The normalized texture is approximated by the texture modes, producing the image $x$ (d); The residual image $r$ is computed (e). The images $j$ and $x$ are more similar the better the model adaptation is.

## 2.2 Finding the Optimal parameters

With this formulation, our goal is to find the parameter vector $p$ that for a given input image $i$ minimizes $e(p)$. We do that by using the Active Appearance Algorithm (AAA) in the following way. For a starting value of $p$, supposed to be close optimum, we compute $r(p)$ and $e(p)$, and find the update vector $\Delta p$ by multiplying the residual image with an update matrix:

$$\Delta p = Ur(p)$$

The vector $\Delta p$ gives us a probable direction in the search space, and we the compute the SSE

$$e_k = e(p + k\Delta p)$$

for $k = 1$. If $e_k < e(p)$, we let $k^* = k$, otherwise we try $k = 0.5$ and $0.25$. Finally, we update $p$ accordingly, that is,

$$p + k^* \Delta p \rightarrow p$$

and iterate until convergence. The magic in this is the update matrix $U$, that we create in advance by training from example images with models correctly adapted.

## 2.3 Creating the Update Matrix

Taylor-expanding $r$ around $p + \Delta p$, we can write

$$r(p + \Delta p) = r(p) + R\Delta p + O(\Delta p^2)$$

where

$$R = \frac{\partial}{\partial p} r(p).$$

Given a $p$ (and thus an $r(p)$), we want to find the $\Delta p$ that minimizes

$$e(p + \Delta p) \approx \left\| r(p) + R\Delta p \right\|^2.$$

Minimizing this is a least squares problem with the solution

$$\Delta p = -(R^T R)^{-1} R r(p),$$

which gives us the update matrix $U$ as the negative pseudo-inverse of the gradient matrix $R$:

$$U = -R^+ = -(R^T R)^{-1} R.$$

To be able to use the AAA, we should consequently estimate the gradient matrix $R$. We do that by perturbating $p$ from a set of (manually) adapted models, parameter by parameter, step by step. The $j$:th row in $R$ can thus be estimated as

$$R_j = \sum_k (r(p + \Delta p_{jk}) - r(p)),$$

where $\Delta p_{jk}$ is a vector that perturbs $p$ in the $j$:th component to the amount of $k \cdot c$ for some suitable constant $c$.

## 3. Training the model

To try out this scheme, the Candide-3 [1] model has been adapted (manually in the beginning, then semi-automatically) to 330 images of six different persons from different angles and with different facial expressions. As Animation Units, the following Action Units from Candide-3 have been chosen:

1. Jaw drop
2. Lip stretcher
3. Lip corner depressor
4. Upper lip raiser
5. Eyebrow lowerer
6. Outer eyebrow raiser

The adapted model has for each image been normalized to a standard shape with the size 40x42 pixels, as Figure 1(c), left, and the resulting training textures collected in a matrix. A PCA has been performed on this matrix, to compute the texture modes. The mean texture $\bar{x}$ and the DC-level have been subtracted from the training textures prior to the PCA.

With the texture modes available, all the parameters have been perturbed, one by one and for each image, in steps of 0.01 in the range [-0.1, 0.1], and the matrix $R$ estimated. From $R$, the update matrix $U$ has been computed.

## 4. Implementation and Results

The algorithm has been implemented in Visual C++. The images have been captured using a Sony EVI-D31 camera and digitised with an Asus V3800 Ultra graphics card with video input. The tests have been performed on a PC with a 500 MHz Pentium III processor. Except for the high-quality camera, this is quite typical consumer hardware that can be found in many homes.

Resulting face model adaptations in a video sequence are show in Figure 2. As can be seen, the global adaptation (rotation, scale, translation) is good and the mouth and eyebrow parameters behave well. All parameters suffer from the algorithm being greedy and easily getting stuck in a local optimum. Typically, this results in the model not being able to follow fast moves, like when the mouth is closed too fast or when the head is moved away quickly. The higher the frame rate, the smaller this problem becomes, and already at a few Hertz normal head motion is handled.

When running the algorithm on live video, input directly from a camera, the computation time is very critical. If the time for each iteration could be reduced somewhat, the frame rate would, of course, be higher. Noticing that the algorithm needs fewer iterations if there is smaller motion between each frame, it is clear that higher frame rate implies that fewer iterations would be needed each frame, which would improve the frame rate even more.

Studying the algorithm, there are three time consuming parts; the image warping, the texture analysis-synthesis, and the update vector computation. Regarding computation time, the image warping is very dominant, but can be sped up by using the graphics hardware of the computer (modern graphics card have specialized hardware for texture mapping on 3D wireframes) or by using pre-computation of barycentric coordinates and vector instructions (modern CPUs have special multimedia instructions). For details, see [2].

Using such tricks, the time for each iteration can be brought down to a few milliseconds. Typically, less than 10 iterations are needed per frame, and the systems runs at more than 15 Hz.

The model parameters for each frame are easily converted to MPEG-4 Facial Animation Parameters (FAPs). The MPEG-4 reference software compressing FAPs has been integrated with the tracking system, and resulting the FAPs can be streamed over the network or stored on a file. Those parameters can then be read into any player compliant to MPEG-4 Face Animation (for example the one by Lavagetto an Pockaj [5]).

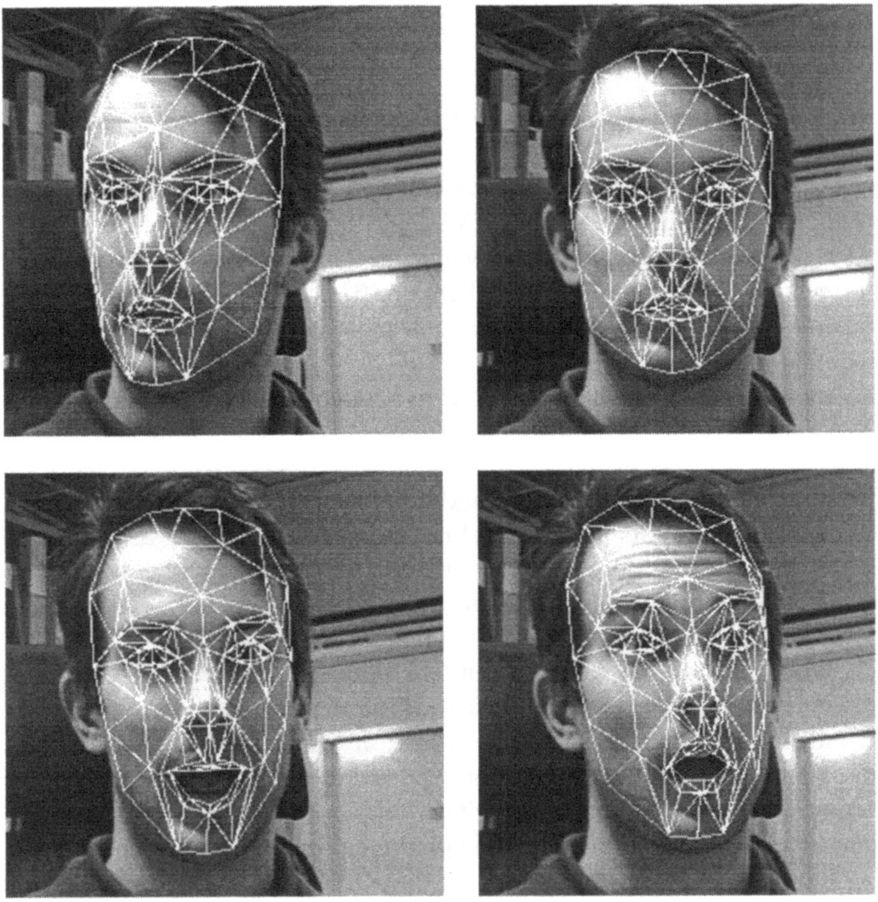

*Figure 2.* The model adapted to four frames of a video sequence.

## 5. Future Work

There are several possible improvements that should be implemented and evaluated. A few of those are mentioned below:

- The AAA should be combined with an algorithm that quickly finds a rough estimate of the face location in the first frame.
- When used for tracking in a video sequence, the initial estimate in a frame should be better predicted than just the adaptation from the previous frame. This could be done, for example, with a simple motion estimation in a few points, a Kalman filter, or a combination thereof.

- No Shape Units are estimated in this example; they are assumed to be known à priori.
- Sooner or later, the tracking will fail. Thus, a re-initialisation algorithm (such as in [7]) should be added to the system.

# 6. Conclusions

The described approach looks promising, with many improvements within reach.

By using features of the hardware (the graphics card or the CPU) in an efficient way or algorithm specific solutions, we can perform an iteration of the Active Appearance Algorithm in a few milliseconds on consumer hardware. Typically, about 10 iterations per frame are needed. Since the video capture and transfer requires some time as well, this allows the system to run at more than 15 Hz on the experiment system.

The main goal, to show that it is possible to implement a real-time model based coder on consumer hardware, is achieved.

# Acknowledgements

This work is sponsored by the Swedish SSF-project VISIT and the European project InterFace (IST-1999-1036).

# References

[1] J. Ahlberg. (2001). *Candide*-3 – an updated parameterized face, Report No. LiTH-ISY-R-2326, Dept. of Electrical Engineering, Linköping University, Sweden.
[2] J. Ahlberg. (2001). Fast image Warping for Active Models, Report No. LiTH-ISY-R-2355, Dept. of Electrical Engineering, Linköping University, Sweden.
[3] T. F. Cootes and C. J. Taylor. (2000). Statistical Models of Appearance for Computer Vision, Draft report, Wolfson Image Analysis Unit, University of Manchester.
[4] T. F. Cootes, G. J. Edwards, and C. J. Taylor. (1998). "Active Appearance Models," Proc. 5th European Conference on Computer Vision, pp. 484 - 498.
[5] F. Lavagetto, R. Pockaj. (1999). "The Facial Animation Engine: towards a high-level interface for the design of MPEG-4 compliant animated faces," IEEE Trans. on Circuits and Systems for Video Technology, Vol. 9, No. 2.
[6] J. Ström et al. (1997). "Very Low Bit Rate Facial Texture Coding," Proc. International Workshop on Synthetic/Natural Hybrid Coding and 3-D Imaging, Rhodes, Greece, pp. 237 - 240.
[7] J. Ström. (2001). "Reinitialization of a Model-based Face Tracker," Proc. International Conf. on Augmented, Virtual Environments and 3-D Imaging, Mykonos, Greece, pp. 128 - 131.

Chapter 10

# OBJECT TRACKING AND SHOSLIF TREE BASED CLASSIFICATION USING SHAPE AND COLOR FEATURES

Lucio Marcenaro, Franco Oberti and Carlo S. Regazzoni
*DIBE, Department of Biophysical and Electronic Engineering, University of Genoa, Italy*

Abstract: This paper proposes a video-surveillance system that is able to track and automatically classify objects detected in a scene. A common object representation based on corners is used for tracking and for classification. Recognition module is based on a self organizing decision tree (SHOSLIF) that is automatically generated by the system during a learning phase. Both color and shape related features are used for object classification. Presented results confirm the validity of the proposed approach.

Key words: Video-surveillance, Objects classification, SHOSLIF trees

## 1. Introduction

The main functionalities of a video-surveillance system are related with the detection and tracking of objects acting within the guarded environment, while higher modules of the system are responsible for objects and events classification. Different kind of features have been proposed for both tasks in literature, shape, motion and color being most frequent ones [1, 2]. Here, we concentrate on shape and color related features for tracking and recognizing objects. In particular, while shape related features are used in order to classify among different postures, histogram properties represent the basis for the classification between different objects. The shape has been chosen as main feature because in many surveillance applications related to transport shape information usually provides a finer discriminant feature allowing objects within a same general class to be classified in a more precise and domain dependent way. In literature there are several approaches for objects recognition that use similar features as silhouette skeleton [3], B-spline border approximation [4], corner points for rigid object tracking [5]. In these approaches the relationship between features used for tracking and for recognition is not clearly defined.

In this paper, we propose a novel approach for tracking and recognition that is based on the same set of features starting from an initially redundant set of significant shape points (i.e. corners). Corner groups and objects histograms are used as basic features for a multilevel shape representation. The multilevel representation is developed through progressive steps oriented to individuate features most useful for tracking (i.e. to locate and to determine the object pose in time) and for recognizing multiple objects (i.e. to classify objects on the basis of a symbolic label representative of a class). The system uses a short-term memory, associated with the tracking process, and a long-term memory, associated with the recognition process.

Methods for representing models for the tracking and recognition phases that are based on the Generalized Hough Transform (GHT) [7] and on SHOSLIF trees [6], respectively. The Generalized Hough Transform (GHT) [7] is used as a basis to integrate object representations in a temporal sliding window close to the instant where tracking has to be performed. This short-term corner-based representation is kept updated by the learning modality of the tracking module along system functioning and used by the matching modality of the same module.

Long-term memory is given by a self organizing tree that is automatically developed from a pre-classified training set. This method is known as Self Organizing Hierarchical Optimal Subspace Learning and Inference Framework (SHOSLIF) tree [6]: this tool uses the theories of optimal linear projection to generate a hierarchical tessellation of a space defined by the training set of samples. This space is generated using two projections: a Karhunen-Loeve projection to produce a set of Most Expressive Features (MEFs), and a subsequent discriminant analyisis projection to produce a set of Most Discriminating Features (MDFs). The system builds a network that tessellates these MEF/MDF spaces and provides and O(log n) complexity for recognizing objects from images.

In Section 2, a system architecture is described, while the introduced multilevel shape representation is shown in Section 3. In Section 4, aspects are discussed related to the use of learned models in the matching phase. In Section 5 results are shown by considering reference sequences of the parking area of the Performance evaluation workshop PETS 2000 [8]. Finally, conclusions are given to highlight major merits and current limitations of the proposed approach.

## 2. System Architecture

Main modules of the surveillance system considered in this paper are shown in Fig. 1.

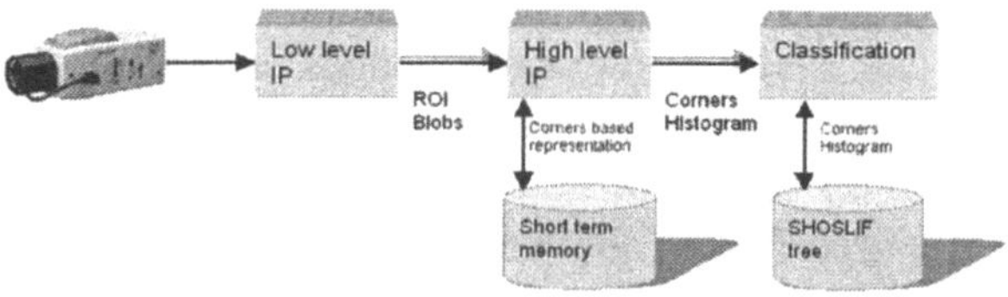


*Figure 1.* Considered Surveillance system modules

The set of tasks performed within the Low Level Image Processing module performs a first stage of abstraction from the sequence acquired from the sensor to the representation that is used for tracking and classification [9]. The acquired frame is first differenced with respect to a background reference scene; the change detection algorithm is followed by a focus of attention step consisting in application specific filtering of changed areas, together with identification of Regions of Interest (ROI) and their minimum boundary rectangles (blobs). An history of blobs is maintained in terms of a temporal graph [9]. Detected ROIs and blob history are considered for further processing by higher level modules.

The High Level Image Processing extracts high-curvature points (corners) and histograms from each detected object by using the algorithm described in [10]. Corners represent the basic common representation used to build up short-term and long-term object models.

Tracking and recognition modules perform their task on the basis of two main working modalities: learning and matching. The tracking module enters in this modality periodically when the object is not overlapped in order to update the short-term object model: this model is used by the object tracker when partial occlusions occur in the image plane.

The recognition module builds up a self organizing tree during the learning modality: a set of human classified samples are presented to the tree that automatically organizes them in such a way to maximize the inter-class distances, minimizing the intra-class variances. During the standard functioning of the system, the SHOSLIF tree is used for objects classification: each object that has been detected from the lower levels of the system is presented to the classification tree that outputs the estimated class for that object and the nearest training sample.

By comparison with previous architectures e.g. [9] one can see that the proposed system does not use pre-defined high-level models, but it tries to develop automatically from a set of supervised examples a joint representation of observed objects. This representation can be obtained in an incremental way and it is structured in such a way to allow tracking activities to be continued during learning. Moreover, the representation can be developed by requiring a low interactivity with teacher, this making it possible learning of sensor specific models in acceptable times. In fact, if it still remains true that recognition performances as well as related decision capabilities are expected to improve with training time, a reasonable classification behavior in some cases of interest can be obtained after training with relatively few examples, as shown in this paper. The easier learning modality with respect to other approaches, e.g. training set collection for neural networks, represent an advantage of the proposed system to be explored in more detail in the future. On the other hand, portability of object models can be also chosen without expected high loss of performances.

## 3. Corner-Based Tracker

Corners are used as high curvature feature points for describing the shape of observed objects. A simple method is used to extract a set of corners from each blob in frames of the sequence that has been described in [10]. In Fig. 4 corners associated with blobs in a short windowed sequence are shown. A list of $N(t)$ corners, $C_k(t)=\{c_i(t):i=1...N(t)\}$ associated with *k-th* blob is then passed to the tracking and recognition modules to allow them to obtain their representations. Each corner $c_i(t)$ is characterized by his relative position with respect to the center of the bounding rectangle of the ROI to which it belongs, as well as by corner orientation as estimated by the corner extraction module. Therefore, each corner can be considered as a triplet $c_i(t)=(\omega,dx,dy)_i(t)$, where $\omega$ is the angle direction of the gradient vector at the corner position with respect to the original image, while $dx$ and $dy$ represent the shift of the corner with respect to the center of mass of the ROI.

### 3.1 Tracking oriented shape representation

The selection of a subset of corners from those extracted for each frame is based on the observation that spatial-temporal regularization is necessary to select robust features to be tracked in order to obtain a good estimation of object position and pose. Spatial grouping and temporal accumulation of evidence have been chosen as basic criteria to develop a short-term object representation from well isolated blobs.

A computationally simpler version of the Generalized Hough Transform (GHT) [7] is used as spatial and temporal grouping tool. A look-up table, called R-table is used to model the template shape of the object. The R-table is used as a voting mechanism during the matching phase to detect objects and to estimate their position in a scene. In the used version of the GHT, a 3 x M(t) cumulative R-table is used for each object, where 3 refers to the triplet $(\omega, dx, dy)_i(t)$ associated with each object and M(t) represents the number of corners used as the short-term model of a generic object. Using gradient information allows saving computational time in the voting phase. In figure 2 an example of voting corner and of the related GHT table is given.

In our approach, the cumulative R-tables from previous frames in the temporal window of size w are used to obtain a more robust model. Then a unique *R*-table, $RC_k(t)$, is constructed by selecting the corners that are present in most of the frames at the same location and with the same orientation (stable corners) allowing small variations it their values.

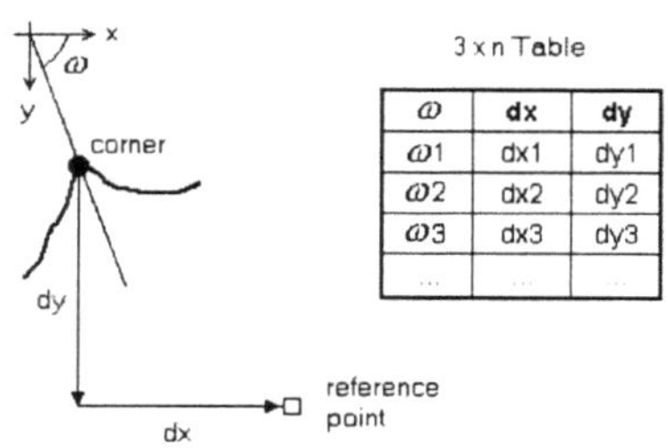


*Figure 2.* GHT: example of voting corner and related R-table at time t.

An example of the considered shape description is shown in figure 3: 3 frames are used to extract the shape model. The extracted corners are shown for each frame, and the superimposed corners over a window of length w=3 at time t are shown for explanation purposes.

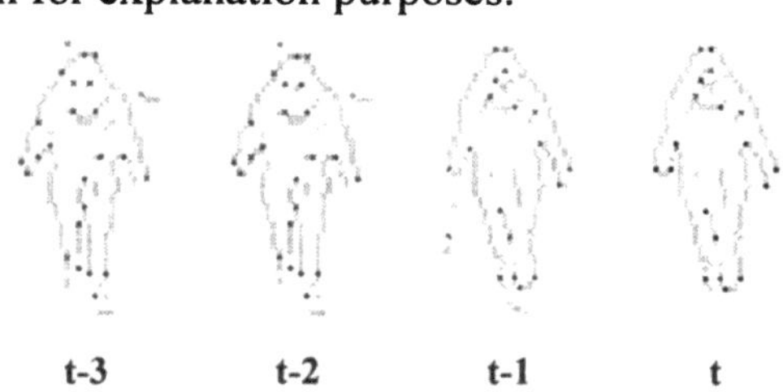


*Figure 3.* Corners extracted for a 3 frames short term memory and resulting shape model (t).

In this way, a simple spatial-temporal regularization is performed. In order to take advantage from such properties, each model is updated only when the related object is isolated on the image plane. When merging between objects occurs the shape model is no longer updated and the models obtained before the merge can be used for separating the objects.

## 4. Object Classification

The long term recognition module uses corner representation and histograms features extracted by the Image Processing modules as a basis for objects classification. While shape related features are useful for classifying objects attitudes and poses, histogram features are used for objects identification. SHOSLIF (Self-Organizing Hierarchical Optimal Subspace Learning and Inference Framework) [6] is the tool that has been chosen for the objects classification. SHOSLIF constitute a supervised self-organizing scheme searching for a tree like organization of a set of pattern in a training set based on most discriminant and most expressive features selected in a distributed way along the multilevel tree structure.

The input of a SHOSLIF is a set of labeled patterns $X=\{(x_n, w_n):n=1..N\}$, i.e. the training set, where $x_n$ is a vector of dimensionality $K$ representing the observed sample and $w_n$ is a class associated with $x_n$ chosen in a set of $C$ classes. The SHOSLIF algorithm produces as output a tree whose nodes contain a decreasing set of samples. The root node contain all samples in $X$. Such samples are distributed to nodes at immediately lower level according to a the following iterative procedure:

1) the Karhunen Loeve expansion is computed [11], i.e. the covariance of the samples in $X$, $S$, and its eigenvalues and eigenvectors are computed; a number of eigenvectors is chosen to obtain $Y=\{y_n\}$ where $y_n$ is a vector of reduced size associated with sample $x_n$. Vector $y_n$ is expressive of $x_n$ in that most important eigenvectors are used to form it, i.e. those eigenvectors associated with higher value (i.e. energy) eigenvalues.

2) Eigenvalues and eigenvectors are computed from $Y$ for the matrix $J=S_w^{-1}S_b$, where $S_w$ is the so called within-class scatter matrix, and $S_b$ is the mixture (between-class) scatter matrix, given by:

$$S_W = \sum_{i=1}^{L} P_i E\{(X-M_i)(X-M_i)^T|\omega_i\} = \sum_{i=1}^{L} P_i \Sigma_i$$

$$S_b = \sum_{i=1}^{L} P_i (M_i - M_0)(M_i - M_0)^T \text{ where } M_0 = E\{X\} = \sum_{i=1}^{L} P_i M_i$$

It is well known [11] that choosing the highest value m eigenvalues and related eigenvectors of J can be a good criterion for ensuring optimal classification under a linear feature selection hypothesis. Therefore, from set Y it is possible to obtain a set $Z=(z_n\}$ each sample $z_n$ being obtained by projecting $y_n$ on a subspace by a linear transformation preserving discriminant properties of Y.

3) New children nodes are created and samples in Z are distributed in sequential way between such nodes: distribution occurs by defining or updating new centers in each node (each center being computed as the average of $z_n$ values of samples associated with the node), and by comparing an appropriately distance between the new sample to be added and previously computed node centers. If the distance is higher than a threshold to the centers of all the children nodes, then a new node is created and the sample $z_n$ is associated with the node. The threshold r(l) decreases with the level, this corresponding to a different classification resolution in the multilevel linear tessellation performed by the SHOSLIF method. Identity and class of samples is preserved along the distribution process in order to allow later classification and retrieval.

4) The process is iterated until each node contains a single sample.

The Principal Component Analysis utilizes the eigenvectors of the sample scatter matrix associated with the largest eigenvalues. These vectors are in the direction of the major variations in the samples, and such can be used as a basis set with which to describe the image samples. Because they capture the major variations in the training data, they can express the samples well and can approximate the samples.

Although the MEF projection is well-suited to object representation, the features produced are not necessarily good for discriminating among classes defined by the set of samples. The MEFs describe some major variations in the set of samples: these variations may be irrelevant to how the classes are divided. This problem can be solved by using discriminant analysis: the between-class scatter is maximized while minimizing the within-class scatter. The samples for each class are projected to a space where each class is clustered more tightly together, and the separation between the class means is increased. The features obtained using a discriminant analysis projection optimally discriminate among the classes represented in the training set, in the sense of the linear transform [12]. The SHOSLIF algorithm has been proven to be a powerful classification and retrieval tool in image retrieval applications. In this context we use it to organize corners extracted by blobs associated with different object classes during a learning phase.

Therefore, the training set *X* is represented by a set of pairs (corners, class) obtained by lower level modules working in the learning modality and partially by human integration of information for blob class identification by visual inspection. However, a problem occurs due to variable dimension of corner vectors at different time instants. As a consequence, a preliminary step is necessary before developing the SHOSLIF tree, i.e. selecting the dimensionality of the corner vector. Moreover, we would like that the recognition system be invariant to scale. A re-scaling step to a common format is also necessary (for being able to compare object shapes at variable distances).

The features selection is performed by partitioning the corner set $C_k(t)$ into $M$ regions, where $M$ is the desired cardinality for the pattern $x$ to be given to the SHOSLIF. The corner set is chosen by iteratively partitioning the blob into two areas, each characterized by the same number of corners (fig 4). The partitioning process is performed alternatively on the two orthogonal axis directions. For each region the vector median corner in the survived local population is chosen as representative sample. As shown in Fig. 5, survived corners are visualized for inspection purposes as two set of connected points: external closed lines connect median corner points in outer regions, while internal lines connect corner points in inner regions.

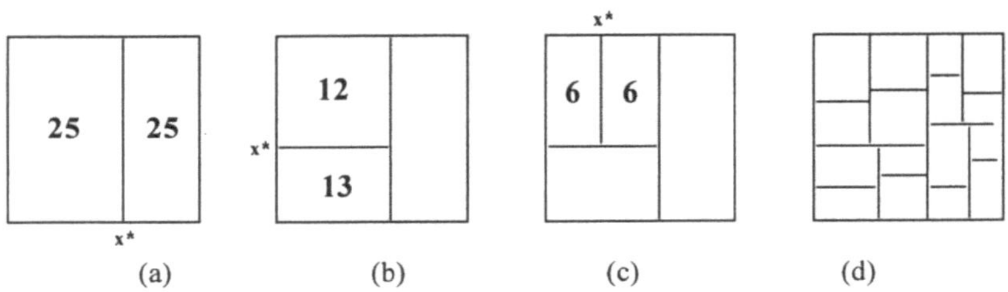


*Figure 4.* Corner partitioning process example: a) first division along x axis, b) second division along y axis, c) third division along x axis, d) final areas with M=16.

In this way, the vector $x_n$ is computed for each sample and a class $w_n$ is associated with it. Note that the class association process can be performed only once and attributed to a whole sequence where the object is well separated by other objects. For what concerns shape, seven different classes have been identified as shown in figure 6.

*Figure 5.* Examples of survived corners: for visualization purposes two set of connected points are designed: external closed lines connect external median corner points, while internal lines connect internal corner points (U=100).

| | |
|---|---|
|  | Class 1: left diagonal car |
|  | Class 2: vertical car |
|  | Class 3: right diagonal car |
|  | Class 4: horizontal car |
|  | Class 5: split legs pedestrian |
| 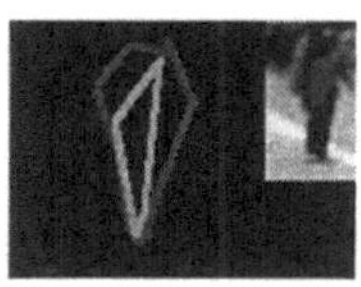 | Class 6: closed legs pedestrian |
|  | Class 7: left diagonal van |

*Figure 6.* Seven different classes of objects has been considered

# 5. Results

Two SHOSLIF trees has been generated by using shape and histogram features respectively for the seven proposed classes. Selected training set is made up by a set of 328 samples distributed over the classes. By using standard parameters for the tree generation, the resulting tree is 8 levels deep and is constituted by 442 nodes, each node having its own projection matrices for MEF and MDF spaces. The system has a null misdetection probability for the training set itself.

The test set is made up by 30 samples: the misdetection probability over the test set is 15% and the system performs a wrong classification between classes 5 and 6 and between classes 1, 2, 3, 4 and 7. The algorithm never classify a car as a pedestrian or viceversa. In particular some examples of correct and wrong classification are shown in figures 7 and 8 respectively.

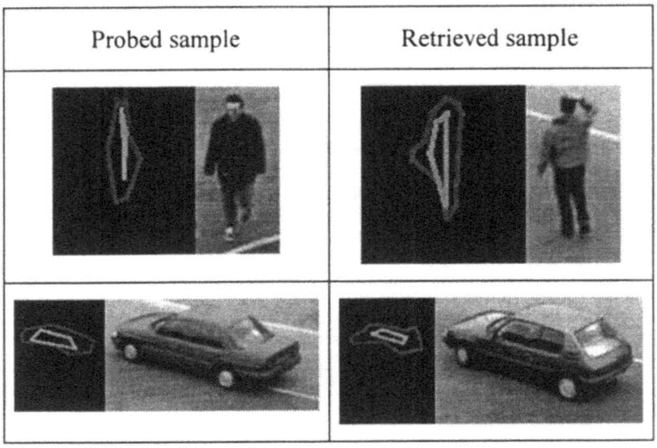


*Figure 7.* Two corrected classified couples of samples.

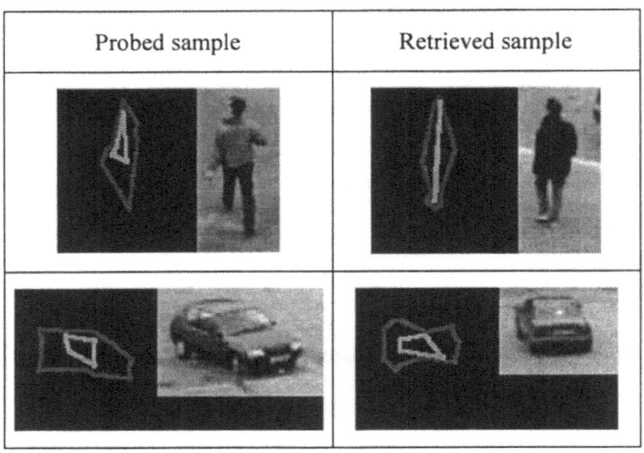


*Figure 8.* Two uncorrected classified couples of samples.

The second test has been done by using histograms for objects identification. In this case seven different classes have been detected depending on the identity of the object (e.g. red car, white van, etc). The probability of misclassification is 8% and the system never misclassify between vehicles because objects belonging to that class are well separated in the color space. There are some classification errors for the pedestrian classes because several samples are badly separated in the selected feature space. Figure 9 shows an example of this kind of misclassification.

| Probed sample | Retrieved sample |
|---|---|
| 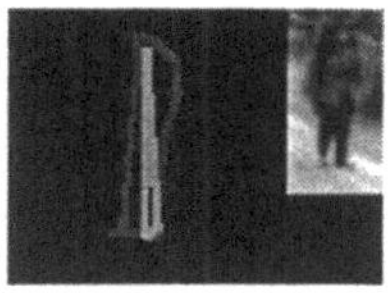 |  |

*Figure 9.* Misclassification between badly separated classes

## 6. Conclusions

A method for tracking and classifying objects in a video-surveillance system has been presented. A corner based shape model is used for tracking and for recognizing an object pose. The classification phase is performed by using self organizing SHOSLIF trees automatically generated during the off-line phase. Computed misdetection probabilities confirm the correctness of the proposed approach.

## Acknowledgements

This work was partially supported by the Ministry of Universities and Scientific Research (MURST) of the Italian Government and by the British Council.

## References

[1] A. Teschioni, F. Oberti and C.S. Regazzoni, "A Neural Network Approach for Moving Objects Recognition in Color Image Sequences for Surveillance Applications", *Nonlinear Signal and Image Processing NSIP'99*, pp 28-32. Antalya, Turkey.

[2] A. Theil, R. Kemp, K. Romeo, L. Kester, E. Bosse, "Classification of Moving Objects in Surveillance Algorithms", *First IEEE International Workshop on Performance Evaluation of Tracking and Surveillance*, PETS'2000, Grenoble, France, March 2000, pp.80-84.

[3] G.L. Foresti, and C.S. Regazzoni, "A new shape representation method for progressive image coding at very low bit rate", *IEEE Signal Processing Letters*, Vol. 4, No.9, September 1997, pp.248-251.

[4] A.M. Baumberg and D.C. Hogg, "An Efficient Method for Contour Tracking Using Active Shape Models", Technical Report 94.11, School of Computer Studies, University of Leeds, Apr. 1994.
[5] M. Haag and H.H. Nagel, "Combination of edge element and optical flows estimates for 3D-Model based vehicle tracking in traffic image sequences", *Int. Journal of Computer Vision*, Vol. 35:3, pp.295-319, 1999.
[6] D.L. Swets and J. Weng, "Hierarchical Discriminant Analysis for Image Retrieval", *IEEE Transactions on Pattern Analysis and Machine Intelligence*, Vol. 21, No. 5, pp. 386-401, May 1999.
[7] D.H. Ballard, "Generalizing the HT to detect arbitrary shapes", *Pattern Recognition*, Vol. 13, No. 2, pp.111-122, 1981
[8] *First IEEE International Workshop on Performance Evaluation of Tracking and Surveillance*, PETS'2000, Grenoble, France, March 2000.
[9] L. Marcenaro, F. Oberti and C.S. Regazzoni, "Short-memory Shape Models for Ground-plane Predictive Object Tracking", *First IEEE International Workshop on Performance Evaluation of Tracking and Surveillance*, PETS'2000, Grenoble, France, March 2000, pp.50-56.
[10] F. Oberti and C.S. Regazzoni, "Real-Time Robust Detection Of Moving Objects In Cluttered Scenes", submitted for publication to the Eusipco2000 Conference.
[11] K. Fukunaga, *"Introduction to Statistical Pattern Recognition"*, Academic Press, London 1990.
[12] S.S. Wilks, *Mathematical Statistics*. Wiley, New York, 1963.

Chapter 11

# An Improved Adaptive Background Mixture Model for Real-time Tracking with Shadow Detection

P. KaewTraKulPong[1,2] and R. Bowden[1]
*[1]Vision and Virtual Reality group, Department of Systems Engineering, Brunel University, Middlesex, UB8 3PH, UK: [2]King Mongkut's University of Technology Thonburi, Bangmod, Toong-kru, Bangkok, 10140 Thailand.*

**Abstract:** Real-time segmentation of moving regions in image sequences is a fundamental step in many vision systems including automated visual surveillance, human-machine interface, and very low-bandwidth telecommunications. A typical method is background subtraction. Many background models have been introduced to deal with different problems. One of the successful solutions to these problems is to use a multi-colour background model per pixel proposed by Grimson *et al* [1,2,3]. However, the method suffers from slow learning at the beginning, especially in busy environments. In addition, it can not distinguish between moving shadows and moving objects. This paper presents a method which improves this adaptive background mixture model. By reinvestigating the update equations, we utilise different equations at different phases. This allows our system learn faster and more accurately as well as adapts effectively to changing environment. A shadow detection scheme is also introduced in this paper. It is based on a computational colour space that makes use of our background model. A comparison has been made between the two algorithms. The results show the speed of learning and the accuracy of the model using our update algorithm over the Grimson *et al*'s tracker. When incorporate with the shadow detection, our method results in far better segmentation than that of Grimson *et al.*

**Key words:** Background Subtraction, Shadow Suppression, Expectation-Maximisation Algorithm, Gaussian Mixture Model

## 1. Introduction

Background subtraction involves calculating a reference image, subtracting each new frame from this image and thresholding the result. What results is a binary segmentation of the image which highlights regions of non-stationary objects. The simplest form of the reference image is a time-averaged background image. This method suffers from many problems and requires a training period absent of foreground objects. The motion of background objects after the training period and foreground objects motionless during the training period would be considered as permanent foreground objects. In addition, the approach cannot cope with gradual illumination changes in the scene. These problems lead to the requirement that any solution must constantly reestimate the background model. Many adaptive background-modelling methods have been proposed to deal with these slowly-changing stationary signals. Friedman and Russell modelled each pixel in a camera scene by an adaptive parametric mixture model of three Gaussian distributions [4]. They also provide some brief discussion on the online update equations based on *sufficient statistics*. Koller *et al* used a Kalman filter to track the changes in background illumination for every pixel [5]. They applied a selective update scheme to include only the probable background values into the estimate of the background. The methods can cope well with the illumination changes; however, can not handle the problem of objects being introduced or removed from the scene. One solution to these is to use a multiple-colour background model per pixel. Grimson *et al* employed an adaptive nonparametric Gaussian mixture model to solve these problems [1,2,3]. Their model can also lessen the effect of small repetitive motions; for example, moving vegetation like trees and bushes as well as small camera displacement. Elgammal *et al* used a *kernel* estimator for each pixel [6]. Kernel exemplars were taken from a moving window. They also introduced a method to reduce the result of small motions by employing a *spatial coherence*. This was done by comparing simply connected components to the background model of its circular neighbourhood. Although the authors presented a number of speed-up routines, the approach was still of high computational complexity. Other techniques using high level processing to assist the background modelling have been proposed; for instance, the Wallflower tracker [7] which circumvents some of these problems using high level processing rather than tackling the inadequacies of the background model. Our method is based on Grimson *et al*'s framework [1,2,3], the differences lie in the update equations, initialisation method and the introduction of a shadow detection algorithm.

A common optimisation scheme used to fit a Gaussian mixture model is the *Expectation Maximisation* (EM) algorithm. The EM algorithm is an iterative method that guarantees to converge to a local maximum in a search space. Due to the space-time requirements in modelling each pixel for the background image, an online EM algorithm is required. Many online EM algorithms have been introduced. They can be classified into two groups.

The first group was in the realm of parametric estimation of *probability density functions* (pdf's). In other words, to use new data in updating the previous estimate without modifying the structure of the previous model. The procedure was introduced by Nowlan [8] and explained in terms of the results by Neal and Hinton [9]. Traven derived an $N$ most recent window version of the procedure [10]. McKenna *et al* [11,12,13] extended the result of Traven [10] to an $L$ most recent window of the results from $L$ batch EM runs and used it for tracking a multi-colour foreground object. This parametric estimation approach can not run effectively without a good initial estimate (normally found by running the batch EM algorithm). The second group is that of non-parametric approaches. Priebe *et al* introduced an adaptive mixture model with stochastic thresholding for generating new Gaussian kernels to the existing mixture model [14,15]. Grimson and Stauffer [2,3], however, applied the same scheme with deterministic thresholding.

In addition to Grimson *et al*, many other authors have applied in applying mixture models to model every pixel in camera scenes. Rowe and Blake applied the batch EM algorithm for off-line training in their *virtual* image plane [16]. However, the model does not update with time and therefore leads to failure for external environments where the scene lighting changes with time. Friedman and Russell modelled road, shadow and vehicle distribution for each pixel using an adaptive mixture of three Gaussian distributions [4]. The classification was based on heuristic method of relative distances in the intensity space. They reported a good segmentation using the expected sufficient statistics formulas. However, this still requires a preprocessing initialisation to learn an initial model using batch EM algorithm.

We explain the background model by Grimson and Stauffer [2,3] and its deficiencies in section 2.1. Our proposed solution to the problem is presented in section 2.2. Section 2.3 explains our shadow detection. Results from each method are shown and compared in section 3 and concluded in section 4.

## 2. Background Modelling

In this section, we discuss the work of Grimson and Stauffer [2,3] and its shortcomings. The authors introduces a way to model each background pixel by a mixture of $K$ Gaussian distributions ($K$ is a small number from 3 to 5). Different Gaussians are assumed to represent different colours. The weight parameters of the mixture represent the time proportions that those colours stay in the scene. Unlike Friedman *et al*'s work, the background components are determined by assuming that the background contains $B$ highest probable colours. The probable background colours are the ones which stay longer and more static. Static single-colour objects trend to form tight clusters in the colour space while moving ones form widen clusters due to different reflecting surfaces during the movement. The measure of this was called the *fitness* value in their papers. To allow the model to adapt to changes in illumination and run in real-time, an update scheme was applied. It is based upon selective updating. Every new pixel value is checked against existing model components in order of fitness. The first matched model component will be updated. If it finds no match, a new Gaussian component will be added with the mean at that point and a large covariance matrix and a small value of weighting parameter.

## 2.1 Adaptive Gaussian Mixture Model

Each pixel in the scene is modelled by a mixture of $K$ Gaussian distributions. The probability that a certain pixel has a value of $\mathbf{x}_N$ at time $N$ can be written as

$$p(\mathbf{x}_N) = \sum_{j=1}^{K} w_j \eta(\mathbf{x}_N; \boldsymbol{\theta}_j) \tag{1}$$

where $w_k$ is the weight parameter of the k$^{th}$ Gaussian component. $\eta(\mathbf{x}; \boldsymbol{\theta}_k)$ is the Normal distribution of k$^{th}$ component represented by

$$\eta(\mathbf{x}; \boldsymbol{\theta}_k) = \eta(\mathbf{x}; \boldsymbol{\mu}_k, \boldsymbol{\Sigma}_k) = \frac{1}{|2\pi\boldsymbol{\Sigma}_k|^{\frac{1}{2}}} e^{-\frac{1}{2}(\mathbf{x}-\boldsymbol{\mu}_k)^T \boldsymbol{\Sigma}_k^{-1}(\mathbf{x}-\boldsymbol{\mu}_k)} \tag{2}$$

where $\boldsymbol{\mu}_k$ is the mean and $\boldsymbol{\Sigma}_k = \sigma_k^2 \mathbf{I}$ is the covariance of the k$^{th}$ component.

The $K$ distributions are ordered based on the fitness value $w_k/\sigma_k$ and the first $B$ distributions are used as a model of the background of the scene where $B$ is estimated as

$$B = \arg\min_{b}\left(\sum_{j=1}^{b} w_j > T\right) \tag{3}$$

The threshold $T$ is the minimum fraction of the background model. In other words, it is the minimum prior probability that the background is in the scene. Background subtraction is performed by marking a foreground pixel any pixel that is more than 2.5 standard deviations away from any of the $B$ distributions. The first Gaussian component that matches the test value will be updated by the following update equations,

$$\hat{w}_k^{N+1} = (1-\alpha)\hat{w}_k^N + \alpha\hat{p}(\omega_k \mid \mathbf{x}_{N+1}) \tag{4}$$

$$\hat{\boldsymbol{\mu}}_k^{N+1} = (1-\rho)\hat{\boldsymbol{\mu}}_k^N + \rho\mathbf{x}_{N+1} \tag{5}$$

$$\hat{\boldsymbol{\Sigma}}_k^{N+1} = (1-\rho)\hat{\boldsymbol{\Sigma}}_k^N + \rho(\mathbf{x}_{N+1} - \hat{\boldsymbol{\mu}}_k^{N+1})(\mathbf{x}_{N+1} - \hat{\boldsymbol{\mu}}_k^{N+1})^T \tag{6}$$

$$\rho = \alpha\eta(\mathbf{x}_{N+1}; \hat{\boldsymbol{\mu}}_k^N, \hat{\boldsymbol{\Sigma}}_k^N) \tag{7}$$

$$\hat{p}(\omega_k \mid \mathbf{x}_{N+1}) = \begin{cases} 1 & \text{;if } \omega_k \text{ is the first match Gaussian component} \\ 0 & \text{;otherwise} \end{cases} \tag{8}$$

where $\omega_k$ is the $k^{th}$ Gaussian component. $1/\alpha$ defines the time constant which determines change.

If none of the $K$ distributions match that pixel value, the least probable component is replaced by a distribution with the current value as its mean, an initially high variance, and a low weight parameter.

According to their papers [1,2,3], only two parameters, $\alpha$ and $T$, needed to be set for the system. The details of its robustness were explained in their papers [1,2,3]; however, with a simple discussion, we can see its incapability. Firstly, if the first value of a given pixel is a foreground object, there is only one Gaussian where its weight equals unity. With only one-colour subsequent background values, it will take $\log_{(1-\alpha)}(T)$ frames until the genuine background can be considered as a background and $\log_{(1-\alpha)}(0.5)$

frames until it will be the dominant background component. For example, if we assume that at least 60% of the time the background is present and $\alpha$ is 0.002 (500 recent frames), it would take 255 frames and 346 frames for the component to be included as part of the background and the dominant background component, respectively. The situation can be worse in busy environments where a clean background is rare. This paper presents a solution to the problem in the next section. Secondly, $\rho$ is too small due to the likelihood factor. This leads to too slow adaptations in the means and the covariance matrices, therefore the tracker can fail within a few seconds after initialisation. One solution to this is to simply cut out the likelihood term from $\rho$.

## 2.2 Online EM Algorithms

We begin our estimating of the Gaussian mixture model by expected sufficient statistics update equations then switch to $L$-recent window version when the first $L$ samples are processed. The expected sufficient statistics update equations provide a good estimate at the beginning before all $L$ samples can be collected. This initial estimate improves the accuracy of the estimate and also the performance of the tracker allowing fast convergence on a stable background model. The $L$-recent window update equations gives priority over recent data therefore the tracker can adapt to changes in the environment.

The online EM algorithms by expected sufficient statistics are shown in the left column while the by $L$-recent window version in the right.

$$\hat{w}_k^{N+1} = \hat{w}_k^N + \frac{1}{N+1}\left(\hat{p}(\omega_k \mid \mathbf{x}_{N+1}) - \hat{w}_k^N\right) \tag{9}$$

$$\hat{\boldsymbol{\mu}}_k^{N+1} = \hat{\boldsymbol{\mu}}_k^N + \frac{\hat{p}(\omega_k \mid \mathbf{x}_{N+1})}{\sum_{i=1}^{N+1} \hat{p}(\omega_k \mid \mathbf{x}_i)}\left(\mathbf{x}_{N+1} - \hat{\boldsymbol{\mu}}_k^N\right) \tag{10}$$

$$\hat{\boldsymbol{\Sigma}}_k^{N+1} = \hat{\boldsymbol{\Sigma}}_k^N + \frac{\hat{p}(\omega_k \mid \mathbf{x}_{N+1})}{\sum_{i=1}^{N+1} \hat{p}(\omega_k \mid \mathbf{x}_i)}\left((\mathbf{x}_{N+1} - \hat{\boldsymbol{\mu}}_k^N)(\mathbf{x}_{N+1} - \hat{\boldsymbol{\mu}}_k^N)^T - \hat{\boldsymbol{\Sigma}}_k^N\right) \tag{11}$$

$$\hat{w}_k^{N+1} = \hat{w}_k^N + \frac{1}{L}\left(\hat{p}(\omega_k \mid \mathbf{x}_{N+1}) - \hat{w}_k^N\right) \tag{12}$$

$$\hat{\boldsymbol{\mu}}_k^{N+1} = \hat{\boldsymbol{\mu}}_k^N + \frac{1}{L}\left(\frac{\hat{p}(\omega_k \mid \mathbf{x}_{N+1})\mathbf{x}_{N+1}}{\hat{w}_k^{N+1}} - \hat{\boldsymbol{\mu}}_k^N\right) \tag{13}$$

$$\hat{\boldsymbol{\Sigma}}_k^{N+1} = \hat{\boldsymbol{\Sigma}}_k^N + \frac{1}{L}\left(\frac{\hat{p}(\omega_k \mid \mathbf{x}_{N+1})(\mathbf{x}_{N+1} - \hat{\boldsymbol{\mu}}_k^N)(\mathbf{x}_{N+1} - \hat{\boldsymbol{\mu}}_k^N)^T}{\hat{w}_k^{N+1}} - \hat{\boldsymbol{\Sigma}}_k^N\right) \tag{14}$$

## 2.3 Shadow Detection and Colour Model

As it is evidence in their papers [1,2,3], Grimson *et al*'s tracker can not identify moving shadows from the objects casting them. The reason behind this is that no heuristic exists to label Gaussian components as moving shadows. One solution is to use a chromatic colour space representation which reduces susceptibility. As many colour spaces can separate chromatic and illumination components, maintaining a chromatic model regardless of the brightness can lead to an unstable model especially for very bright or dark objects. This conversion also requires computational resources particularly in large images. The idea of preserving intensity component and saving computational costs lead us back to the RGB space. As the requirement to identify moving shadows, we need to consider a colour model that can separate chromatic and brightness components. It should be compatible and make use of our mixture model. This is done by comparing non-background pixel against the current background components. If the difference in both chromatic and brightness components are within some thresholds, the pixel is considered as a shadow. We use an effective computational colour model similar to the one proposed by Horprasert *et al* [17] to fulfil these needs. It consists of a position vector at the RGB mean of the pixel background, $E$, an *expected chromaticity line*, $\|E\|$, a *chromatic distortion*, $d$, and a brightness threshold, $\tau$. For a given observed pixel value, $I$, a brightness distortion, $a$, and a colour distortion, $c$, from the background model can be calculated as

$$a = \arg\min_z (I - zE)^2 \text{ and } c = \|I - aE\| \tag{15}$$

With the assumption of spherical Gaussian distribution in each mixture component, the standard deviation of the k$^{th}$ component $\sigma_k$ can be set equal to $d$. The calculation of $a$ and $c$ are trivial using vector dot product. A non-background observed sample is considered a moving shadow if $a$ is within, in our case, 2.5 standard deviations and $\tau < c < 1$.

## 3. Experiment

This section demonstrates the performance of the Grimson model [2,3] and our proposed algorithms on an image sequence. The sequence shown here is 192x144 images. We used an adaptive mixture of five Gaussian components. The $L$ was set at 500 frames ($\alpha$=0.002 in Grimson *et al*'s) and the threshold $T$ was set at 0.6. In the shadow detection module, the brightness threshold, $\tau$ of 0.7 was used. To show the performance of the background models, higher level processes such as noise cleaning or connected component analysis algorithms were not introduced to the results of background subtractions. Figure 1 shows a sequence of busy outdoor scene containing people walking in a public pathway. The sequence includes strong sunshine, large shaded area, tree, reflections from glasses and long moving shadows. We have presented images of the initial phase, busy scene and a long run. Because of no clean images at the beginning, an artefact of the initial image left in Grimson *et al*'s tracker lasted for over a hundred frames. Better segmentation can be seen from our method. The performance enhances dramatically with the shadow detection module.

## 4. Conclusion

We have presented new update algorithms for learning adaptive mixture models of background scene for the real-time tracking of moving objects. The algorithm run under the framework of the real-time robust tracker proposed by Grimson *et al.* A comparison has been made between the two algorithms. The results show the speed of learning and the accuracy of the model using our update algorithm over the Grimson *et al*'s tracker. We have proposed a method to detect moving shadows using our existing mixture model. This significantly reduces additional computational burdens. Shadow detection need only be performed upon pixels labelled as foreground and therefore with a negligible computational overheads. The moving shadows can be detected successfully. The shadow detection also reduces the effect of small repetitive motions in the background scene.

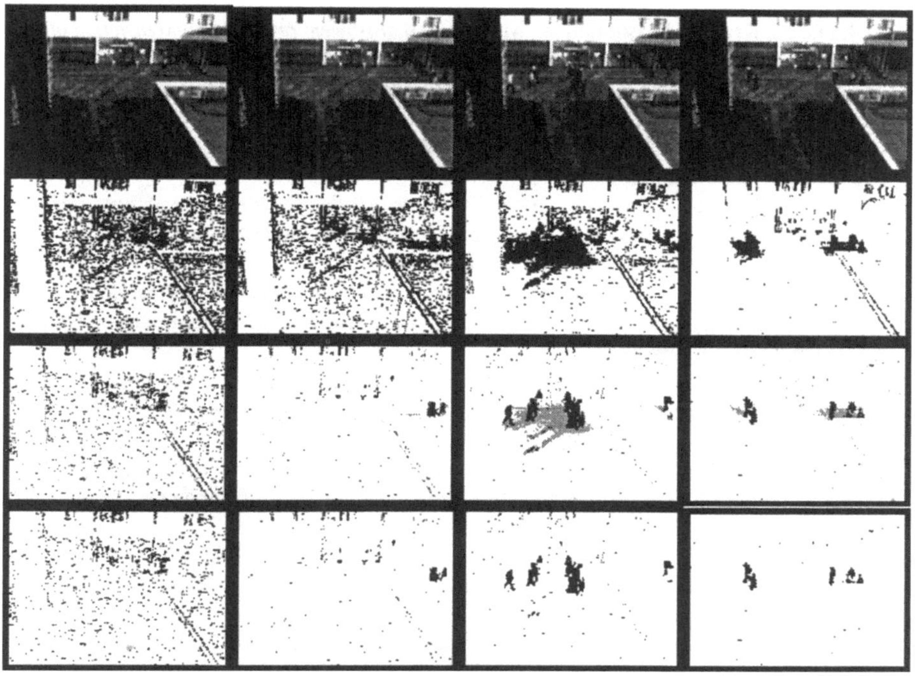

*Figure 1.* The top row displays the original sequence at frames 15, 105, 290 and 1200 respectively. The second row shows the results from Grimson *et al*'s. The last two rows are the results of our proposed method with and without moving shadows displayed in the images. The shadows are shown in grey.

# References


[1] Grimson, W. E. L., Stauffer, C., Romano, R., Lee, L. Using adaptive tracking to classify and monitor activities in a site. in Proceedings. *1998 IEEE Computer Society Conference on Computer Vision and Pattern Recognition.* IEEE Computer. Soc. 1998. 1998.

[2] Stauffer, C., Grimson, W. E. L. *Adaptive background mixture models for real-time tracking.* in *Proceedings. 1999 IEEE Computer Society Conference on Computer Vision and Pattern Recognition.* IEEE Computer. Soc. Part Vol. 2, 1999.

[3] Stauffer, C., Grimson, W. E. L., *Learning patterns of activity using real-time tracking.* IEEE Transactions on Pattern Analysis & Machine Intelligence, 2000. 22(8): p. 747-57.

[4] Friedman, N., Russell, S. *Image Segmentation in Video Sequences: A Probabilistic Approach.* in *The Thirteenth Conference on Uncertainty in Artificial Intelligence.* 1997. Brown University, Providence, Rhode Island, USA: Morgan Kaufmann Publishers, Inc., San Francisco, 1997.

[5] Koller, D., Weber, J., Huang, T., Malik, J., Ogasawara, G., Rao, B., Russell, S. *Towards robust automatic traffic scene analysis in real-time.* in *Proceedings of the 33rd IEEE Conference on Decision and Control. IEEE. Part vol.4, 1994.* 1994.

[6] Elgammal, A., Harwood, D., Davis, L. *non-parametric model for background subtraction.* in *IEEE ICCV'99 FRAME-RATE WORKSHOP.* 1999.

[7] Toyama, K., Krumm, J., Brumitt, B., Meyers B. *Wallflower: principles and practice of background maintenance*. in *Proceedings of the Seventh IEEE International Conference on Computer IEEE Computer. Soc. Part vol.1, 1999*. 1999.

[8] Nowlan, S. J., *Soft Competitive Adaptation: Neural Network Learning Algorithms based on Fitting Statistical Mixtures*, in *School of Computer Science*. 1991, Carnegie Mellon University: Pittsburgh, PA.

[9] Neal, R. M., Hinton, G. E., *A view of the EM algorithm that justifies incremental, sparse, and other variants*, in *Learning in Graphical Models*, M. I. Jordan, Editor. 1998, Dordrecht: Kluwer Academic Publishers. p. 355-368.

[10] Traven, H. G. C., *A neural network approach to statistical pattern classification by 'semiparametric' estimation of probability density functions.* IEEE Transactions on Neural Networks, 1991. 2(3): p. 366-77.

[11] McKenna, S. J., Raja, Y., Gong, S., *Object tracking using adaptive colour mixture models.* Computer Vision - ACCV '98. Third Asian Conference on Computer Vision. Proceedings. Springer-Verlag. Part vol.1, 1997, 1998: p. 615-22 vol.

[12] Raja, Y., McKenna, S. J., Gong, S., *Color model selection and adaptation in dynamic scenes.* Computer Vision - ECCV'98. 5th European Conference on Computer Vision. Proceedings. Springer-Verlag. Part vol.1, 1998, 1998: p. 460-74 vol.

[13] Raja, Y., McKenna, S. J., Gong, S., *Segmentation and tracking using colour mixture models.* Computer Vision - ACCV '98. Third Asian Conference on Computer Vision. Proceedings. Springer-Verlag. Part vol.1, 1997, 1998: p. 607-14 vol.

[14] Priebe, C. E., Marchette, D. J., *Adaptive mixtures: recursive nonparametric pattern recognition.* Pattern Recognition, 1991. 24(12): p. 1197-209.

[15] Priebe, C. E., Marchette, D. J., *Adaptive mixture density estimation.* Pattern Recognition, 1993. 26(5): p. 771-85.

[16] Rowe, S., Blake, A., *Statistical background modelling for tracking with a virtual camera.* in *BMVC `95 Proceedings of the 6th British Machine Vision Conference. BMVA Press. Part vol.2, 1995*. 1995.

[17] Horprasert, T., Harwood, D., Davis, L.S., a statistical approach for real-time robust background subtraction and shadow detection. in IEEE ICCV'99 FRAME-RATE WORKSHOP. 1999.

Chapter 12

# The Sakbot System for Moving Object Detection and Tracking

R. Cucchiara[1], C. Grana[1], G. Neri[2], M. Piccardi[3], A. Prati[1]
*1 D.S.I. - University of Modena and Reggio Emilia, via Vignolese 905 – 41100 Modena, Italy*

*2 D.E.I.S. - University of Bologna, viale Risorgimento 2 – 40136 Bologna, Italy*

*3 Dip. Ingegneria - University of Ferrara, via Saragat 1 – 44100 Ferrara, Italy*

**Abstract**: This paper presents Sakbot, a system for moving object detection in traffic monitoring and video surveillance applications. The system is endowed with robust and efficient detection techniques, which main features are the statistical and knowledge-based background update and the use of HSV color information for shadow suppression. Tracking is provided by a symbolic reasoning module allowing flexible object tracking over a variety of different applications. This system proves effective on many different situations, both from the point of view of the scene appearance and the purpose of the application.

**Key words**: motion detection and segmentation, shadow suppression, background differencing, ghost detection

## 1. Introduction

In the last decade, many approaches have been proposed for moving object detection and tracking from videos, mainly dedicated to traffic monitoring and visual surveillance. Detection of moving objects in videos (moving visual objects, or MVOs for short hereafter) is based on the realistic assumption that in almost each frame MVOs are perceivable as different from background and from recent previous frames. In addition, models of target MVOs can be used for improving detection in those cases where target objects allow us a model-based description and models are known a priori. Instead, model-based detection cannot be performed in cases where precise geometrical models are not possible (for instance, for nonrigid objects like pedestrians) or, simply, object models are not exhaustively known in advance.

In this work we aim to detect different and possibly unknown objects, including vehicles and pedestrians in outdoor traffic scenes but also casual walkers in parking or private areas. The approach we propose (Sakbot, Statistic and Knowledge-based Object Tracker) is meant to be general-purpose and thus adequate for application ranging from vehicle tracking to visual surveillance. Unlike other works in the field, we do not address the relatively simpler case of road surveillance, where one class of moving objects (i.e., vehicles) can be detected along some fixed motion directions. We consider instead more complex situations with moving people and vehicles, with different shape, speed, trajectory, presence of infrastructures and possible occlusions.

Sakbot is conceived as a general-purpose system adopting a single fixed camera, able to deal with several different operational situations:

a) luminance condition changes, due to atmospheric conditions, day hours, shadows.

b) limited, high-frequency camera motions, due to vibrations and the wind, which are compensated in a pre-processing phase;

c) background changes; the background of the scene could change either because objects are stopped (e.g. a stopped vehicle) or because still objects (previously considered background objects) start their motion.

d) MVOs can move with whatever trajectory and speed; thus our approach cannot be based on frame difference, a method that should calibrate difference in dependency on object speed. At the same time, motion models cannot be exploited.

In this context the process must be featured with a general and robust MVO detection phase. For the detection phase, we use a background suppression technique with selective adaptation of the background, improved by the knowledge of motion at object level rather then at pixel level.

The rest of the paper is organized as follows: Section 2 presents the main recent related works. Section 3 describes the Sakbot system with sub-sections for the architecture, the background estimation, and the shadow suppression techniques. Section 4 describes the Sakbot implementation and main application. Conclusive remarks addresses the main results and applications of Sakbot.

## 2. Related Work

Many systems have been proposed recently in the literature for moving object detection and tracking. These systems obviously differ in the approach proposed, but also in various assumptions about the operational environment. One first main distinction is between systems adopting a single, fixed camera [1-6] with respect to systems adopting either multiple cameras [7] or an airborne camera [8]. In this work, we focus on a single fixed camera scenario, since it still captures a wide spread of applications. Another relevant distinction is made between systems oriented to monitoring of traffic scenes, where the main targets are vehicles and pedestrians (see for instance [1,5,6,9]), and systems for video surveillance of unattended areas such as metro platforms,parking areas (see [3, 4]) and unmanned railways environments [10]. In the two cases, different a-priori knowledge about objects in the scene can be exploited in order to improve detection and/or tracking. In this work, we assume that both vehicles and pedestrians must be detected.

Many related works on motion detection and segmentation are based on background suppression. Although foreground points are detected using background differencing, these studies differ by means of the way the compute and update the background (reference) image. Several background update functions use statistics for extracting the background from sampled frames: for instance [11] uses a mean function, [12] uses the Mode, whereas the papers presented in [13,14,15] use mixture of Gaussians. Adaptive background update is proposed in [15,16] by way of a Kalman filter, and in other simpler approaches, which average with different weights the previously computed background and a few sampled frames. Selective background update has been proposed in [13], where the problem of deadlock is discussed.

## 3. The SAKBOT System

The main problem to be solved for the moving object detection phase is the definition of a fast and robust background update technique. To this aim, we have defined a specific approach, called S&KB background update, based on a statistical adaptation of the background, together with a "knowledge-based" selective background update [17]. The statistical adaptation copes with the changing appearance of the scene during time, while the knowledge-based selection prevents updating the background with foreground points. Thanks to adaptation, the system is able to integrate long stationary MVOs (e.g. a parked vehicles) or areas abandoned by previously still objects into the background; selectivity allows the system to discriminate apparently moving objects from real ones.

The main features of Sakbot are:

- the use of *color information* (instead of grey levels only), during the entire MVO detection process, which improves sensitivity; in fact, by utilizing color in the background differencing the discrimination efficacy is strongly improved;
- the use of a *median function with adaptivity* that is a good approximation of a mode function and assures high responsiveness to background changes and good accuracy, even if few time-samples are considered;
- the inclusion of a *knowledge-based feedback* from the segmentation module in order not to insert in the background parts of MVOs; this selective update, based on knowledge of whole objects and not of single pixels only, abates false positives (background pixels recognized as MVO pixels) and prevents from deadlock situations [13] in background update;
- the exploitation of a *shadow detection module* for improving both object segmentation and background updating.

## 3.1 The Sakbot architecture

The architecture of Sakbot consists of several steps (see Fig. 1). The processing time for these steps determines the upper-bound of the achievable processing rate.

Each new input frame ($IF_t$) undergoes a camera motion correction step, which uses a calibrated fixed point in the frame and a standard frame-by-frame cross-correlation for adjusting limited but unavoidable camera movements. The corrected image $I_t$ is computed by shifting the correspondent sampled frame $IF_t$ so that the fixed point is juxtaposed with the minimum of the correlation function.

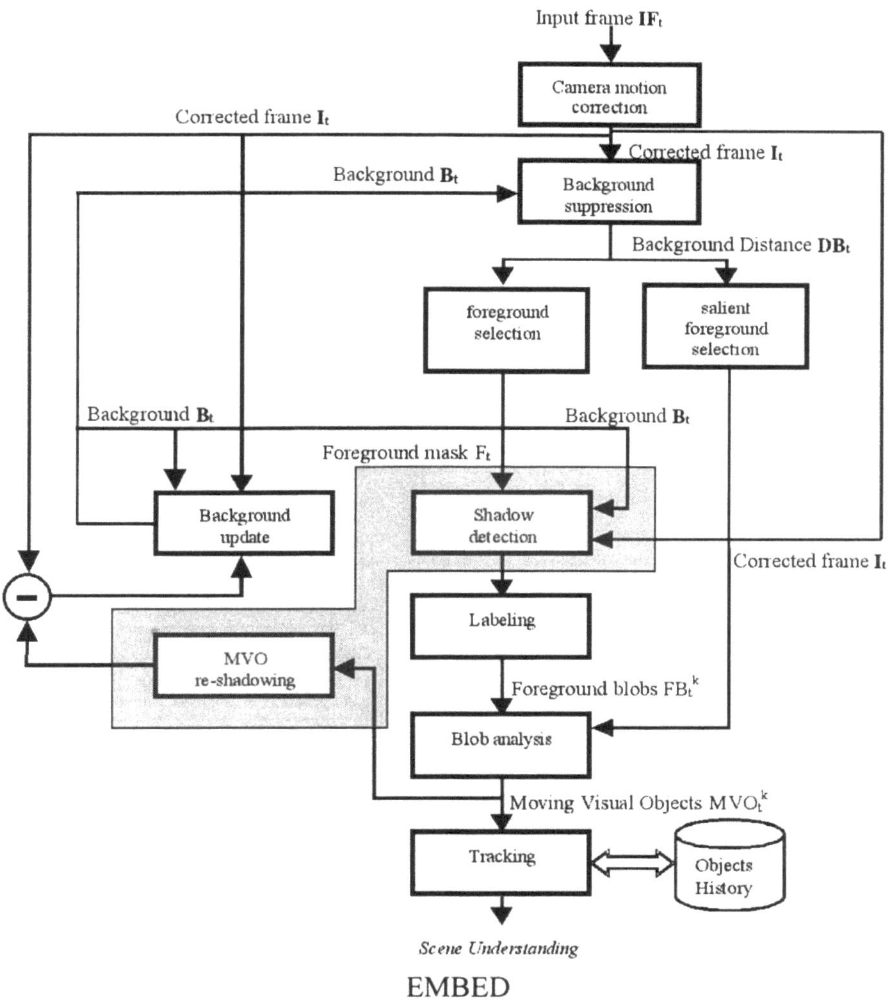


*Figure 1.* The Sakbot architecture

Once the input image has been corrected, the still background is subtracted to detect MVOs. Differently from many other proposals (see for instance [1,6]), we exploit the information given by the pixel chrominance to better discriminate foreground from background.

The difference image $DB_t$ (Difference with Background) is computed as follows:

$$DB_t(x,y)=Distance(I_t(x,y), B_t(x,y))=max(|I_t(x,y).c-B_t(x,y).c|\ c=R,G,B) \quad (1)$$

Therefore the distance image results to be a scalar, grey level-like image containing only candidate moving points.

*Figure 2.* Thresholding with hysteresis

Many background suppression approaches obtain the moving points by simply thresholding the $DB_t$ image. In order to limit the dependency of results from a given threshold, we adopt a threshold with hysteresis, conceptually similar to that exploited in [18]. Moreover, two-level thresholding avoids both the noise inclusion (typical of low threshold choice) and the partial detection of object shape (typical of thresholding based on high values). In [18] points are selected if greater than a low threshold and belonging to a neighbourhood of points higher than a high threshold. In this work, we use threshold with hysteresis added with morphology: we first select moving points with a low threshold Tlow (Fig. 2.b); then, morphological operators (closure and opening) are applied on the image in order to eliminate isolated points or very small spots due to noise. Then labelling is performed, by accepting blobs containing at least one point greater than a high threshold Thigh (see Fig. 2.c). The labelled image at time k thus contains a number of Foreground Blobs ($FB_t$ image, Fig. 2.d). In Fig. 1, the points obtained by these two levels of thresholding are called "foreground points" and "salient foreground points", respectively, and they include the mentioned morphological operations.

On the segmented foreground blobs we perform blob analysis, consisting of two steps:

1) blobs with an area less than a TAREA threshold (which depends on the distance between camera and scene and on the typical size of objects) are discarded;

2) the average optical flow (aOF) is computed for each blob. All blobs with aOF less than a TOF threshold are discarded. They are considered as apparently moving objects or *"ghosts"* (due to a locally incorrect background) and thus not accepted as real MVOs.

## 3.2 The Sakbot background update

The model of background we defined is based on the statistical assumption that the background points should be the most probable points observed in a finite window of observation time. As statistical function we use the median of sampled frames, that we have proven to be effective as the mode function but requiring a limited number of frames, and thus being less time and memory consuming [17]. In the literature, the background is usually computed with an adaptive function which takes into account past background values and the current frame. Accordingly, in order to improve the stability of the background update, we include an adaptive factor, i.e., the previous background weighted with an adequate weight $w_b$. Therefore we define:

$$B_t = median\ (I_t\text{-}\Delta t, \ldots, I_t\text{-}n\Delta t\ , w_b B_t\text{-}\Delta t) \quad (2)$$

The adaptiveness is achieved by including a given number $w_b$ of previously computed background samples in the set of data on which the median filter will be applied. With the use of statistics for background computation, MVO points should not be included in the background, thank to their low occurrence. Nevertheless, many errors occur, especially when objects are large, homogeneous in color, and slow: in this case, points of MVOs could turn out to be included in the statistic.

The statistical and adaptive update could be improved by a selective update, which does not update the background value if the point is marked as moving in the current frame, as in [13]. The drawback is the risk of a deadlock: if for any reason a fixed point is incorrectly detected as moving, it will remain excluded from background update, and therefore detected as moving point forever. In order to avoid this problem we exclude points belonging to detected MVOs after a further validation: MVOs must have a non null average optical flow aOF. Knowledge about the *whole* moving object makes selection more reliable. Thus we define the S&KB background update for each point in $B_k$ as

$$B_t = B_{t-1} \text{ if } I_t \in MVO_t^k \ \& \ aOF_k > TOF \ \text{ for any object } k,$$
$$= median\,(I_t\text{-}\Delta t, \ldots, I_t\text{-}n\Delta t\,, w_b B_t\text{-}\Delta t) \ \text{ otherwise} \qquad (3)$$

where $MVO_t^k$ is the k-th moving visual object detected at frame t.

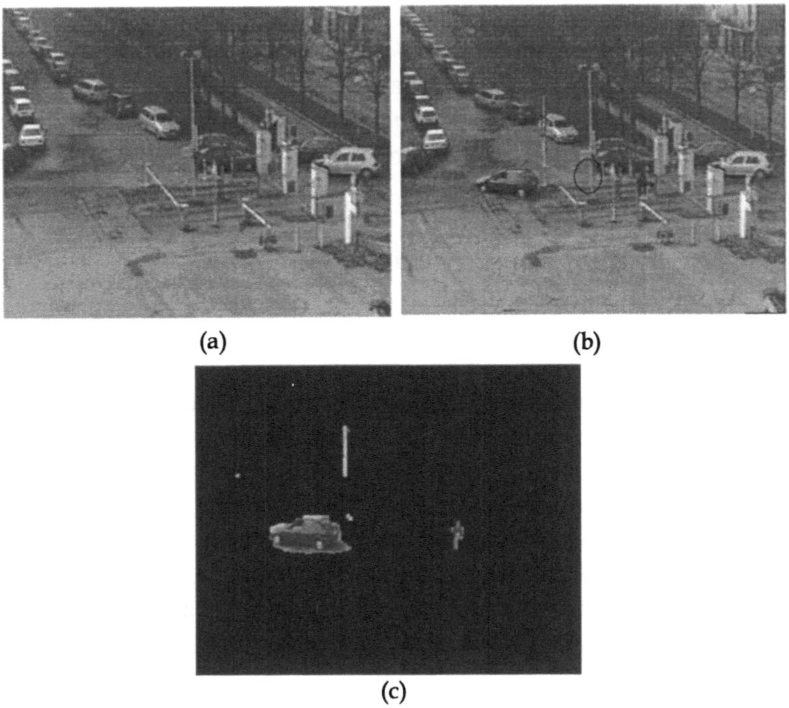


*Figure 3.* MVO detection with S&KB background update

Fig. 3 shows the effect of adding the feedback based on the knowledge of MVOs. Fig. 3.a represents the background with three closed barriers. In Fig. 3.b the bar has risen and a car is passing. In this frame four MVOs are detected: the car, the people, the rising bar and a virtual bar in the horizontal position (in the black circle). This object is a false positive and can be discriminated since its OF is null. Therefore points in the area occupied by the virtual bar will be updated in the computed background with actual background values. Instead, points of the other three MVOs will be masked (and validated as real MVOs, see Fig. 3.c).

## 3.3 Shadow detection in Sakbot

A shadow detection module has been added for improving both MVO detection and background update. Shadow detection is performed based on chrominance analysis, applied to points belonging to moving objects only, in order both to limit the computation and discard fixed shadows belonging to the background.

To detect shadow, we first convert from the color space (R,G,B) to (H,S,V). The color space (H,S,V) better reproduces the human visual behavior and it is more sensitive to brightness changes due to shadows. By exploring many videos with different light conditions we observed that a pixel "covered" by shadows becomes darker (V component), but also exhibits a color saturation (H and S components). Thus, for a shadowed point (x,y) we have:

$$SP_t(x,y) = \begin{cases} 1 \ \text{if} \ \alpha \le \dfrac{I_t(x,y).V}{B_t(x,y).V} \le \beta \wedge \left|I_t(x,y).H - B_t(x,y).H\right| \le \tau_H \\ \quad \wedge \left|I_t(x,y).S - B_t(x,y).S\right| \le \tau_S \wedge FPm_t^{TL} = 1 \\ 0 \ \text{otherwise} \end{cases}$$

with $0 < \alpha, \beta, \tau_H, \tau_S < 1$. Intuitively, this means that shadow darkens a "covered" point, but not too much. Typically $\beta$ ranges from 0.9 to 0.97; higher $\beta$ values cause shadow detection to be too sensitive to noise; instead, $\alpha$ ranges from 0.75 to 0.85, meaning that typically a "shadowed" point becomes darker of about 20%. Typical values for $\tau_H$ and $\tau_S$ are 0.15. The condition on the V component is similar to the shadow detection technique proposed in [13].

**(a)** **(b)**

*Figure 4.* Segmentation with shadow detection

Fig. 4 shows an example of the improvement achievable with the method we use in Sakbot: Fig 4.a shows the original frame; in Fig. 4.b, background pixels are colored in black, foreground pixels in dark grey, shadow pixels detected if using only the condition on the V component in light grey, and eventually shadow pixels detected adding the H and S conditions, in white. In Fig. 4, it is possible to note that the car is partially eroded by light grey pixel, i.e. using color information reduces false positives in shadow detection; moreover, real shadows are detected more effectively. Refer to [19] for more details on shadow detection method and results.

## 3.4 The Sakbot's tracking module

The tracking module has been conceived for managing visual data extracted from the scene at a symbolic level. The adopted approach is a production rule system with forward chaining [20], formalizing the environment knowledge and the relationships between the extracted MVOs. Working at a symbolic level allows the system with flexible object tracking over a variety of applications by simply devising adequate rule sets. The basic symbol is the entity *MVO*, described in terms of a list of attributes:

*MVO =(Id, Extent, Est_Ext, Dir, Displ, Tr_Frames, Lost_n, Stop_n, Status).*

*Id* is the MVO's identifier, *Extent* is the shape extent at the current frame, *Est_Ext* is the estimated extent at the next frame, computed based on the displacement (*Disp*) between the last two MVO's positions, the direction (*Dir*) and the estimated shape variation. *Tr_Frames* states since how many frames the MVO is being tracked. The goal of the system is to track both moving and stationary MVOs, i.e. MVOs in motion in previous frames, but not detected as moving objects at current frames. To this aim, other attributes are added: *Status* (that can be moving or stopped), *Lost_n* stating since how many frames the MVO has been lost by the segmentation module, and *Stop_n* since how many frames the object has been classified as stopped.

At each k frame, the system compares the set $VS=[Vs^{k-1}_1 \dots Vs^{k-1}_n]$ of all n *MVOs* in the scene (computed up to the k-1 frame) with the set $VM=[Vm^k_1 \dots Vm^k_m]$ of the m *MVOs* obtained by segmentation of the current k frame. The basic *match(X,Y)* rule considers several criteria for assessing that *X* and *Y* are the same MVO, being *X* in *VS* set, and *Y* in *VM* set: the *closeness* between the estimated new position of *X* and the computed position of *Y*, the *area-similarity* between extents (corrected by the perspective and the estimated shape variation), and the *pattern-similarity*. If the rule $match(Vs^{k-1}_i, Vm^k_j)$ is verified, the attributes of MVO $Vs^{k-1}_i$ are updated. Furthermore, many other rules cope with object over, under and miss segmentation. Experimental results show that this approach allow for the correction of most segmentation errors, especially due to luminance variation during MVO motion, partial object occlusion, and object cluttering, thus achieving robust tracking.

# 4. SAKBOT Implementation and Application

The current Sakbot prototype has been developed using the Microsoft Visual C++ 4.0 environment and ANSI C libraries as an executable for standard PCs with MS Windows. In the current version, acquisition and processing are performed separately. The input video format supported is AVI. The average frame rate is ten frames per second with reduced (360 x 288) frame size, of course depending directly on the PC's performance.

Test videos have been acquired with a 3-CCD PAL-DV digital camera (Sony DCR-VX9000E), which can be connected to the PC via a standard IEEE 1394 connection. The PAL-DV raw format (720 x 576) does not provide square pixels, but the ratio between height and width is limited (the pixel horizontal size is (4/3)/(5/4) = 1.064 larger than vertical size), thus pixels were assumed square in processing.

Background is updated with a $\Delta T$ of 10 frames, so that a new background is computed every 1 second. All the other operations, including background subtraction, segmentation, shadow suppression and motion validation via optical flow are computed at frame rate. This is achieved by optimizing code performances and by computing time consuming operations (such as optical flow) only when and where necessary (i.e. the optical flow is computed only on MVOs and not on the whole image).

The first Sakbot prototype has been validated in the Campus of the University of Modena, Italy, for people and vehicle surveillance at an access gate and will be used as an experimental system for urban traffic control, installed on traffic-light controllers in Bologna City. In this context, the research has been funded by the "Supervisor of traffic control of Bologna" project co-financed by the Bologna Municipality and Italian Government.

# 5. Conclusions

This paper has presented Sakbot, a general-purpose system for detection of moving objects in traffic monitoring and video surveillance applications. Its architecture is based on an effective and efficient detection module. In this paper, a description is given of the background update and shadow suppression algorithms. Sakbot is currently experimented in the context of a traffic monitoring application at the University of California, San Diego, and will be used in a project for sustainable mobility of the Town and Province administrations of Bologna, Italy, and the local public transportation company (ATC).

## References

[1] Lipton, A., Fujiyoshi, H., Patil, R., "Moving target classification and tracking from real-time video", Proc. of WACV '98, pp. 8-14, 1998.
[2] N. Chleq, M. Thonnat, "Realtime image sequence interpretation for video-surveillance applications", Proc. of ICIP96, pp. 59-68, 1996.
[3] N. Rota, M. Thonnat, "Video sequence interpretation for visual surveillance", Proc. of Third IEEE Int. Workshop on Visual Surveillance 2000, 2000, pp. 59-68.
[4] F. Brémond, M. Thonnat, "Tracking multiple nonrigid objects in video sequences", IEEE Trans. on Circ. and Syst. for Video Tech., v. 8, n. 5, 1998, pp. 585-591.
[5] G. L. Foresti, "Object recognition and tracking for remote video surveillance", IEEE Trans. on Circ. and Syst. for Video Tech., v. 9, n. 7, 1999, pp. 1045-1062.
[6] R. Cucchiara, M. Piccardi, P. Mello, "Image analysis and rule-based reasoning for a traffic monitoring system", IEEE Trans. on Intelligent Transportation Systems, Vol. 1, No. 2, June 2000, pp.119-130.
[7] J. Orwell, P. Remagnino and G.A. Jones "Multi-camera color tracking", Proc. of the 2nd IEEE Workshop on Visual Surveillance, pp. 14-22, 1998.
[8] I. Cohen, G. Medioni, "Detecting and tracking moving objects for video surveillance", Proc. of CVPR 99, 1999, pp. 319-325.
[9] S.J. McKenna, S. Jabri, Z. Duric, A. Rosenfeld, H. Wechsler, "Tracking groups of people", Computer Vision and Image Understanding, vol. 80, no. 1, pp. 42-56, Oct. 2000.
[10] C. Sacchi, C.S. Regazzoni, "A distributed surveillance system for detection of abandoned objects in unmanned railway environments", IEEE Trans. on Vehicular Technology, vol. 49, no. 5, pp. 2013-2026, Sept. 2000
[11] E.L. Dagless, A.T. Ali, J. Bulas Cruz, "Visual Road Traffic Monitoring and Data Collection," Proceedings of the IEEE-IEE Vehicle Navigation and Information Systems Conference (VNIS'93), Ottawa, Canada, pp. 146-149, Oct. 1993
[12] A. Shio, J. Sklansky, "Segmentation of People in Motion," Proc. of IEEE Workshop on Visual Motion, Princeton, NJ, pp. 325-332, Oct. 7-10, 1991.
[13] A. Elgammal, D. Harwood, L. Davis, "Non-parametric model for background subtraction," in Proc. of FRAME-RATE, Corfu, Greece, 1999.
[14] C. Stauffer, W.E.L. Grimson, "Learning Patterns of Activity Using Real-Time Tracking", IEEE Transactions on Pattern Analysis and Machine Intelligence, vol. 22, no. 8, 2000, pp. 747-757.

[15] C.R. Wren, A. Azarbayejani, T. Darrell, A.P. Pentland, "Pfinder: real-time tracking of the human body", IEEE Trans. on Pattern Analysis and Machien Intelligence, vol. 19, no. 7, Jul. 1997, pp. 780-785
[16] D. Koller, K. Daniilidis, H.H. Nagel, "Model-based object tracking in monocular image sequences of road traffic scenes", International Journal of Computer Vision, vol. 10, no. 3, pp. 257-281, 1993
[17] R. Cucchiara, C. Grana, M. Piccardi, A. Prati, "Statistic and knowledge-based moving object detection in traffic scenes", in Proc. of ITSC-2000 - 3rd IEEE Conference on Intelligent Transportation Systems, pp. 27-32, 2000.
[18] M. P., Dubuisson Jolly, S. Lakshmanan, A. K. Jain, "Vehicle segmentation and classification using deformable templates". IEEE Trans. Pattern Anal. Machine Intell., vol. 18, no. 3, 1996, pp. 293-308.
[19] R. Cucchiara, C. Grana, M. Piccardi, A. Prati, S. Sirotti, "Improving shadow suppression in moving object detection with HSV color information", accepted for publication in Proc. of 4th IEEE Conference on Intelligent Transportation Systems (ITSC), 2001
[20] F. Hyes Roth, D. A. Waterman, D. B. Lenat, "Building expert systems". Addison-Wesley, 1983.

Chapter 13

# Assessment of Image Processing Techniques as a means of Improving Personal Security in Public Transport

Luis M. Fuentes and Sergio A. Velastin
*Department of Electronic Engineering, King's College London, London WC2R 2LS, UK*
*{luis.fuentes,sergio.velastin}@kcl.ac.uk*

**Abstract:** This paper presents work that is currently being developed in the Vision and Robotics Laboratory, King's College London concerning the improvement of personal security in public railway transport systems. The application environment is focused on London Underground installations and video sequences have been obtained from standard video recordings. In order to achieve a good segmentation and tracking system working in real time with the specified images, we propose a simple segmentation technique based on the following assumptions: the background does not change; people usually wear dark clothes; colour information is not significant; the perspective of the camera usually results in large and fast changes in the size of the people and the use of low spatial resolution improves segmentation results. These assumptions do not excessively constraint the range of application or reduce performance. On the contrary, analysing reduced-resolution-B&W images without updating background allows real-time processing on incoming video images.

**Key words:** Tracking People, CCTV surveillance

## 1. Introduction

The use of CCTV systems in surveillance has grown exponentially in the last decade. As noted by Norris [1] "the rise of CCTV surveillance system has to be seen in the context of underlying social, political and economic pressures which have led to the intensification of surveillance across a whole range of spheres and by a host of new technologies". In this context, experts predicts an extensive application of CCTV surveillance systems and their integration in a global surveillance network. However, this expansion requires technological aids to overcome the drawbacks associated with the huge amount of visual information generated by these surveillance systems [2]. Anything but reduced human monitoring becomes impossible because of its physical and economic implications, and an advance towards automated surveillance seems to be the only solution. Given these economic and social demands, advanced visual surveillance systems have been developed in the last years. The concept of "advanced visual surveillance" involves not only the interpretation of a sequence of images but also the detection of predefined events that trigger an alarm. The use of constraints obtained from the knowledge of both the task and the environment allows greater computational efficiency [3].

This paper presents work that is currently being developed in the Vision and Robotics Laboratory, King's College London concerning the improvement of personal security in public railway transport systems. The application environment is focused on London Underground installations and video sequences have been obtained from standard video recordings.

## 2. Related work

Intelligent surveillance systems have to deal with the especially difficult task of analysing the behaviour of people. This research domain is currently one of the most active in computer vision. There is, therefore, an extensive literature dealing with the different approaches used to detect and track persons, track hands and detect gestures, to identify faces, etc. as some "surveys" summarise [4,5,6]. The final aim of any surveillance system based on image analysis is to determine whether or not a given event is actually occurring. Although in some cases (presence detection in forbidden areas) there is no need of tracking algorithms, the typical sequence of procedures is detection, tracking and interpretation. There are many different approaches for any of these three stages.

Detection algorithms are normally based on background subtraction algorithms (BSAs) [7,8,9], although some approaches combine this method with a temporal difference [10]. These methods are based on extracting motion information by thresholding the differences between the current image an a reference image (background) or the previous image respectively. BSAs are widely used because they detect not only moving objects but also stationary objects not belonging to the scene. The reference image is defined by assuming a Gaussian model for each pixel. BSAs are normally improved by means of updating their statistical description so as to deal with changing lighting conditions [11,12,13], normally linked with outdoor environments.

Some authors present a different model of background, using pixels' maximum and minimum values and the maximum difference between two consecutive frames [14], a model that can clearly take advantage of the updating process. Pixels of each new frame are then classified belonging to the background or the foreground using the standard deviation to define a threshold. After the segmentation of the foreground pixels, some processing is needed to clean noisy pixels and define foreground objects. The cleaning process usually involves 3x3 median [12] or region-based [9] filtering, although some authors perform a filtering of both images –current and background- before computing the difference [8,11]. Object detection is then achieved through morphological operations [7] or connected component analysis [9]. Once selected, foreground objects (normally called blobs) are characterised with a variety of parameters, the centroid and the size of the surrounding box being the more commonly selected features. As additional features, some authors introduce 3D position [11], a colour-based appearance model [12], colour histograms [14] or features based on the body parts [9].

Once the foreground objects have been segmented, they have to be related with objects/blobs in the previous frame. Tracking algorithms can be very different and use different methods to match the blobs of two consecutive frames. Some make use of trajectory information to predict the position of a blob in the current frame, (e.g. applying Kalman filters [13,15] or recursive least square methods [8]), and use it to find the best match for that blob by simple correspondence or using other statistical methods as Hidden Markov Models [15]. Other approaches use a matching algorithm to compute a "score matrix" with information of all objects [7] or simply match blobs with overlapping bounding boxes [12]. When there is more than one object in the scene, correspondence and tracking algorithm have to take into account the possibility of merging/splitting of blobs [8,9,12,14]. Tracking algorithms also have to deal with occlusion problems.

Within all these processes, the most important stage is segmentation. Although some tracking algorithms can cope with the possibility of "losing" the object being tracked, a proper segmentation means an accurate calculation of blob features which improves the probabilities of a correct match in the correspondence process. In this paper we propose to apply to the segmentation process a set of constraints for important real surveillance scenarios: halls, aisles and platforms in London Underground stations.

## 3. Constraints: simplification of the problem

In order to achieve a good segmentation and tracking system working in real time with the specified images, a number of assumptions about the images can be made. We propose a simple segmentation technique based on the following assumptions:

- The background does not change. Due to the nature of the scenes under study, the chosen scenarios are indoors or outdoors at night and are not likely to be modified by static objects or illumination changes. Thus, the background is considered static and is characterised by a single image.
- People usually wear dark clothes. It is known in illumination engineering that the typical reflectance of a pedestrian is around 0.3. Although this number can change according to people habits, nationality and seasons, it is basically correct in northwestern European countries.
- Colour information is not significant. According to the previous assumptions, the average person will not wear colourful clothes and the low illumination level present in indoors scenarios will not allow colour detectors to work properly. Moreover, outdoors lighting systems very frequently use high –or even worse, low- pressure sodium lamps, with very poor colour reproduction (very low colour rendering index).
- Large variations in people's size. The perspective of the camera usually results in large and fast changes in the size of the people when they move within the scenario.
- Low spatial resolution. When a tracking process does not use body parts as features, an accurate silhouette is not necessary. Thus, reducing the size of the image not only reduces processing time but also acts as a noise-cleaning filter.

These assumptions do not excessively constraint the range of application or reduce performance. On the contrary, analysing reduced-resolution-B&W images without updating background allows real-time processing on incoming video images.

## 4. Detection of people

The method to detect people in an image involves subtracting each new frame from a model of the background and thresholding the resulting image to determine foreground pixels. The foreground image is analysed and divided into different blobs or foreground-pixel clusters. A posterior analysis of each blob determines whether it corresponds to a person, a group of persons an object or noise. These stages are described below.

## 4.1 Modelling the background

The foreground detection is based on a change detection algorithm relying on a background subtraction. A static background B can be successfully modelled with a normal distribution using two images, the background image M and the standard deviation image $\sum$ in the YUV space. Let F be a sequence of $n$ images $\{F_i\}$ acquired with a time interval of $t$ milliseconds, each pixel $x$ in the background model B will be characterised by a mean/median pixel $\mu^x(u,u,v)$ of the $n$ values $F_i^x$ (y,u,v) and the standard deviation pixel $\sigma^x(y,u,v)$. Manipulating the time interval between the images, it is possible to take into account low frequency noise like the horizontal scrolling dark zones from mains interference.

Although the foreground detection will not use colour information, the background model will keep it for further use, when necessary, in refining change detection and blob characterisation algorithms. The selection of the YUV space for colour images is based on the possibility of easy access to brightness and colour information without time-consuming co-ordinate transformation. Digitising images into this colour system will allow black and white processing just accessing the $Y$ co-ordinate. On the other hand, as the background is assumed to remain static, no updating algorithm is necessary. A low resolution Black&White copy of this background model (Y,σ) is used in the next stage.

## 4.2 Foreground detection

Foreground detection is based on a multiple threshold process applied to the difference image resulting from subtracting the current image from the background model. The sign of this difference is kept as giving a first indication of people in the image, according to the second assumption. Let Y(x,y) be the $Y$ co-ordinate of the current image of the pixel located at (x,y) and B(x,y) the associated background pixel. The difference image D(x,y) is defined by the difference between Y(x,y) and B(x,y) truncated at ±127 to be able to store the image in a single byte, equation 1.

$$d(x,y) = Y(x,y) - B(x,y)$$
$$D(x,y) = \begin{cases} +127 & d(x,y) > 127 \\ -127 & d(x,y) < -127 \\ d(x,y) & otherwise \end{cases} \quad (1)$$

Foreground pixels are then signalled as such using a multiple threshold technique, although, according to previous assumptions, more than 95% of flagged pixels are darker than background. Factors $k_-$ and $k_+$ are scene dependent and above the minimum threshold of 3 times pixel standard deviation, equation 2.

$$S(x,y)=\begin{cases}255 & \begin{cases}D(x,y)<-k_-\cdot\sigma(x,y)\\ D(x,y)>\ k_+\cdot\sigma(x,y)\end{cases}\\ 0 & otherwise\end{cases} \tag{2}$$

Negative threshold is usually higher to avoid shadows to be signalled (shadows do not disturb processing of single person blobs but is a serious handicap when dealing with complex blobs comprising more than one person). Figure 1 shows how foreground detection algorithm deals with four different backgrounds.

*Figure 1.* Segmentation and location of people in three different scenarios in London and Paris Underground.

## 4.3 Blob grouping and classification

A sequential analysis of projected histograms is used to identify blobs in the binary image of signalled pixels S(x,y). A first approximation to the problem of assign blobs is achieved by dividing the foreground image into regions using empty columns in the x-projected histogram and finding the smallest window comprising all signalled pixels in that region. Boxes in figure 1 were obtained with this method. Typical cameras produce perspective images with high variations in the size of people according to their position in the scene. Nevertheless, a minimum value and a range of high/width ratio can be used to determine whether a blob corresponds to a person, a part of a person or a group of people. Once the tracking process is operating, previous knowledge of what should be comprised in that blob helps with the assignment. Once again, the analysis of x-projected histogram of group-blobs provides sometimes information enough to split them. Figure 2 show the result in both, signalled and original images.

*Figure 2.* Examples of blob splitting using x-projected histograms.

# 5. Tracking and trajectory analysis

After foreground segmentation and classification, each blob detected in the current frame must be matched with another from the temporally consistent list of tracked persons. The centroid of the blob is used to locate the person in a 2D model of the scene in order to analyse its trajectory and be able to determine the relative position between persons. Correspondence is decided based on the proximity of the centroids and the overlapping between blobs in 2D. An ulterior analysis decides over unmatched blobs dealing with people leaving or coming into the scenario or group dynamics. An accumulated history of trajectories, figure 3, could be used to define "regular" paths and detect associated events.

*Figure 3.* Accumulated centroid position.

# 6. Conclusion and future work

The application of people detection and a tracking algorithm in real scenarios with complex structures is an achievable task that can be applied in real-time with satisfactory result just reducing detection algorithms to work on low-resolution black-and-white images and a static background. Colour images under (YUV) format allow easy access to luminance without time-expensive co-ordinate transformations, while the colour information is kept for its use in further stages of the process. After detecting and tracking people, the next stage will be to detect potentially dangerous situations by analysing the trajectories of the people in the scene and theirs velocities (e.g. running). Due to the lack of images showing the kind of behaviour to detect, a conceptual model will be used.

## References


Norris C., Moran J. & Armstrong G., "Surveillance, Closed Circuit Television and Social Control", Ashgate, UK (1998)

Norris C. & Armstrong G., "The Maximum Surveillance Society. The rise of CCTV", Berg Publishers, UK (1999)

Buxton H., Gong S., "Visual surveillance in a dynamic and uncertain world", *Artificial Intelligence* **78**, 431-459, (1995)

Gavrila D.M., "The Visual Analysis of Human Movement: A Survey", *Computer Vision and Image Understanding*, **73**(1), 82-98, (1999).

Aggarwal J.K. , Cai Q., "Human motion analysis: A review", *Computer Vision and Image Understanding*, **73**(3), 428-440, (1999).

Pentland A., "Looking at people: sensing for ubiquitous and wearable computing", *IEEE Transactions on Pattern Analysis and Machine Intelligence*, **22**(1), 107-119,(2000).

S. S. Intille, J. W. Davis and A. F. Bobick, "Real-time Closed-World Tracking", *Proceedings of the IEEE Conference on Computer Vision and Pattern Recognition* (CVPR'97), 697-703, (1997).

F. De la Torre, E. Martinez, M. E. Santamaria and J.A.Moran, "Moving Object Detection and Tracking System: a Real-time Implementation", Proceedings of the Symposium on Signal and Image Processing GRETSI 97, Grenoble, (1997).

I. Haritaoglu, D. Harwood and L.S. Davis, "$W^4$: Real-Time Surveillance of People and Their Activities", *Trans. Pattern Analysis and Machine Intelligence,* **22**(8), 809-822, (2000).

S. Huwer and H. Niemann, "Adaptive Change Detection for Real-time Surveillance Applications", "Video sequence interpretation for visual surveillance", Proceedings of The IEEE Workshop on Visual Surveillance, Dublin, 37-43, (2000).

N. Rota and M. Thonnat, "Video Sequence Interpretation for Visual Surveillance", Proceedings of The IEEE Workshop on Visual Surveillance, Dublin, 59-68, (2000).

S. McKenna, S. Jabri, Z. Duric, A. Rosenfeld and H. Wechsler, "Tracking Groups ofP", *Computer Vision and Image Understanding* **80**, 42-56, (2000)

C. R. Wren, A. Azarbayejani, T. Darrel and P. Pentland, "Pfinder: Real-Time Tracking of the Human Body", *Trans. Pattern Analysis and Machine Intelligence,* **17**(6), 780-785, 1997.

J.I. Agbinya and D. Rees, "Multi-Object Tracking in Video", *Real-Time Imaging* **5**, 295-304, (1999).

G. Rigoll, B. Winterstein and S. Müller, "Robust Person Tracking in Real Scenarios with Non-Stationary Background using a Statistical Computer Vision Approach", Proceedings of the 1999 International Conference on Image processing, ICIP'99, 242-246, (1999).

Chapter 14

# ON THE USE OF COLOUR FILTERING IN AN INTEGRATED REAL-TIME PEOPLE TRACKING SYSTEM

N.T. Siebel and S.J. Maybank*
*Computational Vision Group*
*Department of Computer Science*
*The University of Reading, Whiteknights,*
*Reading, Berkshire, RG6 6AY, UK*
http://www.cvg.cs.reading.ac.uk/
{nts,sjm}@isg.cs.reading.ac.uk

**Abstract** We present results from multiple experiments with colour filtering methods in order to improve robustness in an integrated surveillance system to track people in a subway station. The system is designed to operate in real-time in a distributed local network of off-the-shelf computers, resulting in practical constraints not found in developmental systems. We show that the quality of colour information is degraded by electrical interference and image compression to such an extent that it is no longer useful for local edge detection. We give a recommendation as to what methods can be used to filter out most of the image noise influencing local edge detection and show how using these methods increases robustness of tracking.

**Keywords:** Visual Surveillance, People Tracking, Active Shape Model, Colour, JPEG, Local Edge Detection

*Figure 14.1.* View from surveillance camera

## Introduction

With the recent advances of computer technology real-time automated visual surveillance has become a popular area for research and development. Surveillance cameras are installed in many public areas to improve safety, and computer-based image processing is a promising means to handle the image databases generated by large networks of cameras. The task of an integrated surveillance system is to warn an operator when it detects events which may require human intervention, for example to avoid possible accidents. The warnings can only be reliable if the system can detect and understand human behaviour and for this it must locate and track people reliably.

A frequent problem with surveillance systems is the poor quality of the video images. Installed cameras are often old, and images are transmitted down long analogue cables which are vulnerable to interference and signal degradation. There are many unsolved problems when it comes to processing this noisy data by means of a computer.

Many methods to make image analysis more robust involve the introduction of prior knowledge, for example the use of complex statistical models to detect image motion [2], and the use of detailed 3D models of the human body to give more accurate tracking results [4, 7].

However, automated visual surveillance systems have to operate in real-time and with a minimum of hardware requirements, if the system

*This work is funded by the European Union, grants ADVISOR (IST-1999-11287) and Smart II (FMRX-CT96-0052).

is to be economical and scalable. Even with today's computer speeds this limits the complexity of methods for detection and tracking. Haritaoglu et al [5] show that real-time performance can be obtained with a simplified person model in low-resolution images.

In this paper we use simple image filtering techniques to improve a people tracking method in which the person model is of medium complexity. We explore whether it is possible to use the colour information which is available on many systems today. We show image filtering methods which can enhance edge detection, thereby improving the robustness of image analysis software for people tracking.

# 1. People Tracking

The tracking system used for our experiments is an extension of the Leeds People Tracker which was developed by Baumberg and Hogg [1]. We have ported the tracker from an `sgi`™ platform to a PC running GNU/Linux to make economic system integration feasible.

The tracker uses an active shape model [1] for the contour of a person in the image. A space of suitable models is learnt in a training stage using a set of video images containing walking pedestrians. Detected person outline shapes are represented using cubic B-splines, giving a large set of points in a high-dimensional parameter space. Principal Component Analysis (PCA) is applied to the obtained set of shapes to generate a lower dimensional subspace $S$ which explains the most significant modes of shape variation, and which is a suitable state space for the tracker.

## 1.1 Basic detection and tracking

People tracking is performed in multiple stages. The tracker maintains a background image which is automatically updated by median-filtering the sequence of video images over time. To detect new people, a motion detector subtracts the background from the current video image. Thresholding of this difference image yields a binary image containing regions where the pixel values of the current image differ significantly from the pixel values of the background image. The detected foreground regions that match certain criteria for size and shape are examined. Their outline shape is approximated by a B-spline and projected into the PCA space $S$ of trained pedestrian outlines. The new shape obtained in this process is then used as a starting point for further shape analysis. Once people are recognised they are tracked using the trained shape model. This is done with the help of Kalman filtering using second order motion models for their movements. The state of the tracker includes the current outline shape as a point in $S$, which is updated as the observed outline changes during tracking.

*Figure 14.2.* Edge search for shape fitting

## 1.2 Fitting a person shape

In order to adjust the outline shape to each new image of a person we use an iterative optimisation method. The current estimate of the shape in $S$ is projected onto the image. It is then adjusted to the actual shape in an optimisation loop by searching for edges of the person's outline, at each spline control point around the shape. Figure 14.2 illustrates this process showing the original camera image, in this case with a fairly dark person against a light background.

It is in this measurement process that image noise and background clutter severely disturb the search for the best fitting outline.

# 2. Colour Filtering

In order to make edge detection more robust we filter the image in such a way that foreground objects become more salient. Normally the search for edges is done in the pixelwise absolute difference image obtained by subtracting the background image from the video image, to improve contrast [1].

## 2.1 Image Quality

When dealing with colour information contained in video images, their quality is an important issue. Before interpreting colour information obtained from from image measurements it is important to trace the path the image took, from camera to computer. The colour images used in our system originate from interlaced colour cameras installed in subway stations. They are transferred over long cables and affected by interference from electrical equipment in the station. The image sequence used for the experiments shown was recorded in a London Underground station[1], stored on analogue (S-VHS PAL) video tape which degraded the colour information. The image quality was further reduced by JPEG image compression needed for transmission over a local network (Ethernet) in the integrated surveillance system. Even though the human eye does not notice a lot of image degradation, the JPEG compression can create considerable problems for image processing algorithms. The reason is that JPEG compression techniques usually sub-sample colour information (the 2 "chrominance channels", denoted $C_B$ and $C_R$) [6], thereby causing problems in algorithms which rely on the correctness of this colour information. JPEG stores the image brightness ("luminance", $Y$) at a higher resolution than chrominance because this component makes up the main part of the information extracted from images by humans.

## 2.2 Filtering Techniques

Our investigations involved a large number of colour filtering techniques to find out how they influence edge contrast and how they behave in the presence of the image noise described earlier. The filters used project each pixel in the image from the 3-dimensional RGB colour space down to a 1-dimensional subspace, giving a grey level image (see below for examples). The resulting image can then be used by our edge detection algorithm to search for those edges included in the outlines of people. We have applied both linear (weighted sums of RGB values) and non-linear mappings (for example, mapping to the hue component in HSV space). However, given the low colour resolution in our images some filters are not feasible in our system. In the following examinations we focus on the results obtained from mappings that are least prone to colour noise.

Another problem for edge detection algorithms is the low contrast of people against the background. The image in Figure 14.1, chosen as an example, shows a person with a white jacket against a light yellow background. In the colour difference image the jacket appears as dark blue and most edge detection methods will not be able to detect it as the edge contrast at the jacket outline is too low.

## 2.3 Filtering the difference image

Many algorithms for colour edge detection start by taking the absolute difference in each colour channel to get a new colour image. Each pixel value $p = (R, G, B)$ of this difference is then mapped to a scalar using a function of its components. The mappings use a scaled Euclidian norm, $||p||_2 = \sqrt{\frac{1}{3}(R^2 + G^2 + B^2)}$, or a weighted sum, $w_1R + w_2G + w_3B$ with $\sum w_i = 1$.

*Figure 14.3.* Luminance $Y$ of colour difference

The image in Figure 14.3 shows the result for a simple linear mapping from RGB space onto the luminance ($Y$) subspace, using the $Y$ projection from CIE[2] recommendation 709, denoted here as $Y_{709}$ (taken from [3]):

$$Y := Y_{709} = .2125\,R + .7154\,G + .0721\,B. \tag{14.1}$$

The weights are chosen in order to mimic the way humans perceive the brightness of a colour image. However, this weighting only gave a small improvement for edge detection compared to the simpler method of weighting each colour channel equally.

Looking at the resulting image, one can see that detection of the outline of the lady in the front fails because her coat has too low a contrast with respect to the background. For the person to the right, the main parts of his body, especially the trousers, are many different shades of grey. Searching for edges around his shape outline is very sensitive to the parameters used in edge detection, which can cause difficulties and mis-detection.

## 2.4 Differencing the filtered images

In a second stage of experiments we filtered both the background image and the current video image using the same $Y$ projection (14.1) and subtracted the results. You can see the difference image in Figure 14.4.

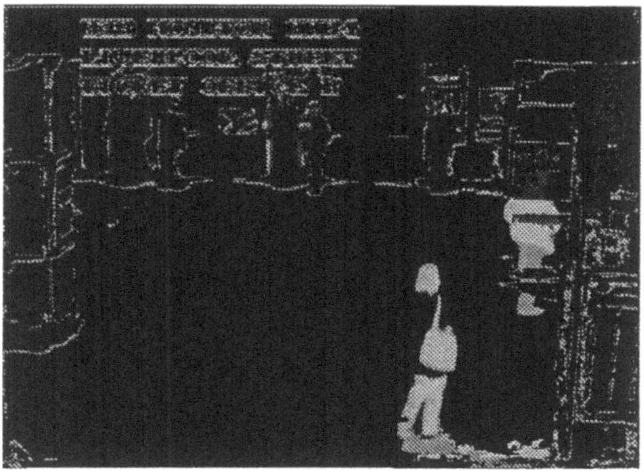

*Figure 14.4.* Absolute difference of luminance $Y$

Compared to the previous image there is not much change when it comes to the woman's coat. Other parts of both people in the front, however, appear much more salient, standing out clearly against the background. For edge detection this gives a significant advantage because it increases robustness by being less dependent on detection parameters such as threshold values. Looking at the right hand person's coat and trousers we see that these are now more uniformly shaded which means that edge detection is also less prone to being distracted by edges located inside the body.

This method of filtering also makes background edges stand out more clearly. For motion detection algorithms this could create difficulties. Our edge detector, however, operates locally in the image so this does not pose a problem.

# 3. Discussion

In this paper we have presented an approach to improve local edge detection for people tracking systems in order to increase robustness in the presence of image noise. The main problem in an integrated system like ours is the noise found in the chrominance channels of the image as seen in Figure 14.5. This is due to electrical interference in a subway station (analogue camera cabling) and image compression techniques (such as JPEG) within our local computer network.

We have shown that using image differencing methods using only luminance gives better results in local edge detection than standard routines, thereby improving local edge detection which is a crucial part of our people detection and tracking system.

It should be noted that even in a system like ours, colour information can still be of good use for the identification of people once they are tracked.

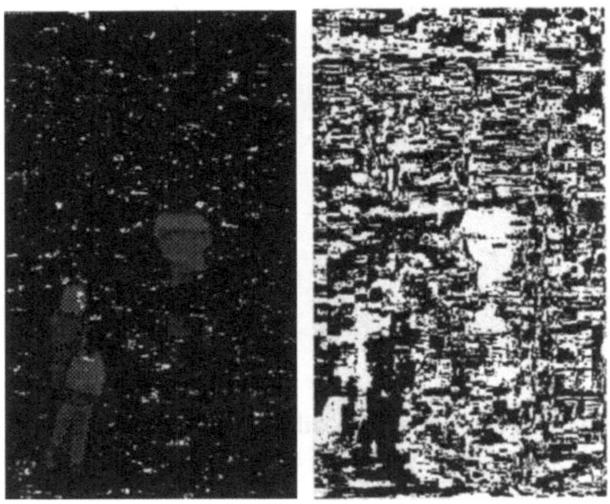

*Figure 14.5.* Chrominance channels $C_B$ and $C_R$

## References


[1] Adam M. Baumberg. *Learning Deformable Models for Tracking Human Motion.* PhD thesis, School of Computer Studies, University of Leeds, October 1995.

[2] Ahmed Elgammal, David Harwood, and Larry Davis. Non-parametric model for background subtraction. In David Vernon, editor, *ECCV 2000, 6th European Conference on Computer Vision*, pages 751–767. Springer Verlag, 2000.

[3] Adrian Ford and Alan Roberts. *Colour Space Conversions*, August 1998.

[4] Dariu M. Gavrila and Larry S. Davis. Tracking of humans in action: A 3-D model-based approach. In *ARPA Image Understanding Workshop*, Palm Springs, February 1996.

[5] Ismail Haritaoglu, David Harwood, and Larry S. Davis. $W^4$: Real-time surveillance of people and their actions. *IEEE Transactions on Pattern Analysis and Machine Intelligence*, 22(8), August 2000.

[6] Thomas G. Lane. *IJG JPEG Library: System Architecture*. Independent JPEG Group, 1991–1995. Part of the Independent JPEG Group's JPEG software documentation.

[7] Hedvig Sidenbladh, Michael J. Black, and David J. Fleet. Stochastic tracking of 3D human figures using 2D image motion. In David Vernon, editor, *ECCV 2000, 6th European Conference on Computer Vision*, pages 702–718. Springer Verlag, 2000.

III

# EVENT DETECTION AND ANALYSIS

Chapter 15

# MODELING AND RECOGNITION OF HUMAN ACTIONS USING A STOCHASTIC APPROACH

Esther B. Koller-Meier and Luc Van Gool

*Communication Technology Lab, Image Science*

*Swiss Federal Institute of Technology (ETH)*

*CH-8092 Zurich, Switzerland*

{ebmeier,vangool}@vision.ee.ethz.ch

**Abstract** This paper describes a self-learning prototype system for the real-time detection of unusual motion patterns. The proposed surveillance system uses a three-step approach consisting of a tracking, a learning and a recognition part. In the first step, an arbitrary, changing number of objects are tracked with an extension of the CONDENSATION algorithm. From the history of the tracked object states, temporal trajectories are formed which describe the motion paths of these objects. Secondly, characteristic motion patterns are learned by clustering these trajectories into prototype curves. In the final step, motion recognition is then tackled by tracking the position within these prototype curves based on the same method, the extended CONDENSATION algorithm, used for the object tracking.

**Keywords:** Tracking, Condensation algorithm, spatio-temporal grouping, human motion recognition

## 1. Introduction

The analysis of human movements or activities from image sequences is an important problem especially for surveillance applications. This paper focuses on the surveillance of bank lobbies where security staff is required to observe human activities on monitors to detect the occurrence of unusual events. The support of automatic video surveillance systems should relieve the operators by directing their attention to the

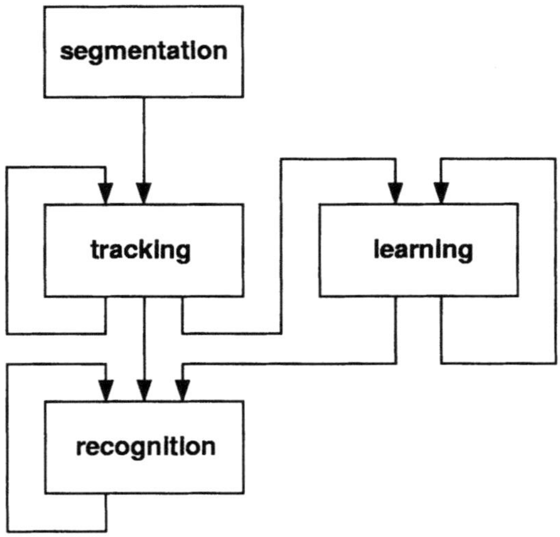


*Figure 15.1. Information flow of the proposed surveillance system.*

unusual cases. We are primarily interested in detecting unusual events such as vandalism, panic or overcrowded areas including *unusual movements* as well as *unlikely object positions*. For this purpose we have developed a self-learning framework for the real-time detection of such unusual motion patterns consisting of a tracking, a learning and a recognition part (see Fig. 15.1).

There is an extensive amount of computer vision work in the area of motion models. Such models are generally first constructed for the motion paths and the detected motions are then classified on the basis of these models. The most frequent attempt based on a state-space approach is the Hidden Markov Model (HMM) [3, 10, 11]. In [2] the CONDENSATION-based Trajectory Recognition (CTR) is proposed which can be seen as a generalization of HMM's. Alternative methods based on template techniques convert an image sequence into a static shape pattern. The most commonly used features for this technique are 2D meshes [8]. Furthermore, these approaches also comprise motion-energy images (MEI) and motion-history images (MHI) which are presented in [4]. A detailed review of human motion analysis can be found in [1].

The outline of this paper is given as follows. In the next section we introduce the tracking approach which is based on the CONDENSATION algorithm [5] to trace an arbitrary, changing number of objects. In Section 3 we propose a self-learning clustering method using the tracking results to build prototype curves as models for the motion paths. The recognition part where the tracking results are matched to the learned prototype curves is then explained in Section 4. Finally, we present some experimental results in Section 5 for the surveillance of bank lobbies.

## 2. Tracking

First of all we focus on the object tracking in video sequences. We use the extension of the CONDENSATION algorithm which we have developed to track an arbitrary, changing number of objects for mobile robots [7].

In principle, we are interested in recursively constructing the *a posteriori* probability density of an object state conditioned on all measurements $\mathcal{Z} = \{z_1, \ldots z_t\}$ which are obtained from a segmentation. The key idea of the CONDENSATION algorithm [5] is to approximate the probability distribution by a weighted sample set $S = \{(s^{(j)}, \pi^{(j)}) | j = 1 \ldots N\}$. Each sample consists of an element $s$ which represents the state of an object and a corresponding discrete sampling probability $\pi$. The evolution of the sample set is described by first propagating each sample according to a system model. Secondly, each element of the set is weighted in terms of the measurements and finally, a particular sample is drawn with replacement, by choosing it with probability $\pi$.

In our application it is necessary to track *multiple* objects simultaneously and to handle *newly appearing* objects. For that purpose, we use a single CONDENSATION tracker for multiple objects such that several object states are represented simultaneously with one sample distribution. Furthermore, we apply an initialization at every iteration step to incorporate newly appearing objects into the tracking process. The programming details of the extended CONDENSATION algorithm for one iteration step are given in Fig. 15.2.

The measurements for our application are obtained by subtracting two successive images. From the resultant motion blobs we extract the central points described by the $x$ and $y$ image coordinates. Accordingly, we define the state of an object blob at time $t$ by

$$s_t = (x_t, y_t, u_t, v_t) \tag{15.1}$$

where $u$ and $v$ describe the velocities in the corresponding coordinate directions. The propagation of each sample is then given as

$$\begin{aligned} x_t &= x_{t-1} + u_{t-1}\Delta t + n_{t-1}^{(x)} \\ y_t &= y_{t-1} + v_{t-1}\Delta t + n_{t-1}^{(y)} \\ u_t &= u_{t-1} + n_{t-1}^{(u)} \\ v_t &= v_{t-1} + n_{t-1}^{(v)} \end{aligned} \tag{15.2}$$

1 **Initialize** the sample set $S_{t-1}$:
  (a) draw $M$ elements $i_{t-1}^{(k)}$ from the initialization density $p(s_{t-1}|z_{t-1})$ where $s$ represents the state and $z$ the measurement
  (b) combine the samples from the previous time step with the initial samples
  $s_{t-1}^{(j)} = s_{t-1}^{\prime\prime(l)} \cup i_{t-1}^{(k)}$ $\quad j = 1 \dots N$, $l = 1 \dots N - M$ and $k = 1 \dots M$

2 **Propagate** each sample from the set $S_{t-1}$ to build a new set $S_t^{\prime}$ according to the system model:
$s_t^{\prime(j)} = f_{t-1}(s_{t-1}^{(j)}, n_{t-1}^{(j)})$
where $n_{t-1}^{(j)}$ is a vector of random variables with a known distribution

3 **Observe** the measurements:
  (a) weight each sample of the set $S_t^{\prime}$ with $\pi_t^{(j)} = p(z_t|s_t^{\prime(j)})$
  (b) calculate the normalized cumulative probabilities
  $c_t^{(0)} = 0$
  $c_t^{(j)} = c_t^{(j-1)} + \pi_t^{(j)}$ $\quad$ for $j = 1 \dots N$
  $c_t^{\prime(j)} = \frac{c_t^{(j)}}{c_t^{(N)}}$ $\quad$ for $j = 1 \dots N$

4 **Select** $N - M$ samples from the set $S_t^{\prime}$ with probability $\pi_t^{(j)}$ to generate the sample set $S_t^{\prime\prime}$:
  (a) generate a uniformly distributed random number $h \in [0, 1]$
  (b) find, by binary search, the smallest $i$ for which $c_t^{\prime(i)} \geq h$
  (c) set $s_t^{\prime\prime(l)} = s_t^{\prime(i)}$ with $l = 1 \dots N - M$

*Figure 15.2. An iteration step of the extended* CONDENSATION *algorithm.*

where $n = (n^{(x)}, n^{(y)}, n^{(u)}, n^{(v)})$ is a vector of random variables with a known distribution and $\Delta t$ is the time interval between two consecutive measurements. Currently, a first order system model is applied for computational reasons, but we plan to expand this model to second order by additionally incorporating the acceleration into the object state. We expect that for example the queuing in front of the cash machines can then be modeled more precisely. By keeping the history of each sample a trajectory is formed as a sequence of states.

*Figure 15.3. The sample distribution is shown as bright grey spots while in darker grey the corresponding motion trajectories are illustrated.*

An example where two persons are tracked simultaneously with a single sample distribution is shown in Fig. 15.3. A bank lobby is observed where three cash machines are located on the left side. The bright grey spots show the current sample distribution while in darker grey the corresponding trajectories are illustrated. The size of the sample set is dependent on the number of objects, for our application we choose $N = 400$ and $M = 40$. The computational cost for the tracking is 0.09 seconds on a PC with a 450 MHz Pentium II processor.

## 3. Learning

The learning phase starts by building prototypes for the trajectories of the tracked object states. The trajectories received from the tracking process are typically short-lived, partially overlapping in space and time and are possibly produced by different objects. Consequently, we cluster these trajectories into long-lived prototype curves each corresponding to a characteristic movement.

The clustering is done in the following steps:

1 All trajectories that are sufficiently long are added as new prototype curves.

2 Prototype curves that are overlapping and sufficiently close are merged into a new curve.

In each iteration step the clustering method is applied to all new prototype curves as well as to the already existing curves. As a result we obtain a set of prototype curves of the monitored environment. A re-

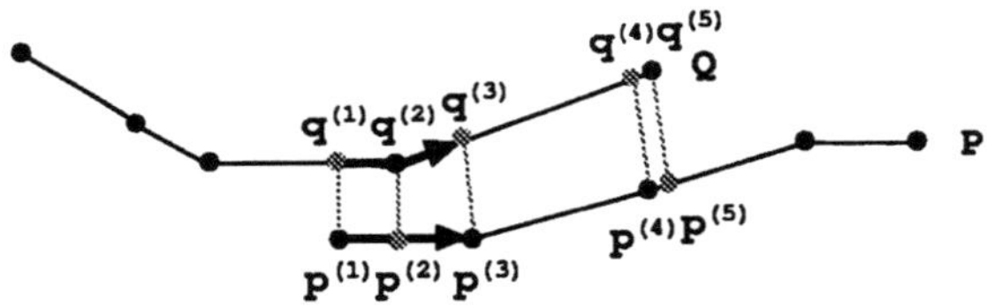


*Figure 15.4. Finding point pairs with equal distance in arc length to the initial points $p^{(1)}$ and $q^{(1)}$.*

lated clustering method, where feature points are grouped to form object trajectories is described in [9].

Each prototype curve describes the spatio-temporal path of an object and is modeled by a sequence of nodes

$$b = (x, y, u, v, w) \tag{15.3}$$

where $x$, $y$ describe the spatial localization and $u$, $v$ the motion. The parameter $w$ indicates the specific weight of this node and is increased with every trajectory that contributes to the node. This weight is used when two curves are merged, to interpolate the nodes of the resulting curve. The initial value is given by $w = 1$.

For the comparison of two prototype curves it is necessary to analyze their distance and their spatial overlap. Let us assume that we have two prototype curves $P$ and $Q$. We first look at the starting point of curve $P$ and find the closest point on curve $Q$. From these *initial points*, we now find more corresponding point pairs on the two curves. A point pair consists of a node on one curve and a corresponding point on the other curve. Such pairs have always the same distance in arc length to the initial points (see Fig. 15.4). We continue to find pairs until we reach the end of one curve.

Two prototype curves are merged if the mean distance of the point pairs

$$d = \frac{1}{K} \sum_{k=1}^{K} (x_q^{(k)} - x_p^{(k)})^2 + (y_q^{(k)} - y_p^{(k)})^2 + \lambda \left( (u_q^{(k)} - u_p^{(k)})^2 + (v_q^{(k)} - v_p^{(k)})^2 \right) \tag{15.4}$$

is sufficiently small and if they have a large enough overlap given by the distance in arc length between the first and the last overlapping point found on the curves. The factor $\lambda$ is used to normalize the velocities and the parameter $K$ indicates the number of point pairs. The merged points on curve R are then calculated by a weighted interpolation of the points on curve P and Q in the overlapping area (see Fig. 15.5).

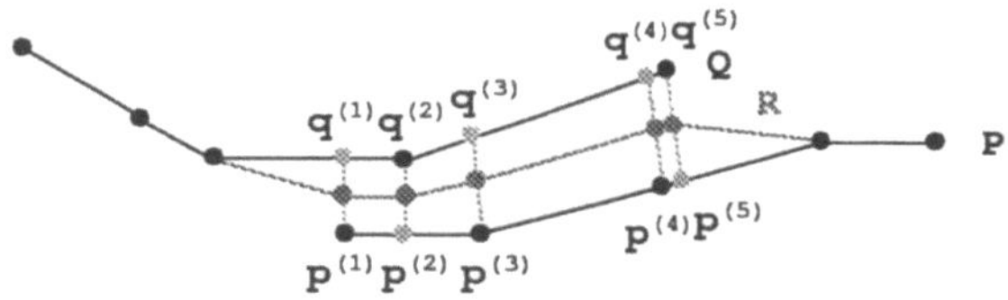


*Figure 15.5. Generate an interpolated prototype curve R from the curves P and Q.*

In order for the merged prototype curve $R$ not to acquire too many nodes, we check for nearby nodes with low curvature values that can be merged into one node without significantly changing the shape of the prototype curve. Furthermore, as relatively unlikely trajectories can also provoke prototype curves we replace weak curves by more dominant ones when a maximal number of prototype curves has been reached.

In Fig. 15.6 a new prototype curve is formed by the person crossing the bank lobby. In grey we see as before the tracking results while in white the prototype curves are shown. The computation power for the learning is dependent on the number of prototype curves. By the chosen maximal number of 30 curves the processing of the learning can temporarily reach up to 1.0 second.

# 4. Recognition

In the real-time recognition mode movements are identified on the basis of the prototype curves. In [2] the authors describe how the basic CONDENSATION algorithm [5] was successfully used to solve different

*Figure 15.6. The development of a new prototype curve caused by the person crossing the bank lobby is illustrated in white.*

recognition tasks. We adapted the idea of this CONDENSATION-based Trajectory Recognition to our extended CONDENSATION method specified in Fig. 15.2.

Instead of the segmentation results, the output of the tracking is now used as the measurement input. However, we usually use less samples for the motion recognition than for the tracking. Therefore, the tracking results are first clustered according to their distance. For each cluster, a mean sample is calculated that consists of a position and a velocity. This sample is then used as measurement input $z$ for the motion recognition.

As we want to detect if and where our objects are positioned on one of the learned prototype curves, the state of each sample is given as

$$s_t = (\mu_t, \phi_t) \tag{15.5}$$

where $\mu$ indicates the prototype curve and $\phi$ specifies the current position on this curve.

The object is assumed to move along the prototype curve, so the state evolution of each sample is described by

$$\begin{aligned} \mu_t &= \mu_{t-1} \\ \phi_t &= \phi_{t-1} + \Delta t + n_{t-1} \end{aligned} \tag{15.6}$$

where $n$ is a random variable with a known distribution and $\Delta t$ is the time interval between two consecutive measurements.

In the first step of the extended CONDENSATION algorithm, we initialize new samples from the measurements. Such a new sample is specified by finding the closest point on a prototype curve $\mu$ and by calculating the corresponding temporal position $\phi$. If the distance to the closest curve is too far, we assign this sample to a *residual curve* with $\mu = 0$. So, to this virtual curve are assigned all movements at unusual places.

In Fig. 15.7 the recognition results for an example sequence are indicated. The tracking results are shown in grey while in white the learned set of prototype curves are illustrated. From the sample distribution of the recognition we have recovered the corresponding object positions which are shown as black spots on the prototype curves. The number of samples required for the recognition is dependent on the number of prototype curves. For our application we choose $N = 300$ and $M = 30$. The recognition can be performed within 0.04 seconds.

## 5. Experimental Results

We give three examples, one for an unlikely object position and two for unusual movements. For the recognition of movements at unusual places we consider the samples of the residual prototype curve. An alert

*Figure 15.7. Recovered object positions of the recognition samples are shown in black on a learned set of prototype curves.*

is triggered when a minimal percentage of samples belong to the residual curve. In the left image of Fig. 15.8 such a situation is illustrated where two unauthorized people improperly occupy the bank lobby using it as a dormitory. The corresponding alert seen by the security staff is shown in the right image. Of course, this case could even be detected by a static analysis that does not use tracking, however it is convenient that the result is also provided by our framework.

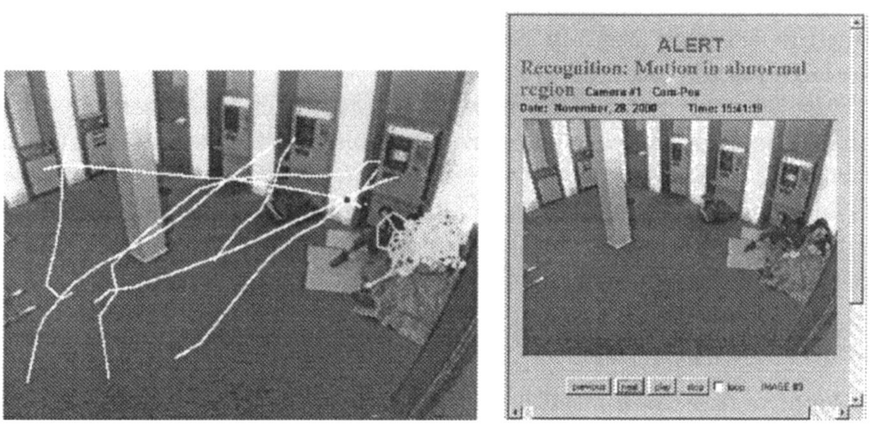


*Figure 15.8. Detection of an unusual motion event caused by the two unauthorized people which improperly occupy the bank lobby using it as a dormitory. In the left image, the current situation of the surveillance framework is illustrated while in the right image the alert seen by the security staff is shown.*

*Figure 15.9. Detection of an unusual motion event caused by rapid movements of an act of vandalism.*

When a person stays in a common place but his behavior is not equivalent to one of the learned prototype curves, e.g. because he has an unusually high speed, we call this an unusual motion event. In that case the tracked position of the person will be too far away from the expected position on the prototype curve. This results in a lower weight for the sample, so that we can detect this event by analyzing the proportion of the recognition samples assigned to one prototype curve before and after the selection step. In Fig. 15.9 such an event caused by an act of vandalism is illustrated.

False alarms produced for example by playing children (see Fig. 15.10) can not be excluded as the system is unable to differentiate between not learned or unusual behavior patters on the one hand and because some distinctions between normal and abnormal behaviors simply require too high a level of semantic interpretation. Of course, in the end it is the operator who will assess the situation.

## 6. Conclusion

We have described a self-learning framework consisting of a tracking, a learning and a recognition part for detecting unusual motion events in real-time. Such a surveillance system is primarily intended to be used in security-sensitive areas such as airports, railway stations, banks or public building lobbies. The tracking approach is solved with an extension of the CONDENSATION algorithm which can deal with an arbitrary, changing number of objects. Motion trajectories are formed by keeping

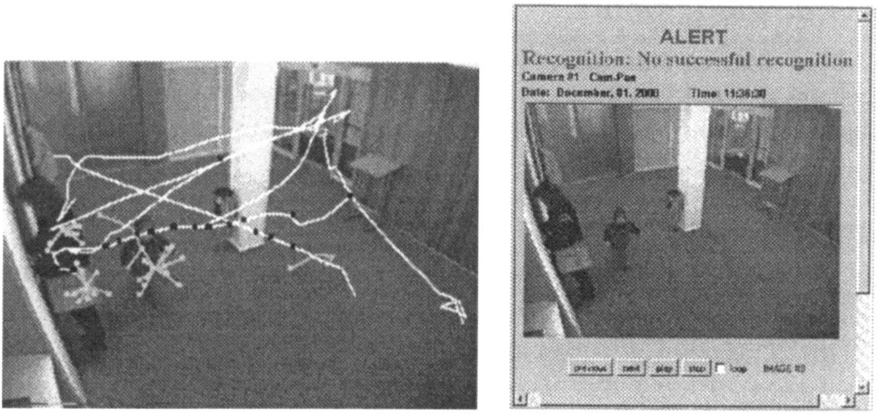


*Figure 15.10. Detection of an unusual motion event caused by the child running around the pillar.*

the history of the tracked object states. Using the clustering we can generate long-lived prototype curves by grouping the trajectories. This processing step is generally the most time-consuming part as the computation power increases with the number of prototype curves. But since the clustering is only done during the training phase, it does not add to the computational cost during normal operation. For the recognition task the tracking results are then assigned to the prototype curves. The idea of the CONDENSATION-based Trajectory Recognition is applied to the extended CONDENSATION tracker so that the recognition and tracking are both based on the same probabilistic framework. Finally, the detection of unusual movements is done by analyzing the sample distribution.

If we have more than one person in the scene, the identification and localization of the object which is responsible for the alert is not directly provided by the sample distribution. On the other hand, we know which prototype curves are active and where in these curves the samples are located.

To improve the accuracy, tracking could be considered across multiple cameras. In addition, the prototype curves which represent the usual events could evolve. Motion patterns could possibly be included or removed from the set of prototype curves over time, for example when the bank closes its counter at 5pm, the corresponding curve could be deleted.

## Acknowledgments

This research was partially supported by the Swiss Commission for Technology and Innovation (KTI). We would also like to thank our industrial partner ASCOM Systec AG for providing us with a video surveillance system including their software [6].

## References

[1] J. K. Aggarwal and Q. Cai. Human Motion Analysis: A Review. *Journal of Computer Vision and Image Understanding*, Vol. 73, No. 3, March, pp. 428-440, 1999.

[2] M. J. Black and A. D. Jepson. A Probabilistic Framework for Matching Temporal Trajectories: Condensation-Based Recognition of Gestures and Expressions. *5th European Conference on Computer Vision*, Vol. 1, pp. 909-924, 1998.

[3] Ch. Bregler. Learning and Recognizing Human Dynamics in Video Sequences. *Computer Vision and Pattern Recognition*, pp. 568-574, 1997.

[4] J. W. Davis and A. F. Bobick. The Representation and Recognition of Human Movement Using Temporal Templates. *Computer Vision and Pattern Recognition*, pp. 928-934, 1997.

[5] M. Isard and A. Blake. Contour Tracking by Stochastic Propagation of Conditional Density. *4th European Conference on Computer Vision*, Vol. 1, pp. 343-356, 1996.

[6] R. Mattone, A. Glaeser, and B. Bumann. SEDOR: A Self-learning Event DetectOR for video-based surveillance. *10th International Conference on Image Analysis and Processing*, 1999.

[7] E. B. Meier and F. Ade. Using the Condensation Algorithm to Implement Tracking for Mobile Robots. *3rd European Workshop on Advanced Mobile Robots*, pp. 73-80, 1999.

[8] R. Polana and R. Nelson. Low Level Recognition of Human Motion (Or How to Get Your Man Without Finding his Body Parts). *IEEE Computer Society Workshop on Motion of Non-Rigid and Articulated Objects*, pp. 77-82, 1994.

[9] J. Segen and S. Pingali. A Camera-Based System for Tracking People in Real Time. *13th International Conference on Pattern Recognition*, Vol. 3, Track C, pp. 63-67, 1996.

[10] T. Starner and A. Pentland. Visual Recognition of American Sign Language Using Hidden Markov Models. *International Workshop on Automatic Face and Gesture Recognition*, pp. 189-194, 1995.

[11] J. Yamato, J. Ohya, and K. Ishii. Recognizing Human Action in Time-Sequential Images using Hidden Markov Model. *Computer Vision and Pattern Recognition*, pp. 379-385, 1992.

# Chapter 16

# VIGILANT

## *Content-Querying of Video Surveillance Streams*

Darrel Greenhill, Paolo Remagnino and Graeme A. Jones
*Digital Imaging Research Centre, Kingston University, Penrhyn Road, Kingston upon Thames, Surrey, KT1 2EE, UK*

Abstract: The VIGILANT concept presented in this paper aims to generate in real-time, a database of tracked scene events richly annotated with low-level chromatic data, 2D and 3D trajectories, as well as object classifications. Rather than generate high-level interpretations online, the creation of a human-oriented graphical user interface will enable untrained security operators to build highly sophisticated queries of complex events and object interactions in typical car-park scenes. We report on how rich annotations may be constructed cheaply directly from tracked event data. In particular, we report on promising classification results. In the last section we briefly present the GUI and demonstrate the complexity of query its is capable of supporting.

Key words: Visual Surveillance, Image Retrieval

# 1. Introduction

The VIGILANT project presented here is essentially a query-driven scene understanding engine capable of generating retrospective video reports of past events. Using a push-button GUI, security operators can describe target events in as human-oriented a fashion as possible. Many state-of-the-art visual surveillance systems aim to automate the scene-understanding process[[2],[3],[4],[5],[6]]. Such an `online' paradigm holds the promise of low false-alarm alarm-generation for site protection. However, despite the considerable success of these demonstrations systems, intelligent monitoring tools face considerable obstacles in their migration to commercial systems both in terms of robust live performance as well as industrial cynicism. In contrast, VIGILANT plans to develop a robust commercial prototype by restricting the online component of the system to real-time event tracking. A database of tracked scene events is richly annotated with chromatic, trajectory and object classification information by low-level off-line processes scheduled to run in idle periods between scene activity. Much of the intelligence required to capture complex scene descriptions is in fact embedded in the query tool (see section 2) which allows the operators to embed complex relations of the low-level chromatic, trajectory and classification information in queries of surprising sophistication.

## 1.1 System Requirements

In order to ensure commercial uptake within current visual security market of the technologies embedded within the VIGILANT system, two primary requirements were defined. First VIGILANT should provide efficient real-time storage of temporally extended events (*i.e.* tracked objects) occurring within a monitored scene. Second, VIGILANT should provide a touch-screen driven GUI to enable untrained security operators to generate sophisticated human-centric queries of the video data *i.e.* queries based on *content*. The first requirement was particularly stringent in that a single high-spec PC must be able to process a video input stream in real-time, and that the daily terabyte of digital pixel data generated by a camera each day should be transformed onto a swappable one gigabyte disk. The second requirement envisaged operators being able to construct sophisticated queries such as *"display all red cars speeding through the car-park"* or *"locate events in which two people abandoned a large vehicle"*. Three crucial issues need to be addressed to provide functionality. First, a query tool must be capable of generating complex queries from simple button presses. Many of the interaction relationships between events must be embedded within the menu choices. Second, the attribute information associated with the detected events must be sufficiently rich to support such queries *e.g.* colour descriptions, time and location of events, object classification. Finally, fuzzy matching of queries is vital to handle both the inherent capture noise and the typical ambiguity of human questions.

## 1.2 System Architecture

In reality the queries described above are essentially retrospective – designed to recover video footage after some incident has occurred. By exploiting this delay between event and query and also the fact that typical car-park environments enjoy periods of busy and idle activity, the system architecture illustrated in *Figure 1* has been developed. First, only the event detection and tracking component need actually be required to operate in real-time. The sub-images associated with the moving event and the estimated 2D-trajectory information are committed to the disk. In addition to a very low-bandwidth background image stream used to reconstruct *video reports*, this pixel data represents the vast majority of disk usage.

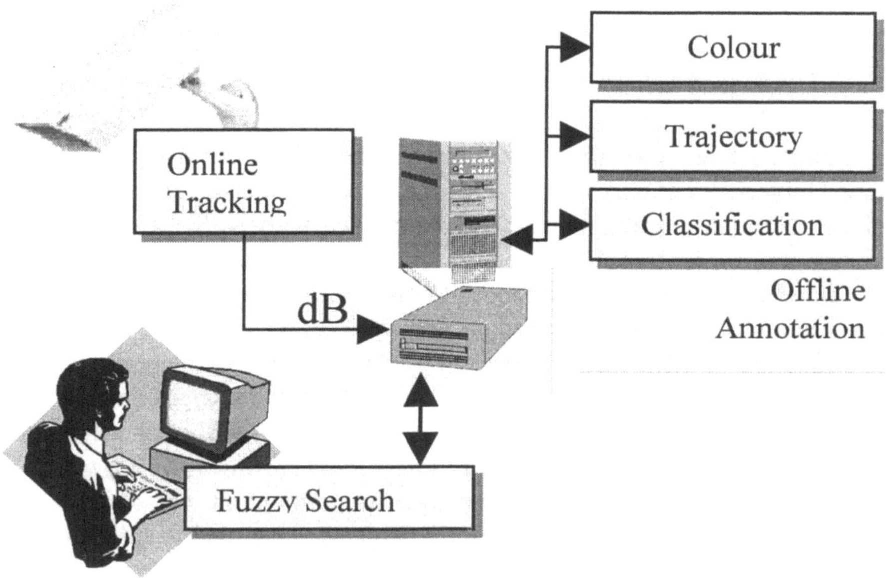


*Figure 1.* System Architecture illustrating the online event-tracking component, the offline event annotation components, and the asynchronous user query tool.

The *offline* event colour, trajectory and classification annotation components are designed to operate during lulls in activity. Unprocessed tracks are recovered from the database and processed by each component in parallel. First the *Colour* component extracts a simple list-based temporal representation of the mode colour "trajectory" of the event's pixels in HSV colour space. Second the *Trajectory* component generates 3D velocity and positional estimates using the *ground plane constraint* [[7]]. In addition, the 2D visual trajectory is combined with a manually generated segmentation of the scene into named *boundaries* and *regions* to generate a simple yet rich description of the progress of an object through the scene *i.e.*

```
08:34:56 Event Id3434 enters "gate"
08:35:03 Event Id3434 enters "approach"
08:35:12 Event Id3434 enters "carpark"
08:35:31 Event Id3434 enters "zone3"
08:36:05 Event Id3434 stops-in "zone3"
```

Finally, the *Classification* stage attempts to classify events into event categories. Currently, we simply wish to differentiate between *Vehicle*, *Person* and *Other* types on the basis of readily available 2D tracking data. Both the algorithm and the level of accuracy available is described in greater detail in section 2.

## 2. Object Classification

A multi-camera surveillance test-bed has been installed overlooking one of the University car parks. The pan, tilt and zoom cameras are pre-set with default positions monitoring car park entrances with wide fields of view. One of these cameras has been used to capture data to validate the approaches described in this paper. In order to evaluate both the object classification algorithm described in section 2.1, a very large data set of 320,000 video frames of a car park was captured during busy arrival and departure periods over four days. This data set contains approximately 400 *Person* and 200 *Vehicle* events all entering, originating within or leaving the car park. Typical image sequences of these events are shown in *Figure 2*. In addition to these common events the data set contains 50 or so *Other* less clear-cut events such as cyclists and big vehicles. This data is split into two equal *Training* and *Testing* data-sets.

Once instantiated, each event must be classified into its object type and specific object behaviour from a limited number of domain-specific activities. This knowledge is derived from the trajectory position and velocity estimates and pixel information provided by the *camera tracker*. In addition, the projected width and height of an object can be temporally tracked by combining the region dimensions extracted from the event map with the position of an object on the ground plane. Classification and behavioural analysis is performed by the following intelligent algorithms.

### 2.1 Object Classification

People and vehicles enjoy distinct velocity width-to-height-ratio characteristics. These are illustrated in Figure 3 by plotting the projected width-to-height-ratio of tracker observations against their estimated image velocity. The velocity estimates need to be normalised by the vertical image position of the observation to compensate for the fact that objects closer to the camera have larger visual velocities. These two class conditional probability density functions for the vehicle and people classes are in fact simply modelled by Normal distributions. Since to some extent these distributions are overlapping, it is necessary to integrate velocity and width-to-height observations over the history of a tracked object to avoid false classification.

(a) Frame 16225

(b) Frame 16250

(c) Frame 16275

(d) Frame 16300

(e) Frame 16325

(f) Frame 16350

*Figure 2.* Example of vehicle entering and manoeuvring through car park. This ten-second event generated nearly 200 frames at a capture rate of 20 frames per second.

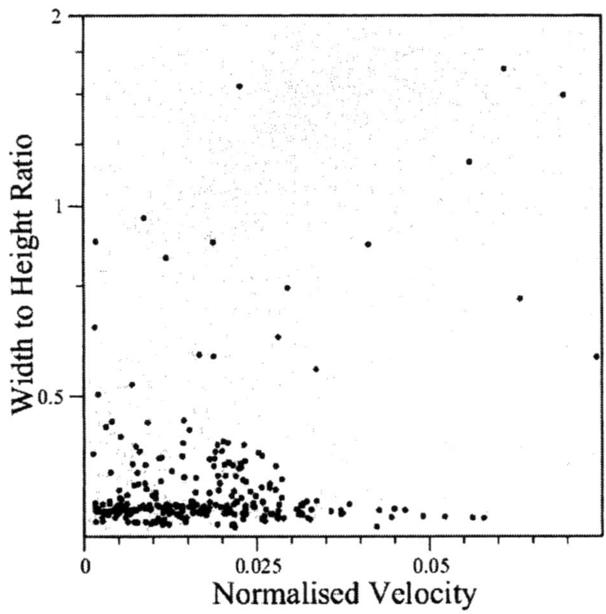


*Figure 3.* Scatter plots in the Width-to-Height Ratio versus Normalised Velocity classification space for Person (black) and Vehicle (grey) training data. Note separate though overlapping distributions.

This is illustrated in *Figure 4* by overlaying the object class PDFs with *trajectories* of a typical person and vehicle event. A simple *maximum a posteriori* decision rule is employed to update the probability of classification given each new observation $\boldsymbol{a}_t = (v_t, w_t)$ of the velocity $v_t$ and width to height ratio $w_t$

$$\omega^* = \arg \max_{\omega \in \Omega} P(\omega \mid \mathbf{a}_t, \ldots, \mathbf{a}_{t_0})$$

where $\Omega$ is the set of possible classifications $\Omega = \{$ *person, vehicle* $\}$, and $t_0$ is the time at which the event started. Assuming each new observation $a_t$ is independent of all previous observations $\boldsymbol{a}_{t-1,\ldots,}\boldsymbol{a}_{t0}$, then the posterior probabilities $P(\omega \mid \mathbf{a}_t, \mathbf{a}_{t-1}, \ldots, \mathbf{a}_{t_0})$ may be expressed temporally as

$$P(\omega \mid \mathbf{a}_t, \ldots, \mathbf{a}_{t_0}) \propto p(\mathbf{a}_t \mid \omega) P(\omega \mid \mathbf{a}_{t-1}, \ldots, \mathbf{a}_{t_0})$$
$$P(\omega \mid \mathbf{a}_{t_0}) \propto p(\mathbf{a}_{t_0} \mid \omega) P(\omega)$$

The class conditional probabilities p(**a**|$\omega$) for vehicle and person events, and associated prior probabilities $P(\omega)$ are derived from training data such as that shown in *Figure 3*.

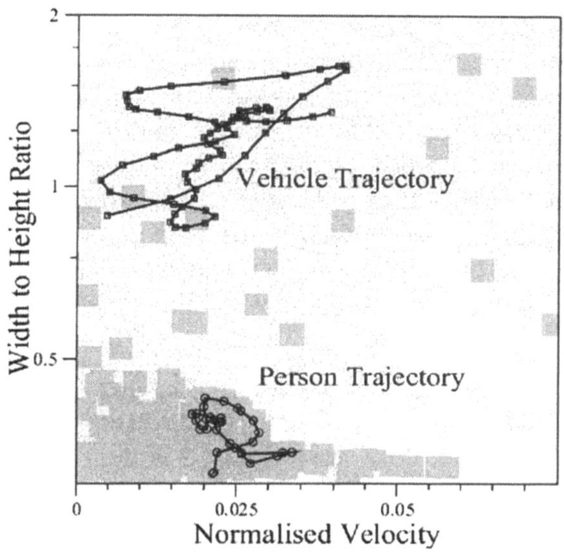


*Figure 4*. 'Trajectories' in Width-to-Height Ratio versus Normalised Velocity classification space for two typical events in which a person and vehicle enter and pass through the car park. For the majority of time, the trajectories of each event remain within the sub-space regions associated with the respective class.

In addition to these two common classes, a number of atypical (in our dataset) event types exist including *cyclists*, and *trucks*. Indeed car, van and truck events are not easily separable. Currently, the training data (for example Figure 3) has been manually separated into vehicle (cars and vans) and person classes, with all other events collectively represented as *Other*, and the classification set $\Omega$ extended to include this new label *i.e.* $\Omega$ = {*Person, Vehicle, Other*}. To account for this *other* class in the classification process specified by equations 1 and 2, a uniform PDF has been assumed. Its prior probability $P(\omega)$ is derived from the training data (examples shown in *Table 1*) while the constant p(**a**|$\omega$) of the uniform PDF is computed experimentally as that value yielding the best classification results on the unseen data set.

*Table 1*. Frequency of Event Types in Training Dataset

| Prior Probabilities | Vehicle | Person | Other |
|---|---|---|---|
| | 0.63 | 0.30 | 0.07 |

## 2.2 Evaluating Object Classification

To evaluate the effectiveness of the classification algorithm, events extracted from the *Testing* training set are classified and compared with the correct manually determined classification. The results are presented in the *scatter matrix* in *Table 2*. These results indicate that for *Vehicle* events, approximately nine-tenths of the events are correctly classified. The remaining incorrectly classified *Vehicle* events are as likely to be classified *Person* as *Other*. The classification of *Person* and *Other* events is somewhat less successful with nearly four fifths correctly classified. In both cases, the incorrect classification is most likely to be *Vehicle*. Nonetheless, the 84% correct classification of the *Testing* dataset is significantly better than the 63% that would result using the prior probabilities alone.

*Table 2.* Object Classification Results (Each row refers to correct manually derived classification of events, while columns refer to the computed classification of the event. Thus the top leftmost cell indicates that 89% of the Vehicle events have been correctly classified as Vehicle, while the top rightmost cell indicates that 5% of the Vehicle events have been incorrectly classified as Other).

| Scatter Matrix | Classification | | |
|---|---|---|---|
| | Vehicle | Person | Other |
| Vehicle Event | 89% | 6% | 5% |
| Person Event | 17% | 79% | 5% |
| Other Event | 28% | 9% | 63% |

While the above results are impressive given the fact that only visual velocity and pixel dimensions have been employed, the accuracy is still far from perfect with 16% of the *Testing* database misclassified. Moreover, the training phase is still highly manual in that video data must be partitioned into sequences containing vehicles or people. It is unlikely that this approach on its own will generalise to include a greater range of vehicle types.

# 3. Query Tool

The purpose of our query tool is to locate and display *video reports* of events described by security operators in as human-like manner as possible. The annotation data generated by the *Colour*, *Trajectory* and *Classification* components described above, is exploited through a touch-screen-driven graphical user-interface has been developed which enables relatively untrained security staff to generate sophisticated queries with a minimum of button presses. *Figure 5* shows the *Object Query Window* in which target objects may be described by *colour*, *type*(classification), *velocity*, spatial and temporal *location* and *time*, and interactions with other objects.

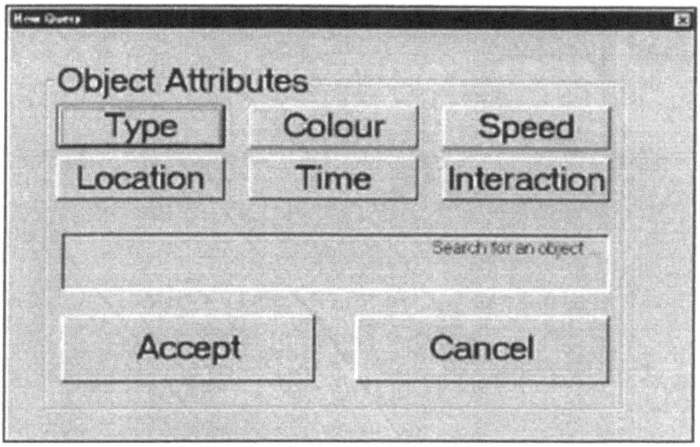


*Figure 5.* Describing an Event (or Object). When specifying an object event in the search, type (classification), colour, velocity, location, time and interaction with other temporally existant objects can be concatenated.

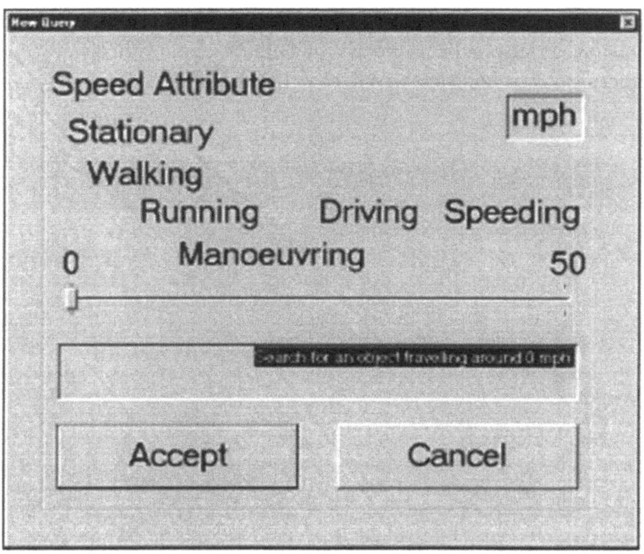


*Figure 6.* Describing Object Velocity. User moves the slider to rough position in previously characterised speed scale.

Three further sub-windows launched from this menu are presented in figures 6 to 8. The first *Object Speed Window* enables the operator to select the speed of an object given some everyday language prompts. The second *Object Colour Window* is based on the Munsell colour space [[1]]. Perceptually significant colour classes are presented as menu of buttons. Finally the third *Object Interaction Window* allows different types of object interaction to be specified.

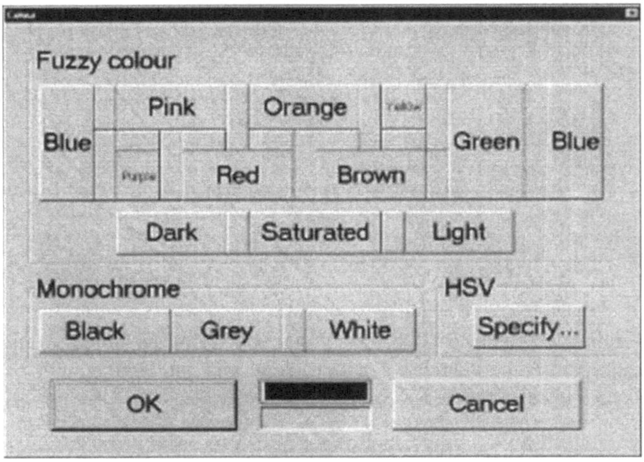


*Figure 7.* Colour space mapped into matrix of buttons each with common language label. Fuzzy colour choice may be adjusted on a dark to light axis.

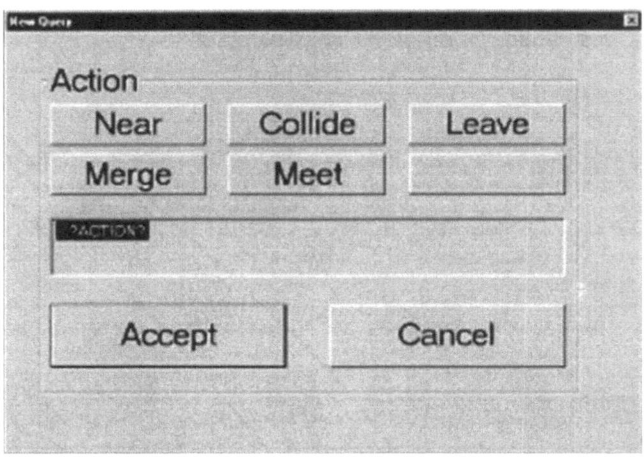


*Figure 8.* Interaction window allows user to specify a range of pair-wise object interactions implying proximity, coincidence, and trajectory genesis or termination

The following SQL-type statement is the result of a typical query building session. Object colour and type are combined with spatio-temporal location as well as a location-coincidence interaction with another attributed object.

```
SELECT   Person.Speed, Person.Locations, Person.tblObject, Person.Time,
         Person.Colour, Vehicle.Speed, Vehicle.tblObject, Vehicle.Time,
         Vehicle.Colour, Vehicle.Locations
FROM     ( Locations INNER JOIN
           Person ON Locations.Locations = Person.Locations )
         INNER JOIN
           Vehicle ON Locations.Locations = Vehicle.Locations
WHERE (  ((Person.Speed)=12mph)
     AND (Person.Locations)="DisabledBay")
     AND ((Person.Time)<11.00am) AND ((Person.Colour)="Red")
     AND ((Vehicle.tblObject)="Van") AND ((Vehicle.Time)<11.00am)
     AND ((tblVehicle.Locations)="DisabledBay") );
```

In practice, such a highly specific SQL statement will not support the fussy matching that will be necessary given the ambiguity and error in human descriptions of events. Nor does the interface syntax yet support queries that specify more than one spatio-temporal location per object *i.e.* entered car-park gate at 9.30am and parked in disabled bay at 2pm. Nonetheless, the prototype demonstrates the ability of a button interface to capture sophisticated descriptions of a scene and incident.

## 4. Conclusion

Currently the VIGILANT system is able to track events within the monitored scene in real-time, and to generate trajectories of high longevity and temporal coherence. From these trajectories (both position and velocity) and the image dimensions of the tracked object, a rich degree of event annotation can be generated by offline *annotation processes*. The operation of the current system was described in necessary brevity in sections 1 and 2. This data describing the object classification, start and end times, the 3D position, semantic region occupancy, colour, 3D velocity implicitly describe the history of the object as it traverses the view volume. Embedded within this data, therefore, is also the relationship between other objects within the scene.

Accurately locating *video reports* of previous events within the visual record depends crucially on accurately describing the target event. Typically, human subjects will specify the objects involved within an incident, describing their appearance and trajectory history, and the active relationships between these objects. Much of this information is available within the low-level annotations generated by VIGILANT. In order to enable matches to be established between operators' descriptions and the database, we need to tackle two main issues. First it is necessary to create a GUI which enable operators to structure queries in as human-oriented style as possible. In section 3, we described our touch-screen button-oriented interface prototype which presents an operator with a series of nested and intuitive menus each designed to obviate the need for keyboard or mouse access. On going human-factors trials are assessing the effectiveness of the tool, and to guide the development of this approach. The second crucial issue which represents the next phase of effort, is the development of fuzzy matching criteria which enables the spirit if not accuracy of a user's description to constrain the search for the correct event match. In addition, evaluation criteria, will need to be developed which order the inevitable lists of partial matches located.

## References


[1] M. Seaborn, L. Hepplewhite, J. Stonham, "Fuzzy Colour Category Map for Content Based Image retrieval", British Machine Vision Conference, volume 1, pp. 103-112, 1999.

[2] I.Haritaoglu, D.Harwood, L.S.Davis "W4: Real-time Surveillance of people and their Activities" PAMI, 22(8), pp.809-830, August 2000.

[3] C.Stauffer and W.E.L.Grimson, "Learning Patterns of Activity using Real-time Tracking", PAMI 22(8), pp.747-757, August 2000.

[4] L.Lee, R.Ròmano and G.Stein, "Monitoring Activites from Multiple Video Streams: Establishing a Common Coordinate Frame, PAMI 22(8), pp.758-767, August 2000.

[5] P.Remagnino, A. Baumberg, T. Grove, D. Hogg, T.Tan, A.Worrall, K.Baker, "An Integrated Traffic and Pedestrian Model_based Vision System", Proceedings of BMVC97, volume 2, Colchester, 8-11th September, University of Essex, UK, pp 380-389, 1997.

[6] M.Bogaert, N.Chleq, P. Cornez, C.S.Regazzoni, A.Teschioni, M.Thonnat, "The PASSWORDS Project", in Proceedings of IEEE International Conference on Image Processing, pp. 675-678, 1996.

[7] G.D. Sullivan, "Visual Interpretation of Known Objects in Constrained Scenes", *Phil. Trans. Royal Society Lon., B*, vol. 337, pp. 361-370, 1992.

Chapter 17

# Evaluation of a Self-learning Event Detector

C. KAAS, J. LUETTIN, R. MATTONE, K. ZAHN
*Ascom Systec AG, Applicable Research & Technology; Gewerbepark – CH 5506 Mägenwil*

**Abstract**: Recently, a novel solution philosophy for complex pattern recognition problems has been proposed (R. Mattone *et al.* in [1]), which is suitable to be applied in the video-surveillance context. It is based on the use of low-level image features and simple, local statistical models for the recognition of complex patterns. For example, the amount of motion in a scene, represented by the number of pixels that changed their intensity over a short time, can be used to recognise "unusual" behaviour of persons like panic or overcrowding. Obviously, this concept calls for some kind of model describing the "normal" situation (e.g., concerning motion) in the field of view. This has been realised by a suitable decomposition of the scene into regions where local models of "normality" can be built. These novel concepts have been implemented in a commercial application called SEDOR, developed at Ascom AR&T. We describe here recent systematic tests of SEDOR, performed on more than 20 hours of video-tapes to demonstrate its performances. We will illustrate that the results can be improved, and point out a possible solution.

**Key words**: Video Surveillance, SEDOR, Topology Map

## 1. Introduction

During the last decade the interest in *intelligent* video-based surveillance has increased tremendously, on the one hand due to the increasing performance of low cost image processing hard- and software and on the other hand due to growing amount of passive – i.e. not program supported – video surveillance systems (see [1] for a recent review). A novel approach to the detection of "abnormal" events in an arbitrary video scene has been proposed by R. Mattone *et. al* [1] and realised in form of a commercial program solution (SEDOR = Self-learning Event DetectOR). The basic idea within SEDOR is that low-level image features and simple statistical models can be used to detect "abnormality" in complex environments (i.e., normally characterised by the presence of several moving objects at the same time), provided that the model of "normality" is decomposed in different local submodels, built in suitable regions of the analysed image. For example, the *amount of motion* [1] can help recognising security-relevant events like panic, fights, overcrowding, if it is properly taken into account that different amounts of motion are "normal" in regions where people normally stay, wait or walk through, and where, due to perspective effects, the observed size of the moving objects is different. In any of these regions, then, the feature behaviour can be approximated with sufficient accuracy by simple functions (e.g., normal distributions of low dimensionality). This map of regions is in SEDOR obtained as result of a novel clustering algorithm (a modified version of the *Growing Neural Gas* algorithm [2], see R. Mattone *et. al* in [1] for details), applied to *motion blobs* extracted from the image [2]. An example of such a *topology map* is shown in Fig. 1. It represents a typical application of SEDOR, where the foyer of a bank with several cash machines is under surveillance.

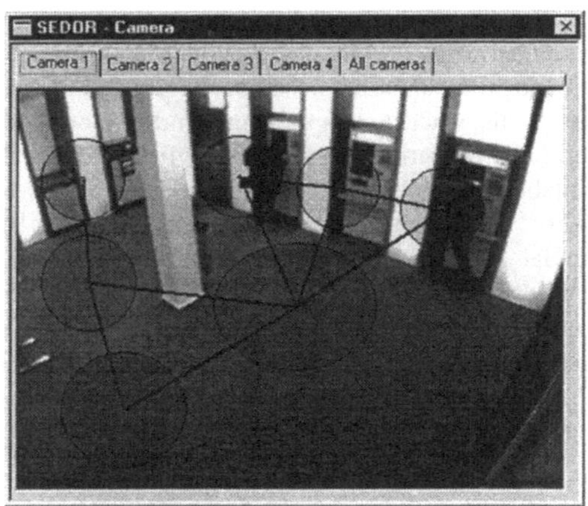


*Figure 1.* Example of a topology map based on the motion blobs.

In any region of the available map, then, a different local model can be built for different kinds of image features. This kind of classification procedure is performed for three different features in the current implementation of SEDOR:

[1] This is defined as the number of pixels that significantly changed in intensity during a short time. They are clustered into so-called *motion blobs*.

[2] The idea is that regions of the image where motion clusters correspond to logically meaningful "places" in the observed scenery.

The first module (SO) classifies the presence of new *objects* (edge-based features [3]) as normal or not, on the basis of their staying time. The alert threshold for the staying time is learned separately for any region of the topological map. An example of an alert generated by the presence of *static objects* (i.e. objects staying still "too long") is illustrated in Fig. 2. This classification module can also be useful in the detection of abandoned and removed objects.

*Figure 2. Example of a "static alert". An unconscious person is lying motionless on the floor*

The second module analyses any single motion blob on the basis of its location w.r.t. the topology map, in order to detect the presence of *motion in abnormal region* (MAR). This module can detect events like the violation of forbidden-access areas.

The third feature classification module, finally, assesses the *total amount of motion in any region*, i.e., the sum of the sizes of all motion blobs detected in any region. An example is given in Fig. 3, showing two persons running around in the foyer. This module is capable to detect abnormalities related to *too much motion* (TMM) in a region, either in absolute value (circular region in Fig. 3), or relatively to the simultaneous motion in some other region (square region in Fig. 3). Note that, according to the above definition of *total amount of motion*, the same "large" amount of motion can be generated by an object moving "fast", by a "large" moving object and by "many" moving objects. This kind of alert can be caused by several security-relevant events, like panic situations, fights, overcrowding.

*Figure 3. Example of a "too much motion alert". Two persons running around in the foyer.*

## 2. Experimental Results

First we will focus on the alert events (AE) which were produced to test the ability of the system to detect relevant events: these scenes have been played by actors and represent some typical alert situation [3] (table 1, top).

*Table 1.* Alarm events (top) and the detection capabilities of the three SEDOR modules (bottom).

| | |
|---|---|
| AE1 | robbery of a customer |
| AE2 | victim remains unconscious on the floor (fig. 2) |
| AE5 | 2 persons painting graffiti on the wall ... |
| AE6 | ... and removing the trash bin [1] |
| AE7 | 2 persons running around (fig. 3) |
| AE9 | "night guests" in the foyer [1] |

| | alarm events | | | | | |
|---|---|---|---|---|---|---|
| Modul | 1 | 2 | 5 | 6 | 7 | 9 |
| SO | | x | x | x | | x |
| TMM | | | x | | x | |
| MAR | | | x | | | x |

The victim of a robbery, which remains unconscious on the floor (AE2) and the two persons running around in the foyer (AE7) are represented in Fig. 2 and 3 respectively. Pictures of the events AE6 and AE9 can be found in [1].

[3] The numbering is chosen to be compatible with our internal notation.

The detection capability of the different modules is shown in table 1 (bottom). The static object module SO reliably detects the unconscious person (AE2) the graffiti (AE5), the missing bin (AE6) and the "night guests" (AE9). The too much motion module TMM creates alerts for the events AE5 and AE7, where the two quickly moving persons represent a large "amount of motion". Finally the events AE5 and AE9 are in addition detected by the motion in abnormal region module MAR as the motion is sometimes located far away from the topology nodes. The only relevant event that is not directly detected (but just through its consequence AE2) is the robbery itself (AE1). It hardly represents any detectable feature (at least for SO, TMM and MAR) apart from a relatively large amount of motion over two to three video-frames, as we will discuss in section 3. In addition to these alert events we tested SEDOR on more than 20 hours of videotape recorded in the same bank foyer with the usual customer flow present and also used to build the necessary models of "normality". In table 2 we present the number of – by definition – false alerts (FA) detected per day (24 hours).

*Table 2. The mean number of false alerts per day (24 h) for the various modules.*

| *modules* | | | *errors* |
|---|---|---|---|
| ***SO*** | ***TMM*** | ***MAR*** | ***FA*** |
| ON *(MTA800)* | *off* | *off* | *5.58* |
| ON *(MTA600)* | *off* | *off* | *23.19* |
| ON *(MTA350)* | *off* | *off* | *35.32* |
| ON *(MTA800)* | *off* | ON | *5.58* |
| ON *(MTA800)* | ON | ON | *27.86* |

First the static module SO has been tested with three different settings of the parameter "minimum time to alert" (MTA) [4]. This is the threshold time after which a still object can cause an alert. Here, 350 cycles correspond to approximately 4 minutes. As obvious from the first three rows of Table 2, the results depend crucially on the choice of MTA. This is due to the fact that many customers remain even up to 10 minutes almost motionless in front of the cash machines. In that sense most of the "false" alerts even demonstrate the functionality of the system. Nevertheless for a time to alert of about 8 minutes the number of false alerts decreases to a value of 5.58 per day. Then, using the setting MTA = 800 for SO, we added the MAR module which did not alter the error rate (forth row). In fact, this module works very reliably in the sense that the relevant alarm events are detected (table 1) without creating false alerts. However, the situation drastically changes

[4] As stated above, this is only a lower bound as in the different topology regions a larger MTA can be learned.

when the TMM module is used (last row) as the error rate increases by more than 22 FA per day: The TMM module cannot distinguish between a real "too much motion" event and several persons that are close together and move "normally" with their individual motion-blobs summing up to a value above the normality threshold. Nevertheless, there exists a simple way of extending the classifier to make it more robust with respect to false alerts and thereby even increasing its reliability in detecting the real alert events. This will be discussed in the following section.

## 3. A modified "too much motion" Classifier

The basic idea is the correlation of motion blob-sizes at different times, thereby introducing the concept of "speed" of the blob. Formally speaking we measure the average time a blob remains in a given node. We also correlate the blob-sizes at different times within different nodes and by that means measure the time necessary for the blob to move between the two nodes. If $A_n(k)$ denotes the size of a motion blob found in node $n$ at processing step $k$ the correlation (between nodes $m$ and $n$) as a function of the time delay $\Delta k$ is calculated according to [5]:

$$\langle A_n(0) \cdot A_m(\Delta k) \rangle = \frac{1}{N} \sum_{k=1}^{N} A_n(k) \cdot A_m(k + \Delta k)$$

The brackets $\langle\rangle$ denote an average over a large number of blobs, typically $N \approx 5000$. In Fig. 4 a typical time correlation function of the blob-sizes in one node has been represented. The curves are normalised to 1 at $\Delta k = 0$ and 0 for large $\Delta k$. The data is well described by an exponential decay and the decay constant $k_0$ is a measure of the presence time of the motion blob in this region. We determine the maximum value of $k_0$ for all nodes $k_{max}$ and will limit our analysis to $0 \leq \Delta k \leq 2 \cdot k_{max}$. In fact above this limit the correlation is too small to give statistically significant results. It is important to determine this limit solely on the basis of the acquired data: one of the important concepts of SEDOR is its self-learning aspect. For values $0 \leq \Delta k \leq 2 \cdot k_{max}$ and all combinations of nodes $(m,n)$ the probability distribution $P(A)$ of the "correlated blob-size" $A = (A_m(0) \cdot A_n(\Delta k)))^{1/2}$ is determined. In Fig. 5 this function is shown for $m = n = 0$ and $\Delta k = 0$. The most interesting feature of this curve is its asymptotic behaviour for large values of $A$, which follows an exponential decay. This is indeed true for all combinations of $m$, $n$ and $\Delta k$.

[5] The time necessary for one processing cycle is usually between 120 to 600 milliseconds. In the following we use the expression time as a synonym for the processing step $k$.

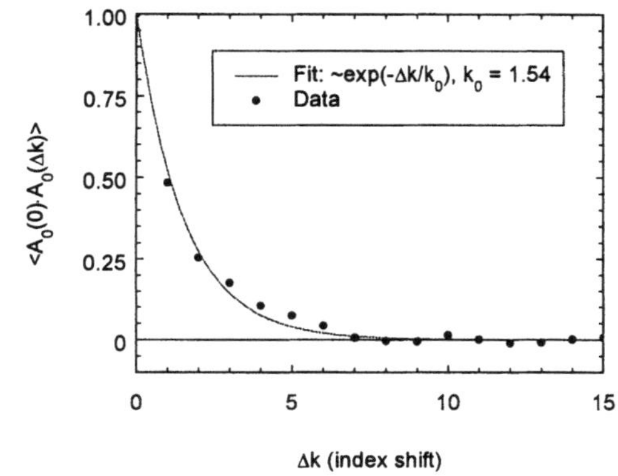


*Figure 4. Time correlation function in node 0*

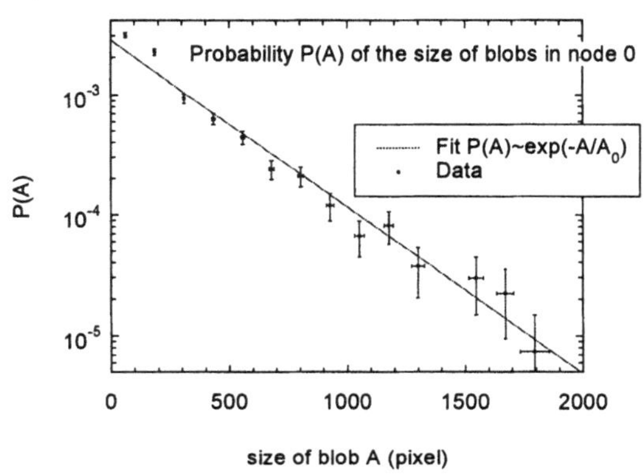


*Figure 5. Probability to find the blob size A in node 0*

This now fits perfectly to the philosophy of SEDOR as we obtained simple models – an exponential distribution – of the "normal" amount of motion within and between nodes. For a typical number of 10 nodes and a value of $2 \cdot k_{max} = 6$, several hundreds of models are available: however, not all of them are statistically independent. The learning time of the classifier is defined by a minimum number of blobs necessary for a good statistics in the probability diagrams (like Fig. 5). Thereafter, for the new detected blobs the quantities $A$ (for all $m$, $n$ and $\Delta k$) are calculated and compared to the models. If the probability for $A$ is less than a defined threshold an *event* is created [6]. The size-threshold $A_{lim}$ can be expressed in terms of a limit probability $P_{lim}$

[6] Note that an *event* does not necessarily correspond to an *alert event*, which is generally characterised by a given number of events detected in a defined interval.

which is the integral of $P(A)$ from $A_{lim}$ to infinity. Tests with two different values of $P_{lim}$ have been performed on the same video-recordings as used above (table 3). The false alerts (FA) and the events detected on the alarm events AE1, AE5 and AE7 are listed, as only these are relevant for the TMM module (a value $\geq 1$ is equivalent to a detection of the event, marked by a cross in table 1). It should be noticed that AE1 could be detected together with a reduction of the total number of false alerts ($P_{lim} = 0.0001$: FA = 12 compared to 22 above). Nevertheless, compared to the alert events AE5 and AE7, which are characterised by a huge amount of events, AE1 cannot be reliably detected. However, if AE1 is discarded, the number of false alerts can even be reduced to zero. This is possible either by further decreasing $P_{lim}$ or by making the alert criteria (on the basis of the detected events) more stringent. For the second alternative a further analysis of the events in table 3 is necessary – so far we did not specify the values of $m$, $n$ and $\Delta k$. First of all it is important to note that out of the 132 events detected in AE5 only 12 result from a value of $\Delta k = 0$. Thus, the concept of correlation at different times is essential for the results. In addition, it turns out that none of the false alerts appeared for different values of $m$ and $n$, i.e. for different nodes. On the contrary, for the alert events AE5 and AE7, one third of the events fulfil $m \neq n$. The reason is that even though several persons can add up to a large motion blob at a given time, it is very unlikely that these persons *together* move on to the same node. They will rather split up and move in different directions resulting in low correlation values between different nodes. In the cases of AE5 and AE7 the two persons quickly moved from one node to the next causing an unusual event. Thus, if an alert is only created in the case of an event with $m \neq n$ it is possible to detect AE5 and AE7 reliably *without any* false alert (table 3, third line).

*Table 3. Results of the improved TMM classifier.*

| | errors | alarm events | | |
|---|---|---|---|---|
| limit probability | FA | AE1 | AE5 | AE7 |
| $P_{lim} = 0.0001$ | 12 | 1 | 132 | 154 |
| $P_{lim} = 0.00025$ | 37 | 2 | 204 | 218 |
| $P_{lim} = 0.0001$ ($m \neq n$) | 0 | 0 | 43 | 49 |

## 4. Conclusions

We presented systematic test results of the SEDOR (Self-learning Event DetectOR) application program developed at Ascom AR&T, performed on a database of more than 20 hours of videotape, and could demonstrate that its philosophy – detection of complex patterns on the basis of low-level image features – works convincingly. The *too-much-motion* module, which detects unusual behaviour like overcrowding or panic, has been improved to guarantee an overall satisfactory performance of the program.

## References

[1] *Multimedia Video-Based Surveillance Systems,* edited by G. Foresti, P. Mähönen and C. Regazzoni, Kluwer Academic Publisher, Boston, 2000.
[2] B. Fritzke, "A growing neural gas network learns topologies," in G. Tesauro, D.S. Touretzky, and T. K. Leen, *Advances in Neural Information Processing Systems 7*, MIT Press, Cambridge MA, 1995.
[3] R. Galli and M. Kiener, *SEDOR, Selbstlernende Ereignis Detektion in Oeffentlichen Räumen—Lösungsansatz: Analyse der Kanten im Bild,* Diplomathesis, Hochschule für Technik und Architektur Burgdorf, CH, 1999.

Chapter 18

# AUTOMATED DETECTION OF LOCALISED VISUAL EVENTS OVER VARYING TEMPORAL SCALES

Jamie Sherrah and Shaogang Gong
*Department of Computer Science*
*Queen Mary, University of London*
*London E1 4NS UK*
jamies@safehouse.com.au, sgg@dcs.qmw.ac.uk

**Abstract** In this work, a general semantics-free method is proposed for the extraction of visual events. The technique is based on the definition that an event is any sort of visual change. It is therefore necessary to determine visual changes that occur at different rates. Wavelet analysis is employed to detect different rates of pixel-wise change in the image, while a Gaussian mixture background model is used to determine absolute temporal change in pixel values. Clustering is then performed in a feature space for these pixel-wise change events. The result is a grouping of low-level events into high-level events. We show results on an artificial shopping scenario.

## 1. Introduction

In general terms, scene understanding for visual surveillance from a fixed camera is ambiguous. Most previous approaches have been based on low-level segmentation or tracking of objects in the scene. For example, in many cases the definition of an event only concerns people in the scene [5, 4]. The tracked trajectories of the objects are then generally used in some high-level knowledge acquisition. Naturally some form of pixel grouping must be performed to obtain high-level knowledge from the scene, but our point is that it should be performed robustly in stages, rather than initially at a low level where insufficient information is available.

Of late, several attempts have circumvented the problem of segmentation and tracking by learning localised or pixel-wise variations [7, 2, 6]. The local models are then used to detect local visual events. Provided only detection is required, this approach overcomes the problems of segmentation and definition of the meaning of behaviour in the scene, which inevitably is problem-specific. However, pixel-wise modelling is computationally expensive, and if high-level knowledge is required then pixel grouping still must be performed.

A further issue is whether learning should be supervised or unsupervised. Supervised learning is generally less difficult since the user can supply examples of the background without any objects (*i.e.* clean frames), and examples of typical behaviours. In this way, semantics are introduced. However there are two difficulties with supervised learning. First, training data or clean frames may be unavailable. Second, it may be infeasible to manually label sufficient training data. That labelling also involves a subjective element, especially when it comes to human behaviour, which means accumulating information about the scene, rather than imposing prior models.

An assumption that most approaches do not explicitly discuss concerns temporal scale. The spatial bounds of the image domain are predetermined by the camera and grabbed image size, however time extends forever. Thus the question arises: over what temporal scale should behaviour or events be defined? We propose a method for unsupervised scene profiling that can learn, classify and detect high-level visual events in an unsupervised manner. The algorithm does not require prior specification of scene semantics. The approach detects pixel-wise temporal change occurring at varying temporal scales.

## 2. Defining an "Event"

The problem at hand is *behaviour* or *scene profiling*, defined here as follows: *Given a fixed camera and long-term video acquisition, learn to detect abnormal behaviour or events in the scene.* Before moving on, let us expand on this definition. Long-term exposure may constitute years, captured at frame rate. Therefore the system may not wantonly accumulate data or models. Iterative optimisation algorithms should be avoided for timeliness constraints. The system must learn to detect. The traditional method is to learn models of normal behaviour and events, then test observations for deviations from these models. We assume the definition of "normal" to be "regularly occurring". Therefore the frequency of occurrence of an event may be low so long as it is regular.

People generally have a conceptual understanding of the term "visual event" as relating to some kind of visual change. Indeed, many past surveillance research approaches are based on motion or change detection [8, 5, 4, 3]. However, these approaches immediately impose some semantics, either explicitly or implicitly, by grouping pixels into high-level events. When performed at a low level, the grouping implicitly requires definition of information such as spatial scale. More importantly, the change detected by these approaches is generally defined as the absolute difference from a reference frame. However, many dynamic scenes contain a rich variety of change occurring at different rates. In [9], high-level visual change is classified into ten categories. Considering their list of different events and the manner of visual change they cause, it is clear that if the definition of visual change were restricted to absolute difference, then many events would become indistinguishable or undetectable. For example, a person passing by and a person stopping for a few minutes would not be differentiated at the pixel level by the dynamic background modelling technique of [8].

In order to encompass the broad range of visual events experienced in everyday scenes, we define an event as *any kind of significant visual pixel-wise change.* It presents the challenge of finding a single technique that can cope with these different types of change. We propose a temporal multi-resolution approach to change detection. Each pixel is analysed in isolation over time to distinguish change occurring at different rates.

# 3. Temporal Multi-resolution Analysis of Visual Change

General low-level event detection requires a method to detect change occurring at different temporal scales. The method must represent different super-imposed frequencies localised in time. It must also be computationally efficient enough so be applied to the huge space-time volume of image data. Clearly a time-frequency analysis algorithm is required. The Fourier transform would be inappropriate since it gives no localisation in time. The Short-Term Fourier Transform (STFT) would be more appropriate, but it is highly redundant which would contravene the requirement for computational efficiency. The STFT also requires the selection of a time window, again restricting temporal scale. Wavelets were designed to overcome these problems [1, 11], and are used in our approach. Wavelets are essentially time-local band-pass filters that yield a multi-resolution time-frequency representation of a signal.

The approach taken is to form a Discrete Wavelet Transform (DWT) at each pixel over time. Consider the intensity values of a single pixel over time. The input time series $x(t)$ can be decomposed into a set of basis functions called *wavelets* [10]:

$$x(t) = \int \int \gamma(s, \tau) \psi_{s,\tau}(t) d\tau ds \tag{18.1}$$

The basis functions $\psi_{s,\tau}(t)$ are scaled and translated versions of the *mother wavelet* $\psi(t)$:

$$\psi_{s,\tau}(t) = \frac{1}{\sqrt{s}} \psi \left( \frac{t - \tau}{s} \right) \tag{18.2}$$

Note that the scale $s$ is inversely proportional to frequency, so that small scale means high frequency. $\gamma(s, \tau)$ is the projection of $x(t)$ onto $\psi_{s,\tau}(t)$:

$$\gamma(s, \tau) = \int x(t) \psi^*_{s,\tau}(t) dt \tag{18.3}$$

The wavelets themselves are localised in time and frequency. The integral of equation 18.1 is infinite, so in practical applications $s$ and $\tau$ are sampled on a dyadic grid, resulting in discrete wavelets. If $x(t)$ itself is discrete, then the discrete wavelet transform is obtained. As the scale increases, the wavelet projections $\gamma(s, \tau)$ have decreasing spatial resolution and highlight features of decreasing frequency. The result is a set of band-bass filters with logarithmic frequency coverage. Since a finite number of scales $s = [1, \ldots, L]$ must be used and the frequency coverage is iteratively halved with scale increase, a low-pass filter called a *scaling function* is used to cover the remaining low-frequency region.

After the DWT is performed, the resulting frequency information can be stored in a time-frequency histogram to characterise events occurring at different instants. For an image containing $N$ pixels, the storage requirement of the histogram is $N.T.L$, which quickly becomes unmanageable. A more feasible approach is to accumulate frequency information at each pixel over time, which requires bins for each pixel and scale level only. This information can subsequently be used to determine which frequencies of change occur over time in different regions of the image.

## 4. Extraction of Events from Visual Data

For an unsupervised scene profiling algorithm, the distribution of events must be learned over time in some feature space. Here we describe a method for learning classes of events in the scene. First, events are detected locally in space and time. These local events are accumulated over time in an unsupervised learning algorithm, resulting in a set of event classes. The class information is subsequently used to detect and label local events, and for establishing high-level assocations between events.

## 4.1 Local Event Characterisation

Two methods are described for detecting pixel-wise change events, and representing localised events in a feature space. First a Gaussian mixture background model is used to detect absolute pixel change, and the pixels are clustered spatially to form local events. Second the DWT is used to detect differential pixel change at different rates, and local events remain at the pixel level. These two methods are compared in section 5.

The Gaussian mixture background model of [6] is used to detect pixel events as absolute colour change in RGB space that does not fit the model. Pixel events are then spatially grouped to form higher-level event boxes. We used a connected components algorithm for spatial grouping. Those foreground box events that remain in predominantly the same position for a non-trivial period of time are subsequently defined as local events. These events $v_i$ are then characterised by the feature vector:

$$v_i = \{t, x, y, d, w, h\} \tag{18.4}$$

where $t$ is the start time of $v_i$, $(x, y)$ is the central position of the event box in the image, $d$ is the duration, and $(w, h)$ are the average event box dimensions.

To detect pixel-wise change events at different time scales, the Discrete Wavelet Transform is used in a two-pass fashion. First a wavelet histogram is formed, then on the second pass the actual events are extracted. The purpose of the histogram is to identify regions of the image where change occurs regularly so that spurious detections can be avoided. We use $L = 7$ levels, which means the DWT is performed in blocks of $2^7 = 128$ time samples (frames). Frames are accumulated until 256 frames have been collected, then the DWT is performed over this time for each pixel. A histogram of spectral power is accumulated for each scale or level as follows. For each scale, the output array is visited and those absolute values that are over a noise threshold contribute to the histogram at that level. The final histogram is thresholded on a per-level basis to remove noise. This step is currently manual, but will be automated in the future. The 18-sample Daubechies wavelet was used [11].

The second-pass of the algorithm involves going over the sequence again and computing the DWT. For each pixel, level and time frame, form an event if the DWT output value is greater than the threshold. Only pixels with histogram values above threshold are included. These pixel-wise detections are our local events, and are characterised by:

$$v_i = \{t, s, x, y, h, g\} \tag{18.5}$$

where $s$ is the temporal scale, $h$ is the DWT output value for that scale, and $g$ is the grey level at the pixel $(x, y)$.

Given the set $v_1, \ldots, v_N$ of local events, the final step in training is to perform clustering in feature space to determine the distinct classes of event in the scene. We use k-means clustering with k a manually-determined parameter.

## 4.2 Correlations Between High-Level Events

Suppose we are at the beginning of an event $v_a$ of class $A$. We want to know if $v_a$ causes events of class $B$. What we are essentially saying is that $v_b$ (nearly) always occurs in the not-too-distant future of $v_a$, but not necessarily vice versa. For example, if a shopper takes an item from the shelf, he will pay within the next few minutes, but afterwards the next shelf item may not be taken for another half an hour if it is not a busy day. Using a one-sided Gaussian weighting function centred on the starting time of event $v_a$, the Gaussian-weighted time difference between the beginnings of events $v_a$ and $v_b$ can be determined. The purpose of the weighting function is to give higher weight to events that are temporally proximate. The weighted differences are accumulated in a covariance matrix for all combinations of events:

$$C(A, B) = \sum_{i:v_i \in A} \sum_{j:v_j \in B, t(v_j) > t(v_i)} G(t(v_j); t(v_i), \sigma) \tag{18.6}$$

where $t(v_i)$ is the time at which event $v_i$ occurred. The element $C(A, B)$ is the accumulated weighting of events of class $B$ following events of class $A$. The matrix is upper-triangluar because a one-sided Gaussian was used so that only connections forward in time are sought. By thresholding these correlations, rules can be established such as "if $A$ happens then expect $B$" to happen. These rules are then triggered by detection of the causing event ($A$), and in response an alarm shows the expectation of the occurrence of the caused event ($B$).

# 5. Experimental Results

We have collected a 20 minute test sequence called the *shop* sequence, involving an artificial shopping scenario. A shop keeper sits behind a desk on the right side of the view. An assortment of drink cans is laid out on a display table. Customers enter from the left and either browse the wares without purchasing, or take a can and pay for it. Abnormal

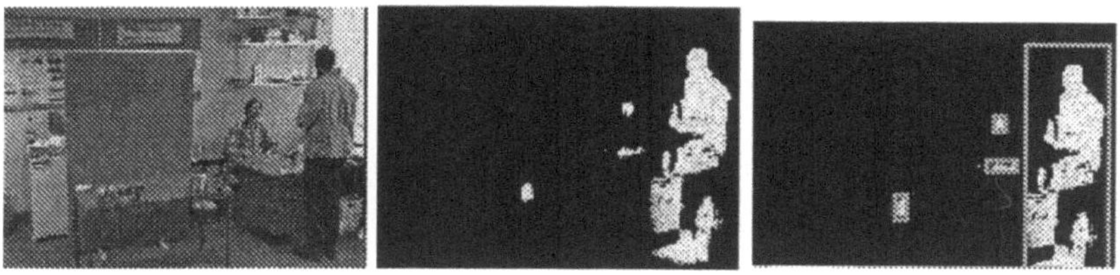

(a) Original Scene (b) Pixel Change Events (c) Grouped Pixel Events

*Figure 18.1.* Example of local event characterisation using Gaussian mixture background model.

behaviour would be to take a can and leave without paying. We will show how the system can (a) detect the events of browsing, taking a can, and paying, and (b) learn the temporal association between these events. The data were sampled at about 8 frames per second.

**Detection of Absolute Change** Absolute change event boxes were extracted from the sequence, an example is shown in Figure 18.1.Using this method, $N = 213$ local events were extracted from the *shop* sequence.

These local events were then clustered using k-means with k=4. The absolute change local events $v_i$ and their respective classes are shown in Figure 18.2. Although clustering is performed in a high-dimensional space, only the x-y co-ordinates of the events are shown here. The four different colours show the four clusters found. These have been manually labelled as being caused by the shop-keeper, browsing, paying and cans taken.

Note that the results are quite noisy due to mis-detections and poor segmentation. In particular, the payment events are on the ground, and segregated from the change caused by the upper-body of the customer. This is due to errors made by the spatial clustering algorithm when forming the local events.

**Detection of Multi-Scale Temporal Change** In contrast, the DWT was then used to extract events from the same sequence. Since the frame rate is about 8 fps, the longest event we can characterise with 7 levels is 8-16 seconds in duration. An example of the wavelet histogram is shown in Figure 18.3. At each scale level and pixel, the thresholded histogram value is true (white) if significant change occured there. It can be seen that different occurrences are highlighted at different scales with continuous variation over scales. The histogram was then used to extract

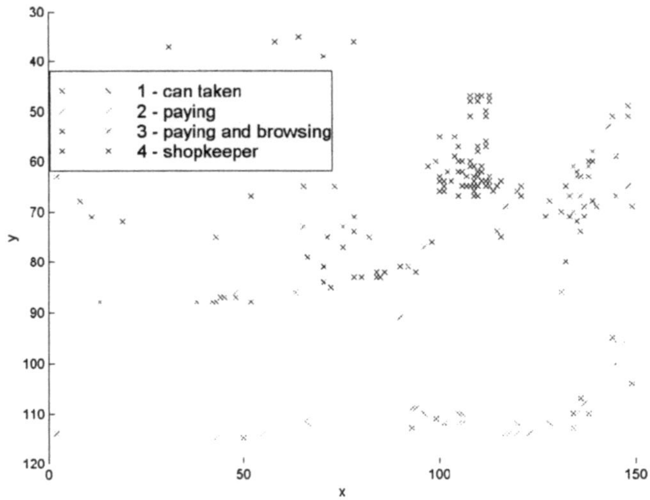


*Figure 18.2.* Clustered events using the Gaussian mixture background model to detect absolute change local events. Only the x-y positions of the events are shown.

pixel-wise events at different levels from the *shop* scenario, the resulting event list contained $N = 250,297$ events.

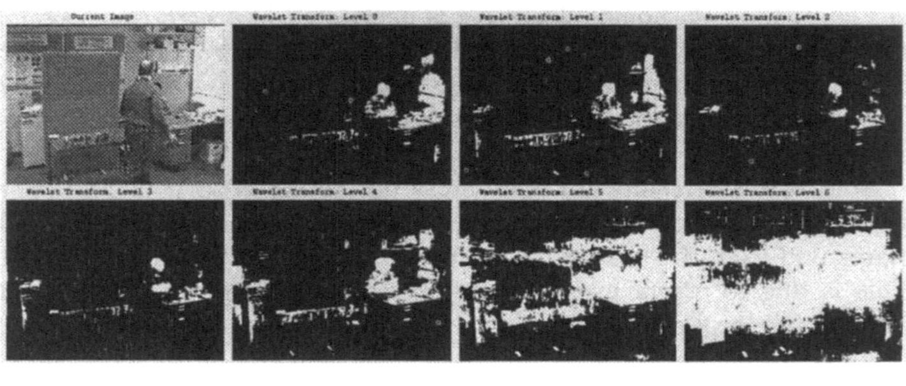

*Figure 18.3.* 7 level wavelet histogram of the shop scenario. From left to right, top to bottom, the figures are: original scene, then thresholded histograms in binary form from level 7 to level 1.

After clustering the multi-scale temporal change local events, the $v_i$ and respective classes are shown in Figure 18.4. Again the 4 different classes, though found using unsupervised learning, have been manually-labelled, this time as shop-keeper, paying, browsing-left and browsing-right. In this case, the events are very clearly distinct from each other, with very dense detection rates. The main difficulty is that the cans being taken have not fallen into a class of their own. In fact they are not very distinct in the histogram of Figure 18.3. The reason is that the can is taken suddenly, with no subsequent change. In terms of temporal change, this would be hard to distinguish from noise. Since the wavelet method is based only on temporal difference and not absolute difference, it cannot detect such events. Rather it needs to be combined with the absolute change method.

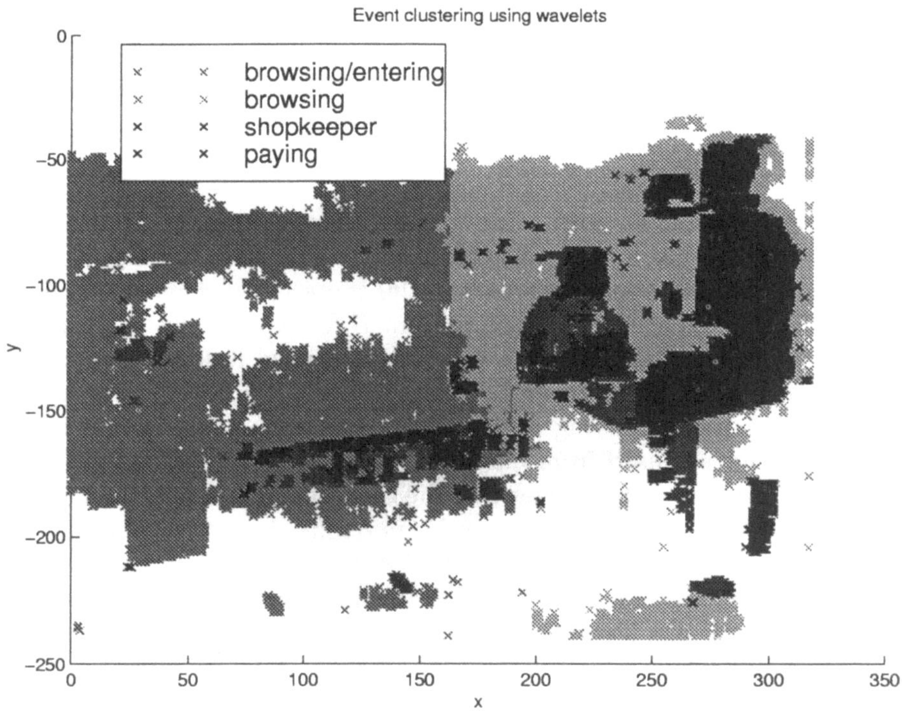


*Figure 18.4.* Clustered events using the wavelet model to detect multi-resolution change local events. Only the x-y positions of the events are shown.

In comparing the two methods, we can see that the wavelet-based approach gives a much better clustering. One advantage of the Gaussian mixture model method is that one ends up with fewer events, making the algorithm computationally less expensive. The trade-off is that some grouping must be performed at a level that is too low to be robust. Note that the absolute change method involves a much more restrictive defini-

| | can | paying | shopkeeper | browsing |
|---|---|---|---|---|
| can | 0.0 | **3.3698** | 0.5837 | 0.6546 |
| paying | 0.0 | 0.0 | 1.3320 | 1.3819 |
| shopkeeper | 0.0 | 0.0 | 0.0 | 0.5691 |
| browsing | 0.0 | 0.0 | 0.0 | 0.0 |

*Table 18.1.* Causal correlation matrix showing causal connections between events.

tion of an event at the low level. Semantics have been imposed through the interpretation of the background modelling results, spatial clustering of pixels into local events, and temporal grouping of instantaneous events at the same position. The multi-resolution method imposes no such semantics, but again suffers from a large number of resulting events. Both of the local event detection methods need to be combined in one final solution.

**Causally Linking Events** Experimentally, the causal correlation matrix $C$ shown in Table 18.1 was established. The standard deviation of the time-weighting Gaussian was $\sigma = 100$ frames. The element $C(A, B)$ shows the accumulated weighting of event $B$ following event $A$. It can be seen in the table that there is a very strong causal link between taking a can and paying, highlighted in bold. There is also a relatively strong connection between paying and the shopkeeper moving, which makes sense since he is roused from his book when receiving the money.

In our case, the system developed the rule "if can is taken, then expect payment". An example is shown in Figure 18.5. A can been stolen (left box) which triggers display of the centre of the payment event cluster (right box). Until a payment event transpires, the box will continue to flash. Although this algorithm is quite simple and naive, the results are still quite powerful.

## 6. Conclusion

A methodology for detecting general visual events in a scene has been presented. Two algorithms for detecting local events have been presented and compared, one based on absolute change and the other on different rates of differential change. The methods were tested on real data.

Future work will involve further investigation of the wavelet approach, including the use of an adapted wavelet analysis to remove noise and reduce the number of detections. The two detection methods need to be merged to detect both absolute and relative change. An on-line implementation of the algorithm would enable experiments conducted over

*Figure 18.5.* Example of the causal rule linking cans taken and payment. A can been stolen (left box) which triggers display of the centre of the payment event cluster (right box).

many days in real situations. Finally, there is great scope for more work on using these visual events for high-level reasoning, such as finding causal connections.

## References

[1] A. Akansu and R. Haddad. *Multiresolution Signal Decomposition: Transforms, Subbands, and Wavelets.* Academic Press, 1992.

[2] O. Chomat, J. Martin, and J. Crowley. A probabilistic sensor for the perception and the recognition of activities. In *Proceedings of the Sixth European Conference on Computer Vision*, volume 1 of *Springer-Verlag Lecture Notes in Computer Science*, pages 487–503, Dublin, Ireland, June 2000.

[3] I. Haritaoglu, D. Harwood, and L. Davis. $W^4$: Real-time surveillance of people and their activities. *IEEE Transactions on Pattern Analysis and Machine Intelligence*, 22(8):809–830, August 2000.

[4] S. McKenna, S. Jabri, Z. Duric, and H. Wechsler. Tracking interacting people. In *IEEE International Conference on Face & Gesture Recognition*, pages 348–353, Grenoble, France, 2000. IEEE Computer Society.

[5] R. J. Morris and D. C. Hogg. Statistical models of object interaction. *IJCV*, 37(2):209–215, 2000.

[6] J Ng and S Gong. Exploiting pixel-wise change for inferring global abnormal events in dynamic scenes. Technical Report QMW RR-00-07, Queen Mary, University of London, 2000.

[7] C. Stauffer and W. Grimson. Adaptive background mixture models for real-time tracking. In *IEEE Conference on Computer Vision and Pattern Recognition*, Colorado, USA, June 1999.

[8] C. Stauffer and W. Grimson. Learning patterns of activity using real-time tracking. *IEEE Transactions on Pattern Analysis and Machine Intelligence*, 22(8):747–758, August 2000.

[9] K. Toyama, J. Krumm, B. brumitt, and B. Meyers. Wallflower: Principles and practice of background maintenance. In *IEEE International Conference on Computer Vision*, volume 1, pages 255–261, Corfu, Greece, September 1999.

[10] C. Valens. A really friendly guide to wavelets. http://perso.wanadoo.fr/polyvalens/clemens/wavelets/wavelets.html, 1999.

[11] M. Wickerhauser. *Adapted Wavelet Analysis: from Theory to Software*. IEEE Press, 1994.

# Chapter 19

# REAL–TIME VISUAL RECOGNITION OF DYNAMIC ARM GESTURES

Héctor Hugo Avilés-Arriaga and Luis Enrique Sucar-Succar
*Instituto Tecnológico y de Estudios Superiores de Monterrey Campus Cuernavaca*
*A.P. 99-C, Cuernavaca, Morelos 62589, México*
00374765@academ01.mor.itesm.mx, esucar@campus.mor.itesm.mx

**Abstract** Visual recognition of arm gestures is an important skill for computers to interact with humans in a more natural form. We propose a real-time monocular visual system to recognize a set of five right-arm dynamic gestures, two of which present depth motion. These gestures were selected for their potential application to command mobile robots. The system employs a radial scan segmentation algorithm combined with a statistical-based skin detection method to find the candidate face of the user and to track his right-hand. It uses simple features to describe the user's right-hand movement and Hidden Markov Models as recognition technique. The system was tested with different users, and is able to recognize these five gestures in real time with an average recognition rate of 82.4%.

**Keywords:** Dynamic gesture recognition, Hidden Markov Models

## Introduction

Gestures are very important for human to human communication and useful for human-machine interaction. Diverse interesting systems have been developed for gesture recognition either, as insolated gestures recognition [10, 6] or studying its relation with spoken language [1]. However, gestures can also be used as a simple and natural form to command service robots [3, 5, 11]. The problem with these systems is that they are still limited in the number of gestures recognized and the form to execute them. This document presents a visual system to recognize a set of five right-arm dynamic gestures using Hidden Markov Models as recognition technique. These gestures were chosen for their potential application in human-mobile robot interfaces. In contrast with other similar systems [11], it recognizes natural gestures which are widely understood around the world [7]. The system employs a radial scan segmentation algorithm combined with a statistical-based skin detection method to find the user's candidate face and track his right-hand. The system uses 4 simple features to describe the user's right-hand movement. These features are applied as motion observations for the Hidden Markov Models. The system was tested with 2 users, and is able to recognize five gestures in real time with an average recognition rate of 82.4%. The main contribution of this work is a novel methodology for dynamic gesture recognition that works in real time (30 images per second), under different illumination conditions and different distances from the camera to the user.

# 1. Methodology

Natural execution of gestures is important to increase the number of potential users and the utility of the system. In this way, our system considers five gestures that are widely understood in many countries: *go-right* (Figure 19.1a), *go-left* (Figure 19.1b), *come* (Figure 19.1c), *stop* (Figure 19.1d) and *attention* (Figure 19.1e). For example, this gestures would tell the mobile robot where to go or when to stop its motion. These gestures are executed with the right-hand of a user placed in front of a videocamera. The methodology employed on this approach is divided in the next 6 sequential stages: (i) *skin pixels classification*, (ii) *skin regions segmentation*, (iii) *user's candidate face segmentation and localization of the right-hand*, (iv) *tracking of the user's right-hand*, (v) *motion feature extraction*, and (vi) *gestures recognition*. Each stage is described in the following sections.

## 1.1 Skin classification

The skin pixels classification method used by our system is based on the method developed by Jones [2]. He suggests the construction of two histogram color models, one for skin pixels and other for non-skin pixels, both in 24 bits RGB color space. Using the histogram models, we built a skin classifier based on Bayes rule:

$$P(skin|rgb) = \frac{P(rgb|skin)P(skin)}{P(rgb|skin)P(skin) + P(rgb|\neg skin)P(\neg skin)} \tag{19.1}$$

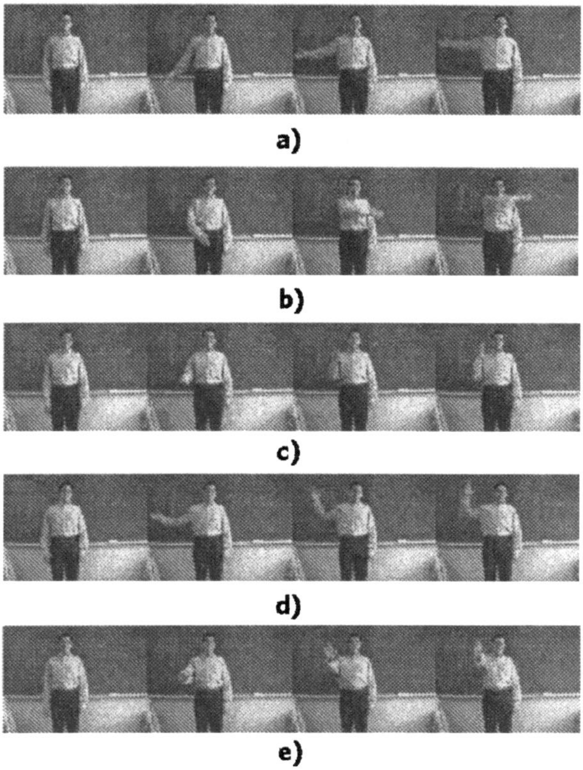

*Figure 19.1.* Gestures considering by our system: a) *go-right*, b) *go-left*, c) *come*, d) *attention* and *stop* gesture.

where $P(skin)$ and $P(\neg skin)$ are *a priori* probabilities, and $P(rgb|skin)$ and $P(rgb|\neg skin)$ are taken directly from the skin and non-skin histograms respectively. Pixels are classified as skin *iff* $P(skin|rgb) > \Theta$, where $0 \leq \Theta \leq 1$ is a threshold. However, to increase the speed of the classification stage, our system employs the rule $P(rgb|skin) \geq P(rgb|\neg skin)$, which is equivalent to use $P(skin|rgb) > \Theta$ with $\Theta = P(skin)$ [Jones 98]. Since $P(skin|rgb)$ and $P(skin|\neg rgb)$ are taken directly from the skin histogram model, other calculations are not necessary.

## 1.2 Skin region segmentation

To segment skin regions, the algorithm developed in our system is based on the radial scan segmentation algorithm proposed by SAVI Group [9]. The algorithm traces lines with certain angular distance among them, from the center of the image to its edges, classifying pixels over these lines, as skin or non-skin pixels (Figure 19.2). At the same time, it uses some segmentation conditions to grow skin regions. The advantage of this algorithm is the fast speed to find skin regions. For example, if we consider an square image $N \times N$ with $N = 480$ pixels, then we have an image of 230,400 pixels. If we use 360 lines to sweep the image, then we have to visit only $360 \times 240 = 86,400$ pixels, i.e., 37.5% of the total pixels in the image. These reductions in the search space of the image are very important when we are working with real-time systems.

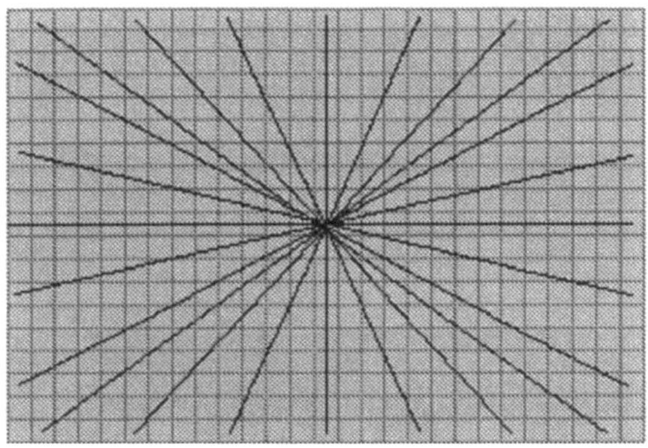

*Figure 19.2.* Scan radial segmentation.

## 1.3 Face segmentation and hand localization

To find the user's candidate face the skin segmentation algorithm is applied only in the upper half of the image, supposing that the user's face is the predominant skin region in this area. Figure 19.3 shows an example of face segmentation. Once the face has been located in the image, to find the user's right hand the skin segmentation module is applied on a small region that contains the right hand, using anthrophometric measures and considering that the arm is at the rest position (see figure 19.3).

## 1.4 Hand tracking

Once the system has localized the hand at the rest position, the hand can be tracked in the image sequence. For tracking, in every image we define a 120x120 search window around the previous position of the hand, using the same radial algorithm for skin segmentation described above. The size of this search window is based on the usual speed of arm movements, obtained experimentally. Some images that show the hand tracking are presented in figure 19.4. With this strategy, hand tracking is performed in real time i.e., 30 images per second.

*Figure 19.3.* Example of face segmentation and localization of the user's right hand.

## 1.5 Motion feature extraction

Motion feature extraction is derived from the method proposed by Starner [10]. Our system uses 4 simple features to describe the hand displacement: $\Delta area$ or changes in area of the hand, $\Delta x$ or changes in hand position on the *x-axis* of the image, $\Delta y$ or changes in hand position on the *y-axis* of the image, and $\Delta form$ or comparison between sides of the square region that segments the hand. To evaluate the hand motion between two images, these features take only one of three possible values, (+), (−) or (0) that indicate increment, decrement or no change, depending on the position and posture of the hand in the previous image.

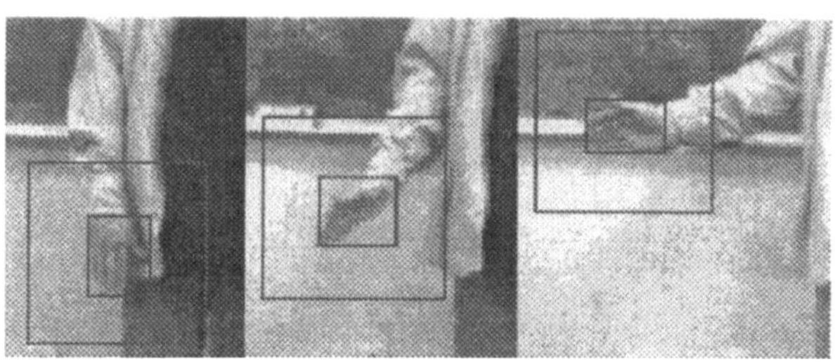

*Figure 19.4.* Tracking of the user's right hand.

Consider the $\Delta x$ feature. If the motion of the hand's centroid is to the user's right side, then $\Delta x = (+)$. If the motion is to the left side of the user, then $\Delta x = (-)$. If the system doesn't detect motion on the *x-axis*, then $\Delta x = (0)$. The $\Delta y$ feature is obtained in a similar way. In addition to the $(\Delta x, \Delta y)$ descriptions, some information about the *form* of the hand is useful to describe the gesture. A simple form's description can be obtained doing comparisons between sides of the square region that segments the hand. If the rectangle's side parallel to the x-axis ($side_x$) of the image is longer than the side parallel to the y-axis ($side_y$), then $\Delta form = (+)$. If $side_y$ is longer than $side_x$, then $\Delta form = (-)$. If $side_x = side_y$, then $\Delta form = (0)$. To estimate depth motion in a simple way, we use the $\Delta area$ feature. If the hand increases its area we suppose its forward motion, and then $\Delta area = (+)$. If the hand decreases its area, we suppose its backward motion, and then $\Delta area = (-)$. If there is no change in the area, then $\Delta area = (0)$.

We use a set of 81 (the product of the 3 values for each of the 4 features) different possible hand motion descriptions, which corresponds to $V = \{v_1, v_2, ..., v_{81}\}$, i.e. the finite set $V$ of different motion observations where $v_1 = (+)(+)(+)(+), v_2 = (+)(+)(+)(-), v_3 = (+)(+)(+)(0), ..., v_{81} = (-)(-)(-)(-)$.

## 1.6 Gesture recognition

Starner [10] introduces Hidden Markov Models (HMM) for visual recognition of dynamic gestures. A HMM can represent the set of motion observations generated by one gesture. These models can take into consideration variations in the way the gesture is performed by different people and at differente velocities. Moreover, the topology of the HMM can be refined for better adaptation to the set of motion observations. A detailed description of HMM can be found in [8].

For the recognition stage we use two different types of models: a 3-state model topology for go-right, go-left and attention gestures (Figure 19.5). A 5-state model is used for come and stop gestures (Figure 19.6), which have more complex motion.

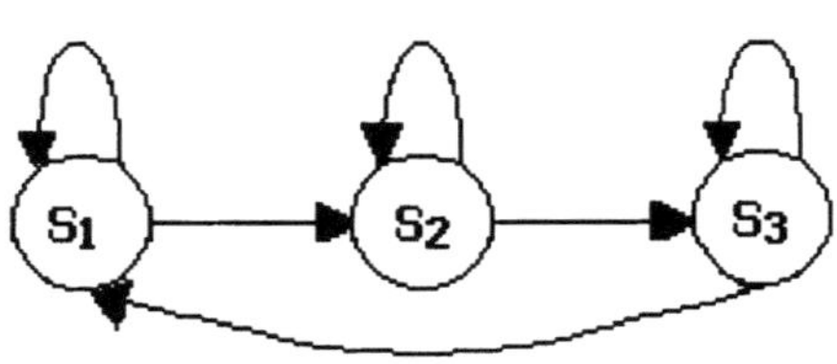


*Figure 19.5.* Topology of the 3-state Hidden Markov Models used to represent the go-left, go-right and attention gestures.

The 3-state topology can be explained as follows. Consider an initial (rest) position of the arm. The state 1 ($S1$) in figure 7 represents this initial position. Doing the go-right gesture, $S2$ represents the motion of the hand to point to the right. After that, $S3$ represents the motion of the hand back to the initial position. The 5-state model, used for the more complex gestures was obtained through sequential refinments of the 3-state topology.

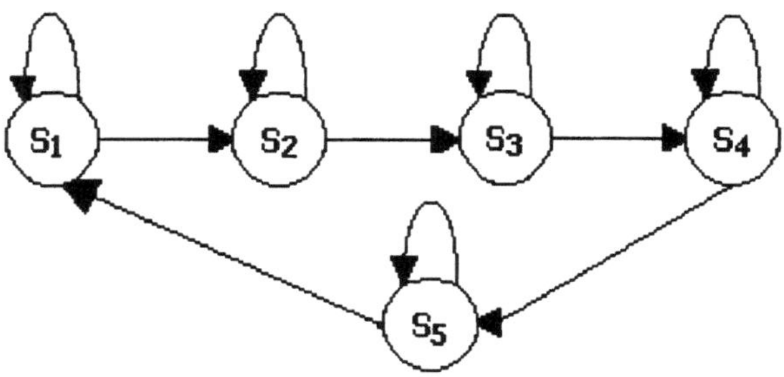


*Figure 19.6.* Topology of the 3-state Hidden Markov Models used to represent the go-left, go-right and attention gestures.

## 2. Experimental results

Initially we built and tested the skin classification and segmentation modules. To construct the histogram color models employed to classify pixels as skin or non-skin, it was necessary to segment $2,101,612$ skin pixels, and $19,552,655$ non-skin pixels by hand. The skin pixels were taken from 170 images of faces, arms and hands of 30 people at different lighting conditions, using 4 videocameras.

To test the tracking and recognition stages, we supposed a laboratory environment. The image resolution employed was $640 \times 480$ pixels. The distance from the user to the camera ranges between 1.5m and 2.5m. The experiments made with HMM for the visual recognition of gestures can be divided in two parts: training and recognition.

For training, the HMM for each gesture, we used an average of 126 sequences of observations, taken from 11 different people. Althought training sequences must have a great amount of possible variations [10], it must be taken into consideration that whenever a gesture is made, this one must be executed with similar form, speed, force and amplitude so that it won't be confused with other gestures. In this manner, the ambiguity is reduced, and clarity in the message is obtained [7].

For testing, 2 users made more than 80 executions of each gesture in front of the videocamera. The sequences of observations we used to test the system were different from those used for training. However, 11 of these sequences were repeated, because the system lost the tracking of the hand. In order to start and end of each gesture, a small region of

| | Executions | Come | Attention | Stop | Right | Left |
|---|---|---|---|---|---|---|
| Come | 81 | 61 | 8 | 1 | 10 | 1 |
| Attention | 82 | 1 | 58 | 22 | 1 | 0 |
| Stop | 85 | 2 | 27 | 56 | 0 | 0 |
| Go-right | 81 | 0 | 0 | 0 | 81 | 0 |
| Go-left | 83 | 0 | 0 | 0 | 0 | 83 |

*Table 19.1.* Results of the recognition system. Each row shows the number of executions to test every gesture, as well as the number of times it was classified erroneously like another gesture.

tolerance was established around the initial position of the hand. The return of the hand to its initial position defines when the gesture is completed. In table 19.1 and table 19.2 we show the recognition results for the 5 gestures. Table 19.1 shows in each line the total number of test executions to recognize each gesture, and the number of times in which this one was classified erroneously like another gesture. Table 19.2 shows the percentage of correct classification of these gestures.

| | Correct clasification rate |
|---|---|
| Come | 75.3% |
| Attention | 70.7% |
| Stop | 65.8% |
| Go-right | 100% |
| Go-left | 100% |
| Average | 82.4% |

*Table 19.2.* Gestures recognition rates. Rows present the percentage of correct classification in the execution of each gesture and the recognition average of the system.

As it can be seen in the results of tables 19.1 and 19.2, there is a great difference in correct recognition for the left and right gestures in comparison to the other gestures. These results are easily explained considering the nature of the necessary movements in one or another case. Whereas the gestures go-left and go-right require displacements that predominate towards the left and the right (parallel to the image plain); the gestures to come, attention and stop involve movements in depth (perpendicular to the image plane). So the gestures that involve depth motion have a lower recognition rate, given that direct depth information is not considered.

Another factor is the variation of the distance between the user and the videocamera. On the one hand, when increasing the distance, we obtained a better recognition of the "attention" gesture; on the other hand, if the distance diminished, it improved the recognition of the "stop" gesture. So it seems that when it is more perceivable the change of the hand area, it is more useful as description for movement in depth.

The system was implemented on a SG O-2 R5000 computer. The recognition process takes aproximately 0.2 sec. after the gesture is completed. Unfortunately, direct comparisons with other similar gestures recognition systems cannot be done because the nature of gestures [6].

# 3. Conclusions and future work

In this document we described a visual system to recognize 5 dynamical gestures. The system operates in real–time (30 images per second) with an average recognize rate of 82.4%. As future work we consider different possibilities. First, the system doesn't consider egomotion or motion of the videocamera. This skill is important to run the system on a mobil robot. Second, to recognize more complex gestures, it will be helpful to obtain other features to describe depth motion. Third, to increase the number of gestures to recognize, we will include the left hand and the face of the user.

# References

[1] Cassell, J. (1998). *A Framework For Gesture Generation And Interpretation.* In Cipolla, R. and Pentland, A. (eds.), Computer Vision in Human-Machine Interaction. New York: Cambridge University Press.

[2] Jones, Michael J., James M. Rehg. (1998). *Statistical Color Models with Application to Skin Detection.* Technical Report CRL 98/11. Cambridge Research Laboratory.

[3] Kahn, Roger E. (1996). *Perseus: An Extensible Vision System for Human-Machine Interaction.* PhD Thesis. The University of Chicago.

[4] Kendon, Adam. (2000). Gesture in conversation: problems in description and interpretation. *Fourth IEEE International Conference on Automatic Face and Gesture Recognition 2000.*

[5] Kortenkamp, David, Eric Huber, and R. Bonasso, Peter. (1996). Recognizing and interpreting gestures on a mobile robot. *Proceedings of the AAAI-96*, pages 915-921, AAAI Press/The MIT Press.

[6] Jerome Martin, Jean-Baptiste Durand, (2000). Automatic Gestures Recognition Using Hidden Markov Models. *Fourth IEEE*

*International Conference on Automatic Face and Gesture Recognition 2000.*

[7] Morris, Desmond. (1977). *El hombre al desnudo.* Vol. 1 Ediciones Orbis. Barcelona.

[8] Rabiner, Lawrence R. (1990). A Tutorial on Hidden Markov Models and Selected Applications in Speech Recognition. In Alexander Waibel and Kai-Fu Lee (Editors), *Readings in Speech Recognition*, págs. 267-296. Morgan Kaufmann Publishers.

[9] Stereo Active Visual Interface Group. (1999). www.cs.toronto.edu/~herpers/projects.html. Univesity of Toronto. May, 1999.

[10] Starner, Thad Eugene. (1995). *Visual Recognition of American Sign Language Using Hidden Markov Models.* Masters Thesis. MIT. Program in Media Arts and Science.

[11] Waldherr, Stefan. (1998). *Gesture Recognition on a Mobile Robot.* Diploma Thesis. Carnegie Mellon University. School of Computer Science.

# IV

# DISTRIBUTED ARCHITECTURES

# Chapter 20

# Distributed Multi-Sensor Surveillance: Issues and Recent Advances*

Pramod K. Varshney and Ioana L. Coman
*Department of Electrical Engineering and Computer Science, Syracuse University, Syracuse, NY 13244-4100, USA*

**Abstract:** Distributed multi-sensor surveillance systems are becoming increasingly important in a variety of military as well as non-military applications. This chapter discusses three important issues regarding design of such systems, namely sensors and sensing technologies, system architecture, and information processing and fusion. Some underlying concepts are illustrated by considering the problem of concealed weapon detection.

**Key words:** Integrated surveillance systems, multi-sensor systems, information fusion, concealed weapon detection.

## 1. Introduction

Conventional surveillance systems of the past were devised to monitor the activities of large military units or entities such as aircraft and ships. These missions were accomplished by using sensors such as radar and sonar that could view large volumes of interest in more or less continuous manner. Times have changed and one may need to monitor the activities of humans and vehicles for a variety of law enforcement, security and defence applications. There has been an increasing interest in the development of surveillance systems for other non-military applications. By surveillance of a particular environment, we mean the monitoring of that environment for the detection and tracking of specified activities. For the design of such a system, we need to develop an integrated architecture and define tasks for efficiently collecting, storing and processing all the data that is relevant to

* This work was supported in part by ARO grant DAAD19-00-1-0352.

the monitored activities within the environment for fast and reliable human or automated decision making. This important and complex problem arises in both public and private sectors, e.g. medicine, environment, defence, communications systems, and building systems. In general, for surveillance applications one has to take into consideration the following: sensing device development, system and network design, and information processing tasks.

As a specific application, one can consider a surveillance system for monitoring changes in the living environment. Important chemical, biological, and physical modifications occur continuously in our environment that can have a short-term and/or long-term impact on our living environment and our quality of life. Many of these modifications, especially the ones that occur in harsh environments, can be monitored through sensor technology. They can be embedded into Intelligent Distributed Systems (IDS) which can provide accurate Intelligence, Surveillance and Reconnaissance (ISR) capabilities. The dependability of ISR systems can be increased, at a reasonable additional cost, by using a large number of small inexpensive sensors with different functionalities that are connected in a network using wired or wireless links. In such Intelligent Real-time Integrated Sensor (IRIS) systems redundant sensors are useful in decreasing system vulnerability to sensor failures. Due to the presence of a large number of sensors, the quantity of information increases and it becomes very important to find methods to efficiently combine this information and store the relevant data. Otherwise, the decision makers (human or automatic) will not be able to make timely decisions and the network might become overloaded because of insufficient bandwidth and/or computing power. An IRIS System not only will inform the human decision-maker about changes/problems in the monitored environment but also will indicate the location and time when the event occurred. Through the analysis performed by the intelligent system, IRIS could provide solutions and even execute a course of action to find a remedy for the problem.

Another interesting example of such a prototype system used for human health monitoring is the Georgia Tech Wearable Motherboard™ (GTWM)[4, 9]. The GTWM is a wearable vest which through sensors mounted on the vest, and using optical and other special fibres meshed with the vest's material is able to sense, process, transmit and monitor different vital signs, e.g. body temperature, and ECG.

Development of systems for surveillance applications relies on new smart sensors, integrated system design, data storage, real-time data analysis, data fusion and system control functions. Research and development in this area is highly interdisciplinary, where electronics, mathematics, physics, signal processing, communication theory, data networks and computer science play major roles. Since the physical implementation and experimentation can be quite costly, it is expected that theoretical approaches and computational modelling and simulations will be widely used for system design.

This chapter is organised as follows. In Section 2, we propose a model of a distributed multi-sensor surveillance system. We briefly discuss the past, present and future research in sensor technology and the main components of a distributed surveillance system. We also present several problems that arise in these systems in terms of data processing, e.g. communication and data fusion. Section 3 contains a case study dealing with a particular multi-sensor surveillance problem, namely Concealed Weapon Detection (CWD). In this particular problem one has to determine whether or not a weapon is concealed under a garment by fusing data sensed through different imaging technologies (infrared and millimetre wave). Final remarks are presented in Section 4.

## 2. Distributed Surveillance Systems

A distributed surveillance system is usually composed of a large collection of simple and inexpensive *smart sensory devices* (infrared (IR), chemical, biological, acoustic, magnetic, motion, etc.). For instance, in a military surveillance application [1], such systems may consist of $10^2$ to $10^7$ sensors, placed on fixed or mobile platforms, able to collect data locally or remotely. Some of these sensory devices are included to provide complementary data, while others are just redundant to minimise data inaccuracy due to potential device failures.

These devices would be of little value unless they are equipped with communication and processing capabilities that would allow them to self-organise into independent subsystems. These subsystems would exhibit emerging behaviours and capabilities to further integrate into a *super-system*. However, when these distributed super-systems contain large numbers of subsystems, issues such as system organisation and topology, protocols for subsystem interactions including communication standards and information processing become important. Systematic consideration of these issues and design of distributed sensor networks are open research problems.

### 2.1 Sensing Devices

Over the past decades, great progress has been made in research and development of sensing devices. Early research and development in this area was especially fuelled by military applications. In the 1960s and 1970s increased effort was devoted to the development and integration of different types of sensors, such as acoustic/sonar, IR, seismic, and magnetic sensors.

An example of an application is the Remote Battlefield Acoustic Sensor System (REMBASS) developed during the 1970s, which incorporated acoustic, seismic, magnetic and IR sensors connected by radio communication links. Unfortunately, the sensor systems at that time had limited signal-processing capabilities that resulted in large amounts of data being sent to a central unit, which eventually overloaded the operators. During the 1980s and 1990s, new advances in material sciences, microelectronics, physics, biology, chemistry and computer science provided new technologies for the sensor industry. The MEMS technology provides substantial information processing capabilities at the device level, which greatly reduces communication requirements. Research groups are continuously working in developing new low cost, low power, and wireless sensors with various sensing capabilities that can withstand harsh environments for long periods of time [2, 6]. Desired features of these sensors include self-monitoring capabilities for reliability, power efficiency for long term operation, autonomy for local control operation, and re-configurability for easy adaptation at different base stations.

## 2.2 Distributed Multi-Sensor System Architecture

The overall architecture of an integrated surveillance system depends greatly on the application. For an airport surveillance application, for example, such a system could include video cameras (to monitor traffic), chemical and biological sensors (to monitor gas/toxic emissions and possible medical/biological outbreaks), and motion sensors (to trigger other more advanced surveillance devices). While data is currently collected for many of the factors mentioned above, it is usually not integrated into an overall system. Due to a lack of information fusion and resulting improvement of diagnosis, decision-making and control action, many factors that might induce short-term and long-term effects in the environment are "missed".

Each device or groups of devices can be viewed a subsystem since many of them are equipped with local processing and communication capabilities. Based on this assumption, the architecture of such a system can be envisioned as having a self-adapting hierarchical structure, where the lower-level subsystems can self-organise based on the requirements and the health of the device (See *Figure 1*). Such a system should allow the following [7]:

1. Operational Independence of the Subsystems: The subsystems must be able to usefully operate independently.
2. Managerial Independence of the Subsystems: The subsystems are separately acquired and integrated but maintain a continuing operational existence independent of the integrated system.

3. Evolutionary Development: The development of the integrated system is evolutionary with new functions added, removed, and modified based on the experience gained.
4. Emergent Behaviour: The integrated system performs functions and carries out tasks that do not reside in any subsystem. They are emergent properties of the entire system and cannot be localised to any subsystem. The principal purposes of the integrated system are fulfilled by these new properties.
5. Geographic Distribution: The subsystems can exchange only information and not substantial quantities of mass or energy.

Next, we describe the architecture of a two-level hierarchical system (see *Figure 1*) that is being considered for the monitoring of indoor environmental quality (IEQ) at Syracuse University [16]. At the lowest level are the *sensor-level intelligent subsystems*, containing a single or several sensing/monitoring devices, which are able to respond rapidly to local conditions and communicate with the *regional* or *local subsystems*. The

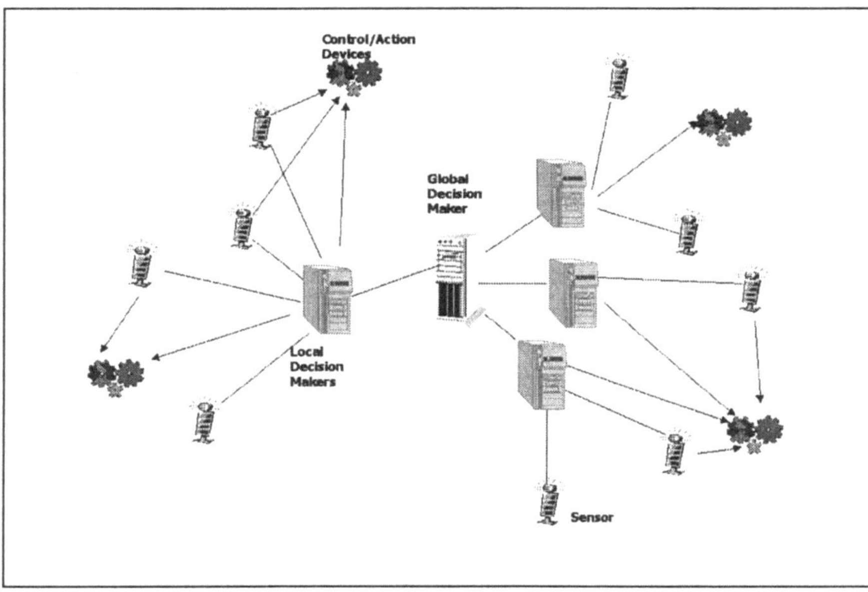


*Figure 1.* Example of an integrated system architecture

regional subsystems are integrated in a distributed environment where data fusion is performed to allow regional-level actions to be taken. They communicate with *central intelligence units* (for large coverage area applications, e.g. military surveillance, urban environment surveillance and

IEQ monitoring for large buildings) as well as with sensor-level systems. The data processing tasks of a regional subsystem may be distributed to other units (regional subsystems or central units), depending on the communication capabilities, the activity of the other regional subsystems and the response time requirements. The central intelligence units communicate with the regional subsystems in order to make complex decisions, including evaluating and overriding decisions and actions taken at a lower level, future decision planning and communication with human operators.

## 2.3 Information Processing

Information processing tasks vary greatly and depend on the type of application, the sensory devices integrated in the system and the type of data provided by these devices (images, numeric, sound, etc.). In general, these systems have to consider the communication network, low-level information processing tasks (at the sensor level if it has computational capabilities), information fusion processing (at different levels in the system architecture), data manipulation in real-time or near-real-time (storage, searching, retrieval and visualisation) and decision making. In a distributed surveillance application, information processing tasks can be divided into two categories: those performed at the sensor subsystem level (the lowest level) and the ones performed at higher levels, namely at the regional or central levels.

Tasks that a sensory subsystem may perform include the following: a). Periodic broadcasts of the current status and location. b). Responding to requests from control centres and/or neighbours. c). Acquisition, pre-processing, compression, encryption and transmission of the data. For imaging sensors, local data processing consists of image processing tasks (e.g. restoration, segmentation, etc.), feature extraction and object recognition. Depending on the communication and processing capabilities of the subsystem, it can also acquire and process network data (e.g. data fusion and system status management), refine and extract knowledge (e.g. scene analysis, target recognition, situation status analysis, information reliability status, etc.), make decisions and execute control protocols (activate alarm, close security doors, etc.), and monitor the activity of its neighbours (e.g. monitoring the health of the neighbouring sensory devices, reduce message redundancy, etc.)

At the higher level (with or without a control centre), the tasks include support of the knowledge base, communication among subsystems, integrated system strategic planning and optimisation of system resource utilisation. Since most of the subsystems (see Section 2.2) are independent, they can collaborate only through information exchange [7]. To allow efficient collaboration, standards and protocols need to be specified for communications among different units. The seven-layer OSI model [12] for network architecture may be considered for network design. However, a subset of these layers may suffice for the design of this special purpose network. Resource optimisation may include strict time restrictions on the data processing and data communication within the system for the applications that require real-time or near real-time responses.

An important aspect of these systems is the ability to integrate or fuse data from different sensors. This is required so that we can utilise the information collected by the system in the most efficient manner and make most effective and intelligent decisions for system control. Information fusion can be done at different levels. At the data level, we can combine raw sensor data. This is the most effective data fusion method but requires commensurate sensors and possibly large communication bandwidth. At the feature level, appropriate features and relevant information can be extracted and fused. Finally, fusion at the decision level involves the fusion of tentative decisions made at the lower level. This requires extensive signal processing capabilities at the lower level units. Bandwidth requirements are reduced but the performance of the overall system also degrades. For additional details on fusion architectures and decision fusion, the reader is referred to [5, 14, 15].

## 3. Case study: concealed weaponS

Over the years, global terrorism and crime have grown. One key step in preventing terrorism and crime is to determine whether or not people are carrying weapons concealed underneath their clothing. Concealed weapons detection (CWD) is, therefore, an increasingly important problem in the area of multi-sensor surveillance. This technological challenge requires innovative solutions. A number of sensors based on different phenomenology are being developed to observe objects underneath people's clothing. However, each sensor has its limitations. For example, infrared (IR) imagery cannot penetrate heavy clothing and millimetre wave (MMW) images have poor resolution. In order to overcome these limitations, multiple sensors that provide complementary information may be employed for CWD. Recently IR and MMW sensors have been considered for CWD purposes [11] since an IR sensor provides a high-resolution image while MMW sensor can penetrate through clothing. By fusing this complementary information, more complete information is obtained which can then be utilised to detect concealed weapons.

A block diagram describing the major tasks in a multi-sensor surveillance system for CWD is shown in *Figure 2*. Once images from different sensors are available, it is necessary to find the corresponding points of those images before employing any fusion algorithm. In the literature, the process of finding the corresponding points from two or more images is called *image registration*. In addition to registration, filtering and noise removal are also carried out in the pre-processing stage. The resulting registered images are fused, features are extracted and then decision is made on whether or not the person under surveillance is carrying a weapon. The registration and fusion tasks for CWD are briefly discussed here. For details, see [13].

Several different registration problems arise in the CWD application. For example, images are taken at the same time from different but almost co-located (adjacent and parallel) sensors; images are taken simultaneously from sensors of the same type but located at different sites; images are taken from different sensors which are located in different places, etc. We have developed an automatic two-stage registration algorithm for images simultaneously taken from two parallel sensors that are of different type [[3]].

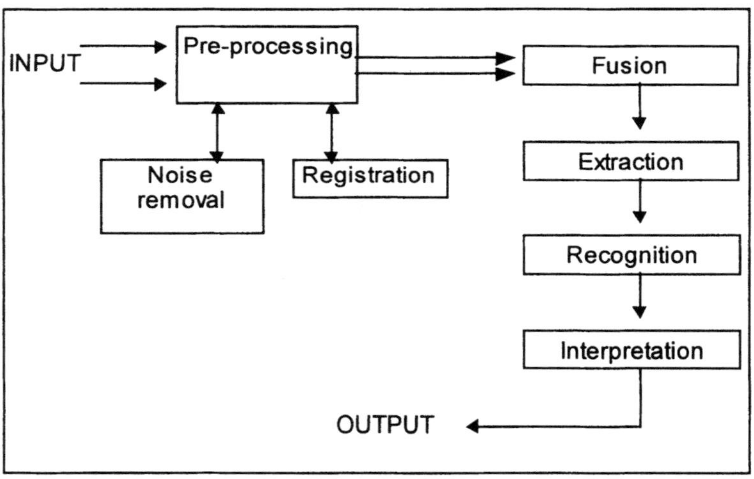


*Figure 2.* Multi-sensor image processing for CWD application

This case considers a simpler geometrical relationship between images, but it is quite difficult due to the use of different sensors, therefore different type of images. *Figure 3* shows a schematic of the set-up under consideration.

A typical image pair taken from IR and MMW sensors before registration can be viewed in *Figure 4*. *Figure 5* presents the registered MMW image. We superimposed the boundary of the body extracted from the registered IR image on the registered MMW image to evaluate the accuracy of the registration task.

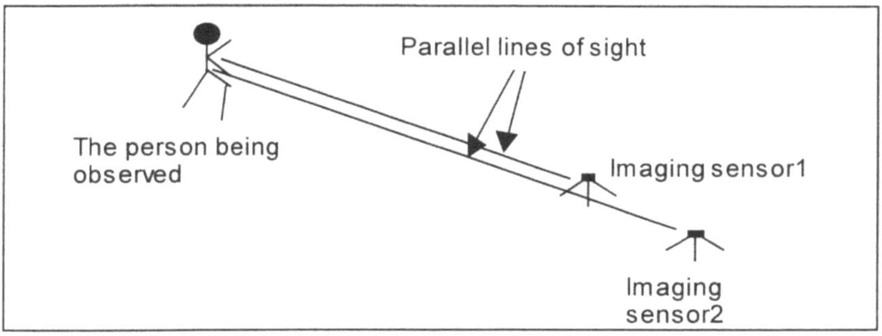


*Figure 3.* The set-up considered in our application.

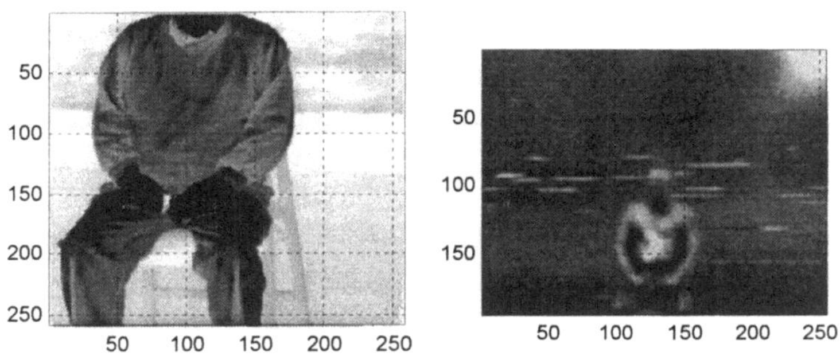


(a) (b)

*Figure 4.* Typical (a) IR and (b) MMW images. Notice the strength of each sensor. (High resolution of IR and better penetration of MMW)

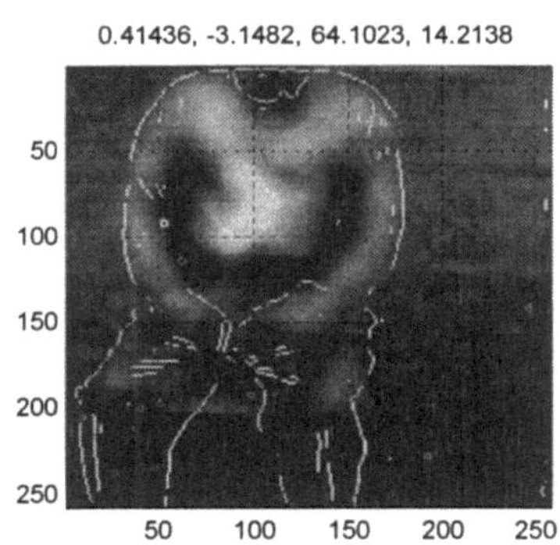


*Figure 5.* Registered MMW Image. The four numbers indicated above the image show the four transformation parameters (scaling, rotation angle in degree, vertical displacement, and horizontal displacement) found by the two-stage automatic registration algorithm.

The registration algorithm we developed consists of two stages. The main tasks involved in the first stage are two image segmentation algorithms and one binary correlation algorithm. The goal of this stage is to register the two images coarsely so that the result can be used as a starting search point of the optimisation process employed at the second stage, registration by maximising mutual information. The details of the registration algorithm developed are available in [3].

After the images taken from different sensors are registered, the next step is to integrate the information contained in each image. This task is referred to as *image fusion*. To accomplish this task we developed a wavelet based image fusion algorithm [10,13]. The examples shown in *Figure 6* are images before and after fusion. Thresholding (using Otsu's method [8]) was used to compare the fused results to the original IR and MMW images. *Figure 7(a)* and *Figure 7(b)* show the thresholded results for the original IR and MMW images. The IR image shows the gun, but it is also connected to wrinkles on the lower part of the image. In addition, several other wrinkles are visible for the IR image. No shape resembling a gun is apparent in the MMW image. *Figure 7(c)* shows the thresholded results for the fused image. It can be easily seen that the gun shape is very clear and completely separated.

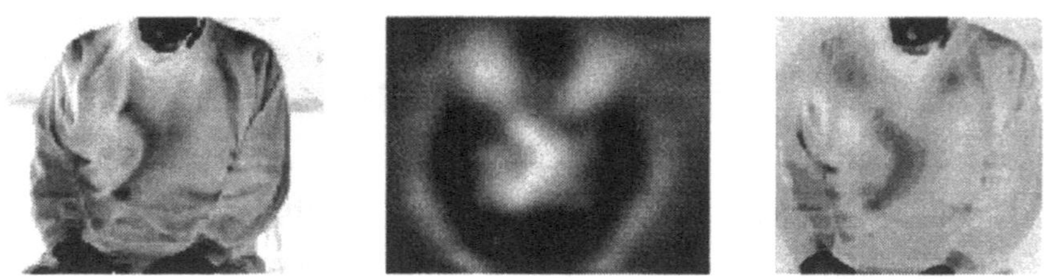

(a) (b) (c)

*Figure 6.* Examples of images before and after the fusion process.

(a) (b) (c)

*Figure 7.* Result images after thresholding was applied on the fused images.

Concealed weapon detection (CWD) technology utilises multiple sensors to improve detection of weapons hidden underneath a person's clothing. This is a typical multi-sensor surveillance task. The processing steps in the CWD task include registration, fusion, noise filtering, data enhancement, partitioning, feature extraction, and weapon identification. Here we have demonstrated how to integrate information from multiple sensors to help in extracting gun shapes from the image, with improved detection of the weapons. Once the result of CWD is available, further actions can be taken. They may include letting the person go or stop them for further search.

## 4. Summary and conclusions

A great deal of effort is being put on the design and development of distributed multi-sensor surveillance systems. This indicates the significance of this field. While research on different aspects of the problem is important and interesting, one of the extremely important tasks is to find a solution on how to integrate all these tools in creating new surveillance systems with superior capabilities. New theories and computational methods will be essential in efficiently processing the data acquired and manipulated through these integrated systems. It is also expected that highly interdisciplinary teams will be one of the keys in finding effective solutions to this increasingly important problem.

## References

[1] Distributed Microsensing: Devices, Networks and Information Processing. Strategic Assessment Workshop, Technical Report, Army Research Laboratory, Jan. 11-12, 1999, Adelphi, Maryland.

[2] Bult K. et. al. "Low Power Systems for Wireless Microsensors", *Proc. of the 1996 Int. Symposium on Low Power Electronics and Design*, Monterey, CA, Aug. 1996, pp.17-22.

[3] Chen H. -M. and Varshney P. K. "An Automatic Two-stage IR and MMW image Registration Algorithm for Concealed Weapons Detection", accepted for publication in *IEE Proceedings of Vision, Image and Signal Processing.*

[4] Firoozbakhsh B., Jayant N., Park S., Jayaraman S., "Wireless Communication of Vital Signs Unsing the Georgia Tech Wearable Motherboard", Proc. ICME2000 IEEE International Conference on Multimedia and Expo. (Latest Advances in the Fast Changing World of Multimedia), vol.3, pp. 1253-6.

[5] Hall D. and Llinas J., "An Introduction to Multisensor Data Fusion", *Proc. IEEE*, vol. 85, Jan. 1997, pp. 6-23.

[6] Lin T.-H., Sanchez H., Kaiser W. J. and Marcy H. O. "Wireless Integrated Network Sensors (WINS) for Tactical Information Systems", *Proc. of the 1998 Government Microcircuit Applications Conference.*

[7] Rechtin E. and Maier M. W. " The Art of Systems Architecting (Systems Engineering Series)", CRC Press, 1996.

[8] Otsu, Noboyuki. "A Threshold Selection Method from Grey-Level Histograms", *IEEE Transactions on SMC,* vol. SMC-9, Jan. 1979, 1: 62-66.

[9] Park S., Gopalsamy C., Rajamanickam R., Jayaraman S., "The Wearable Motherboard: A Flexible Information Infrastructure or Sensate Liner for Medical Applications", *Proc. Medicine Meets Virtual Reality 7*, Jan. 1999.

[10] Ramac L.C., Uner M.K., Varshney P.K., Alford M., Ferris D. "Morphological Filters and Wavelet Based Image Fusion for Concealed Weapons Detection", *Proc. SPIE Aerosense* '98, 1998, 3376: 110-19.

[11] Slamani M. *et. al.*. "Image Processing Tools for the Enhancement of Concealed Weapon Detection," *Proc. 1999 International Conference on Image Processing*, 1999, 3: 518-522.

[12] Tannenbaum A. S., Computer Networks, 2nd Edition, Prentice Hall, Englewood Cliffs, NJ, 1989.

[13] Varshney P.K., Chen H. -M., Ramac L.C., Uner M., Ferris D., Alford M. "Registration and Fusion of Infrared and Millimetre Wave Images for Concealed Weapon Detection", *Proc. 1999 International Conference on Image Processing,* 1999; 3: 532-536.

[14] Varshney P. K. "Multisensor Data Fusion", *Electronics and Comm. Eng.*, Vol. 9 , Dec. 1997, pp. 245-253.

[15] Varshney P.K. "Distributed Detection and Data Fusion", Springer-Verlag, 1997.

[16] "New York Environmental Quality Systems Centre: Intelligent Control and Management Systems", http://nyeqs.syr.edu/ .

# Chapter 21

# Intelligence Distribution of a Third Generation People Counting System Transmitting Information over an Urban Digital Radio Link

Carlo S. Regazzoni, Claudio Sacchi, Gianluca Gera
*University of Genoa, Department of Biophysical and Electronic Engineering (DIBE)*

Abstract: The present work is focused on the performance analysis concerning some feasible strategies for the distribution of modular image processing tasks in new-generation surveillance systems. A brief overview of surveillance systems technical evolution will show the current trend towards distributed and reconfigurable image processing systems. We have considered for test a distributed surveillance system for people counting. The transmission of multimedia information from the different levels of the aforesaid system, i.e. the peripheral level composed by one or more advanced sensors with autonomous processing capabilities, and the remote level usually referred to a central processing architecture close to the operator, is performed by means of a wireless link simulated in a realistic way. Results shown clearly evidenced the relevant effects on the system performances achieved by the different choices of intelligence distribution strategies.

Key words: Wireless surveillance networks, distributed surveillance architectures

## 1. Introduction

Third generation multimedia surveillance systems can exploit the most recent developments concerning embedded image processing and digital transmission techniques, in order to consider the actual implementation of distributed surveillance networks, characterized by adaptive relocation of processing capabilities through the different levels of the network itself [1]. This is made possible now by the technological improvement provided by DSP programmable architectures [2], that allow to embed image-processing tasks (namely: system intelligence) into advanced intelligent sensors. Roughly speaking, a video-based surveillance network can be subdivided into two main processing levels (see Figure 1):

- The peripheral sensor level, characterized by the presence of one or more video sensors with different autonomous processing capabilities (called smart cameras [1]).
- The remote operator level, which is generally, identified with a centralized architecture for multimedia information processing installed close the operator. In a multi-sensor surveillance network, the overall processing capabilities of the remote operator level are conventionally assumed to be loaded in the hub (i.e. the local gathering node of the network). The operators' PCs are regarded as simple consoles devoted at displaying processing results by means of suitable interfaces and at sending commands in remote.

A pictorial scheme of a new-generation multimedia surveillance network with wireless transmission is shown in Figure 1. In [1], a preliminary analysis of intelligence distribution through a video-based surveillance network has been performed with respect to the real-time processing constraints. The optimal intelligence distribution strategy has been investigated from a mathematical point of view by considering the minimization of a cost function based on the processing time required to the intelligent devices in order to execute the assigned tasks.

In [5], a first attempt of joined analysis of intelligence distribution and radio link configurability in wireless video-based surveillance networks has been performed mostly from the point of view related to the allocation of the communication resources in terms of bandwidth. In this paper, we wish to analyze the effects of different choices of intelligence distribution strategies on the performances of an image processing system for video-surveillance. The test application considered here is related to a people counting system originally designed for remote monitoring of tourist sites served by cable-TV (CATV) network infrastructures [6]. In the proposed analysis, we shall consider a video-based surveillance network with a fixed wireless radio link connecting sensor level with remote level and different level of intelligence transferred from the remote level to the sensor level. The performance analysis is performed by considering as quality parameter the counting precision in terms of absolute error and absolute mean error. The paper is structured as follow:, Section 2 is aimed at providing a global overview of the evolutionary trends toward of video-based surveillance systems towards distributed architectures, Section 3 will describe the proposed video-based people counting system considered as a case of study, Section 4 will illustrate the different strategies of distributed decomposition of the people counting system, Section 5 will show the simulation results, and Section 6 will draw the conclusions of the paper.

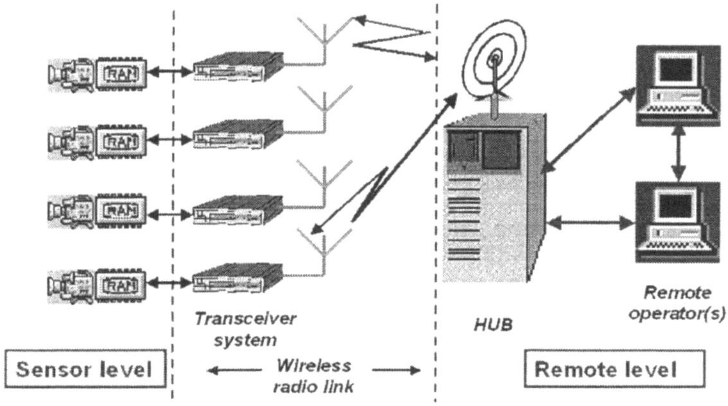


*Fig. 1* A Multimedia surveillance network with wireless transmission (pictorial scheme)

## 2. Technical evolution of video-based surveillance systems towards distributed architectures

Historically, first applications of remote monitoring based on digital signal processing techniques concerned with military surveillance (e.g. spy-satellites) and space research (e.g. planet exploring missions). However, the increasing request of security coming from modern society together with the difficulties of satisfying this request by public institutions quickly lead to the transfer of such technologies to the civic field. At the present time, automatic and semi-automatic monitoring systems can provide reliable technological answer to many actual problems such as: security in public environments, transport safety, tourist management, environmental monitoring etc. A graph explaining the technical evolution of video-based surveillance systems versus time in the civil field is reported in Figure 2.

Since beginning of 60's, 1st generation (1G) analog CCTV systems were exploited mainly in indoor surveillance applications [1] (e.g. surveillance of supermarkets, banks, garages etc.). More recently, CCTV systems tried to exploit digital wireless transmission techniques in order to extend their application range to remote monitoring of outdoor sites and vehicles. The main disadvantage of such systems concerns with the actual impossibility of recognizing in real time the situations of interest. Indeed, 1G systems would require the presence of operators devoted at continuously observing high-complexity scene(s), thus making the efficiency of the system depending on the employed persons' attention level that is proven to decrease very fast in the time [1].

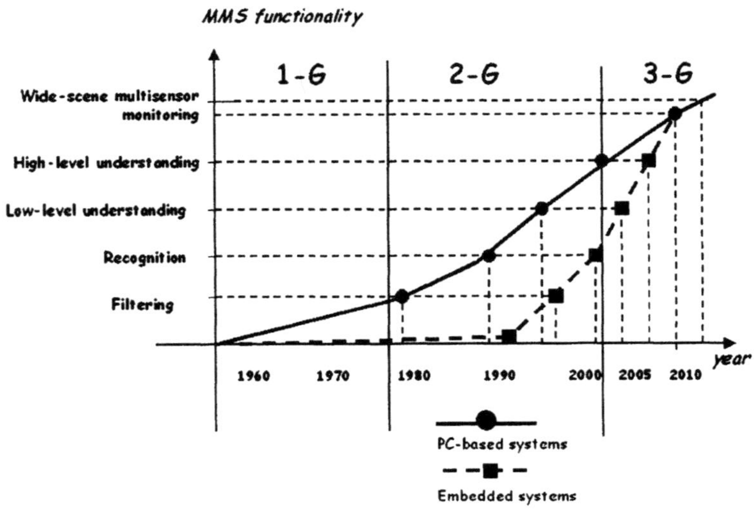


*Fig. 2.* Technical evolution of multimedia surveillance systems versus time

Since 80s digital image processing technologies drastically decreased their costs. This fact led to a significant technological improvement of video-based surveillance systems. A new generation, i.e. the 2nd generation (2G), of surveillance systems arouse at the mid of 80s and grew during the overall 90s. The turning point in such a development was the commercial explosion of low-cost PC-based image processing technologies occurred in these last ten years. As a logical consequence, the market for PC-based video-surveillance systems has increased very much. Referring to the solid-line graph of Figure 1, we can observe that processing capabilities of 2G PC-based video-surveillance architectures shifted form low-level tasks like noise filtering and change detection (early 80s), to middle-level tasks like recognition and tracking of interesting regions of the image (early 90s), up to more complex image understanding tasks (e.g. image interpretation of low-complexity scenes – late 90s) [1]. This lead to the development of semi-automatic video-based surveillance systems able at driving the attention of the human operator by means of suitable messages displayed over graphical interfaces. In parallelism with 2G development phase, the increasing commercial availability of high-speed wireless and wired digital network infrastructures allowed since middle of 90s to consider the effective implementation of remote monitoring applications, where information acquired at peripheral level by video sensors can be transmitted through existing or private network infrastructures to a remote processing center. Digital video sensors are already available on the market, however the integration of existing networks and remote monitoring applications is not a trivial task. The main problems arise from the necessity of transmitting compressed video streams (by using e.g. JPEG or MPEG standards) over band-limited and noisy channels. In remote video-based surveillance applications real-time and quality-of-service constraints are more severe than those ones related to other video-processing applications. In particular, the performances of lower level image-processing tasks that directly process received and source-decoded images can be severely degraded by noisy artifacts due to channel errors. For this reason, real-time error concealment strategies must be introduced in order to improve system performances in 2G remote video-based surveillance systems. This is still an open research field. An example of real-time algorithm for error recovery in remote 2G surveillance applications is discussed in [9]

Considering the favorable market perspectives, the current research on automated surveillance is already working on the new generation systems: the so-called 3rd generation (3G) ones [1]. From the image processing point of view, the basic innovative concept of 3G systems with respect to 2G ones is the distribution of processing capabilities (namely: system intelligence) across the computer network. This is due to the very recent technological push concerning embedded signal processing devices using DSP programmable hardware architectures [2], which has started during second half of 90s. This push is also evidenced by the pattern of the dashed line of Figure 1, depicting the evolution of embedded systems. DSP-based embedded signal processing technology makes it possible to build intelligent cameras with autonomous processing capabilities [3], called also smart cameras. Smart cameras are currently able to perform in real-time MPEG and JPEG compression coding and low-level image processing tasks, such as: noise filtering, change illumination compensation and change detection. Due to the announced launch on the market of very fast DSPs (like the TMS320C64x by Texas Instruments that can support clock frequencies comprised between 600 MHz and 1100 MHz), or general-purpose processors like INTEL STRONGARM® 700MHz, it is foreseen that embedded image processing systems will be able to implement higher-level image processing functions, such as: motion detection and tracking. The crossing point between the solid curve and the dashed curve of Figure 1 is roughly expected by 2010, when embedded architectures should be able at implementing also higher level understanding tasks at commercial level. The upper step of the technical evolution in remote monitoring architectures will be the multi-sensor monitoring of wide scene, where multimedia sensor network will work in multipoint-to-multipoint configuration. Moreover, the use of advanced programming languages like JAVA allows the user to relocate in a dynamic way the software modules among the different intelligent devices present both at peripheral level (i.e. intelligent sensors) and at central level [1][4]. In this sense, we can speak about reconfigurable intelligence distribution in 3G surveillance networks [4]. The choice of the optimal intelligence distribution strategy with respect to real-time processing capabilities and bandwidth constraints can be found by means of formal approaches using mathematical cost functions [4]. In the next sections we shall consider the analysis performed by means of laboratory simulations of the intelligence distribution of the image processing tasks of an actual video-based surveillance system.

## 3. The image processing system considered as a case of study

The proposed image processing system allows estimating the number people by means of advanced image processing methodologies. The modular structure of the system is shown in Figure 3. The Low-Level Image Processing (LLIP) modules directly work on the acquired image sequences performing noise filtering, background updating, adaptive change detection, and focus of attention [6]. The High-Level Image Processing (HLIP) modules work on the binary image of differences fed by the LLIP module. The image processing functions performed at HLIP level are: detection of spatially connected change detection regions (i.e. the blobs [6]) blob tracking, and the feature extraction [6]. The output of the HLIP module is a list of tracked mobile areas, together with some extracted features concerning them. The Image Understanding (IU) module is the intelligent part of the system. Its output is basically a list of classified mobile areas, from which the incremental number of persons entering in the scene from an initial time instant to a final one (namely: people counting) is derived and displayed to the operator by means of a suitable Man-Machine-Interface (MMI). The core of the IU module is a neural network, assisted by other pattern recognition tools like Hough transforms. Finally the performances of the counting system are improved by some time-regularization mechanisms exploiting:

- the knowledge of the history of the dynamic evolution of the scene (this information is returned by the HLIP level);
- the introduction of a linear Kalman filtering stage in order to smooth eventual noisy counting peaks.

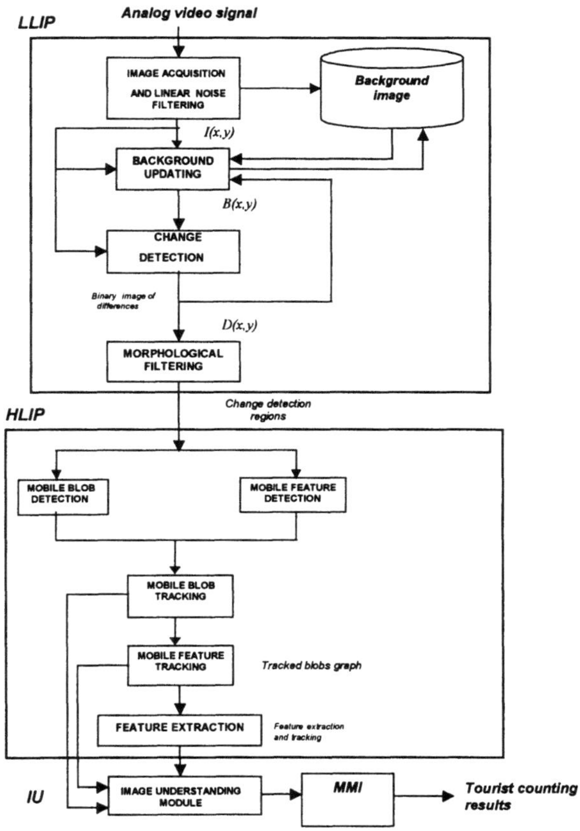


*Fig. 3* Block diagram of the image processing system for people counting

It is worth noting that the three main modules of the system above described are structured into submodules (in Figure 4, the submodules are depicted with boxes receiving some defined inputs and returning some defined outputs). These submodules are complying with the definition of image processing task (or simply task) exposed in [4]. Actually a task is defined as a modular and auto-inclusive instruction set, characterized by structured input and output data flows. A video-based surveillance system can be generally formalized by a sequential cascade of tasks. For this reason, the tasks are typically the software tools candidate for relocation through the different network levels in distributed image processing architectures. The relocation of tasks, called also intelligence distribution, can be actually performed by means of downloading operations exploiting the potentialities of advanced programming languages like JAVA [4]. The video-based people counting system considered here (fully described in [6]) was originally designed for processing image sequences acquired by an analog CATV signal. Of course, the overall image processing tasks were entirely performed at the remote level.

## 4. Distributed decomposition of the people counting system over a wireless surveillance network

Now, let's consider a wireless and distributed implementation of the video-based people counting system described in Section 3. The aim of the proposed analysis is to test an eventual adaptation of the proposed AVS system also for tourist zones not reached by cable networks or for other kind of people counting applications not restricted to the tourist flows estimation. The test of the new distributed image processing architecture was performed by laboratory simulations. The simulated wireless transmission between the sensor level and the remote level, now identified with the hub, is performed by means of a Spread-Spectrum-based transceiver working in the ISM band (2.4-2.4385 GHz), using a QPSK modulation and a rake receiver [7]. The distributed decomposition of the image processing system has been performed into two steps, corresponding to the two different distributed architectures depicted in Figures 4. In Figure 4a a distributed image processing system for people counting is shown with the LLIP modules up to change detection transferred from remote hub level to the sensor level. This kind of architecture is fully compliant with the state-of-the art of DSP-based embedded image processing devices. In Figure 4b a more advanced distributed architecture is shown with some parts of the HLIP modules (i.e. up to blob detection) transferred to the sensor level. This other architecture can represent, at the present time, an upper bound for current smart camera prototypes. The motivations for such a distributed architecture consist in the possibility of reducing the redundancy due to the large black zones of the change detection image. In fact, the output of the blob detection task is a list of localized areas of interest selected from the binary image of difference. Such areas, called blobs, are related to compact and physically connected change detection regions that could be regarded by the image processing system as mobile groups of persons to be counted. The middle level modules enclose each blob inside a rectangle, called also obstruction rectangle, characterized by some geometrical features (horizontal and vertical dimensions, area, etc.), and by a set of 2D Cartesian co-ordinates in the image plan (i.e. co-ordinates of the mass center, co-ordinates of upper and lower points, etc.). From this consideration, one can realize that if an intelligent sensor can process the acquired visual information up to blob detection, only the essential information related to the obstruction rectangles should be transmitted to the remote level. The processed information transmitted by architecture 4a is a binary source-coded image sequence, which is a broadband and unstructured kind of information (i.e. the value of a coded pixel gray level does not influence the gray level values of other pixels close to it). The processed information transmitted by architecture 4b is a set of blob packets. Each blob packet contains:

- *Header information:* 2D co-ordinates of the blob center, blob rectangle dimension, 2D blob rectangle co-ordinates (upper and lower points);
- *Blob pixels:* only the binary source encoded pixels belonging to the obstruction rectangle containing the blob are transmitted to the remote level.

For each frame, some blob packets are transmitted by the digital transceiver placed at the sensor level. It is clear that blob packets are a narrowband and highly structured kind of information, where errors in the header information can involve heavy errors in blob image reconstruction performed at hub level.

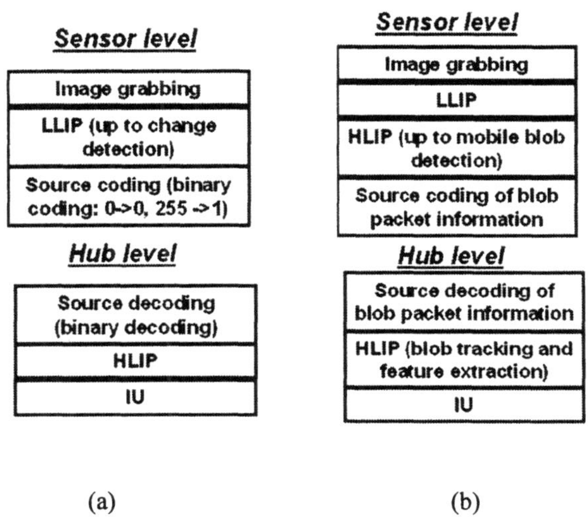


*Fig. 4.* a) Distributed image processing architecture for people counting with low-level intelligence transferred to the sensor level, b) distributed architecture with low and middle-level intelligence transferred to the sensor level

Together with the two advanced architectures shown in Figure 4, we shall also evaluate a state-of-the-art video-based surveillance architecture, using a digital wireless camera with JPEG source coding and fully centralized processing. This last architecture, entirely realizable with commercial components, is introduced in our analysis as a touchstone for system performance evaluation. In the fully centralized architecture, the multimedia information transmitted through the wireless channel is a JPEG-coded image sequence that will be decoded by a devoted module at remote level and processed by a centralized people counting system that is structured in the same way as the one shown in Figure 3.

## 5. Experimental results

Some simulation results related to different intelligence distribution strategies are shown in this section. The image sequence used for test is composed by 15 frames.

An ISM-band frequency-selective Rayleigh-fading channel [8] in pedestrian environment has been simulated by means of a tapped-delay-line model [8]. The parameterization in terms of coherence bandwidth and Doppler spread of the simulated wireless channel has been taken by the experimental measurements exposed in [6]. Accordingly with such measurements, the Doppler spread has been settled to 1.9 Hz. and the coherence bandwidth equal to 3.5 MHz. In order to focus in a better way the effects of the multipath-fading channel on the system performances, we neglect the Gaussian noise term, as often done in the literature. For what concerns the spread spectrum transceiver, we fixed the processing gain at the value of 15 (AO-LSE sequences employed for spreading). Moreover, a convolutional FEC coding [8] with rate 1/2 has been introduced in order to protect the signal against channel noise. As a fixed bit-rate of 1Mb/s has been chosen for all simulations, the bandwidth occupation of the transmitted signal is about 8 MHz In this case, the rake receiver can extract and combine two separated paths [8]. The measured BER after FEC decoding is about 10-3. In order to comply with the real-time constraints concerning the information transmission between sensor level and hub level, we did not consider any kind of retransmission in case of error detection in the received bitstream. In Figure, 5a, 5b, 6a, and 6b examples of the consequences involved by channel errors in the different cases are shown.

(a) (b)

*Fig. 5.* Examples of effects of channel errors over: (a) and (b) JPEG decoded images

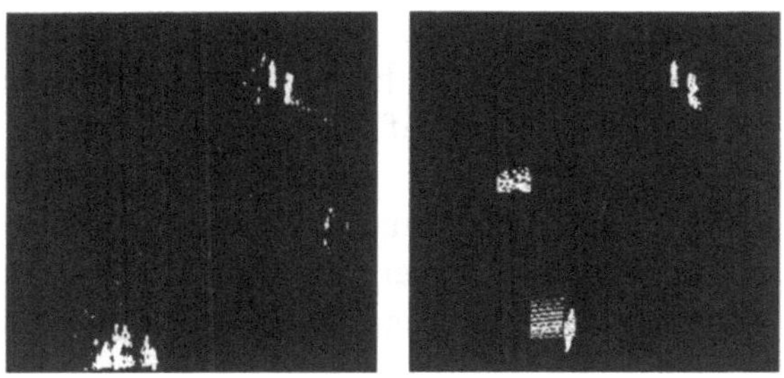

(a) (b)

*Fig. 6.* Examples of channel errors over: (a) binary image of difference, (b) reconstructed blob image from received blob packet information

One can see from Figures 6 aand 6 that the channel errors have different effects over transmitted information. The centralized image processing system should process image sequences with some JPEG decoded frames characterized by heavy degradations such the one shown in Figure 5a and 5b. The counting precision will obviously suffer very much by such kind of errors, as the LLIP modules will erroneously regard noisy artifacts as very big change detection regions. Persons contained inside such regions will be missed. In fact, time regularization of the counting working in the IU module is explicitly devoted at avoiding overcounting involved by incorrect change detection due, to unpredictable luminance variations and artifacts occurring in the image. Thus some consecutives noise-corrupted frames will lower the people counting results in a permanent way. Figure 6a depicts the effects of the channel errors over a received and decoded binary change detection image. In this case, we are referring to the distributed architecture of Figure 4a. The effects encountered are not so heavy as the ones experimented by the JPEG transmission. Indeed, the channel errors only involve the presence of some isolated noisy pixels. In most situations, the morphological operators working in the HLIP module can filter off such isolated noisy points. In Figure 6b, a wrongly reconstructed blob image due to errors in the header information of some blob packets is shown. Of course, we are dealing now with the distributed architecture of Figure 4b. The perceptual effects are related to blob missing and blob splattering. The people counting precision can be severely compromised by such kind of errors. However, the reduced amount of information sent in blob packets should prevent a high occurrence of potentially disastrous header errors (in our simulation only two frames were affected by them).

In Figure 7 a comparison is shown among the results in terms of people counting precision achieved by the different architectures tested in the present work. The curve provided by the distributed architecture of Figure 4a is very close to the ideal curve of the people counting performed by inspection by a human operator who visually counted the persons entering the monitored scene. On the other hand, the curve provided by the state-of-the-art centralized architecture is very far from an acceptable degree of precision. The counting errors are, in this case, due to some consecutives noise-corrupted frames. The results provided by the distributed architecture of Figure 4b are quite good, however some errors in blob packets involve a slight displacement from the exact people counting curve.

Let's consider some mathematical precision indexes for the image processing application chosen for test. A quantitative method for evaluating the counting precision is the absolute counting error (AE), related to the single i-th frame defined as:

$$AE_i = \left| P_i - \hat{P}_i \right| / \mathrm{P_i} \quad i = 1..N_{frame} \quad (5.1)$$

where $P_i$ is the exact people counting value and is the people counting value estimated by the image processing software. Another method for assessing the performances of the people counting system is the mean absolute error (MAE), simply obtained by averaging the AE with respect to the number of frames of the image sequence used for test.

The plot of the absolute counting error (AE) versus the frame number for the different kinds of intelligence distribution strategy analyzed in this paper is shown in Figure 8. As expected, the centralized image processing architecture provides huge errors in people counting even higher than 40%. Apart few initial frames, needed by the system for convergence, the distributed architecture with LLIP modules embedded at sensor level (i.e. architecture depicted in Figure 4a) provide a very good counting precision with an absolute error rapidly converging to zero. Good performances are also provided by the distributed image processing architecture with part of the HLIP module embedded at sensor level (i.e. architecture depicted in Figure 4b). However some channel errors in the transmitted blob packets can involve some peaks in the absolute counting error. The results of counting precision in terms of MAE achieved by the distributed architectures of Figure 4a and Figure 4b are equal to 4% and 5% respectively. On the other hand, the centralized processing architecture provides an unacceptable value of the MAE equal to 28%. It is worth noting that the precision limit usually required by users of people counting systems is equal to 10% [6].

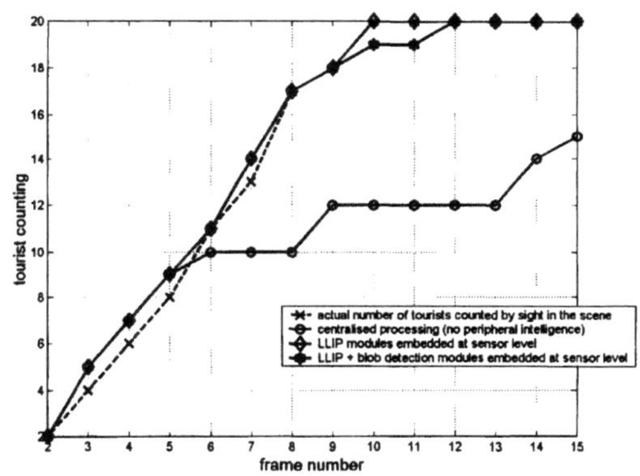


*Fig. 7.* Numerical results achieved in terms of counting precision by the different image processing architectures

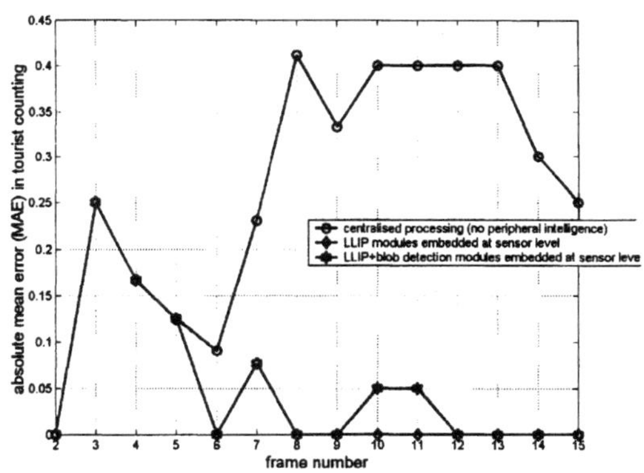


*Fig. 8.* Absolute counting error achieved by the different image processing architectures

# 6. Conclusions

To conclude this paper, it can be said that the distribution of intelligence in advanced video-based surveillance networks is a relevant factor that can influence very much the performances of the considered application (i.e. video-based people counting). The state-of-the-art centralized image processing architecture cannot assure an acceptable precision of the people counting due to the errors affecting the JPEG-coded video transmission. The introduction of effective post-processing error concealment algorithms would be explicitly required in order to recover corrupted frames by channel errors. The transferring of processing capabilities to the sensor level can be suitably exploited for enhancing the performance of the image processing system without adding new modules, particularly when the sensor level transmits to the remote level unstructured information like the binary change detection image. A more efficient distributed architecture, like the one of Figure 4b could be successfully exploited especially when the communication resources in terms of bandwidth are limited. However, channel errors can severely affect highly structured information like blob packets. In this latter case efficient strategies of joint source and channel coding should be considered in order to prevent header errors in blob packet information.

# Acknowledgements

This work was partially supported by the University and Scientific Research Ministry (MURST) of the Italian Government under the National Interest Scientific Research Program.

# References

[1] C. S. Regazzoni G. Foresti, P. Mahonen, (eds.), "Multimedia Video-based Surveillance Systems: Requirements, Issues and Solutions", Kluwer Academic Publishers, Norwell (MA): 2000.

[2] P. G. Paulin, C. Liem, M. Cornero, F. Nacbal, G. Goossens, "Embedded Software in Real-Time Signal Processing Systems: Application and Architecture Trends", Proceedings of IEEE, Vol. 85, No. 3, March 1997, pp. 419-435.

[3] K. Illgner, "DSPs for image and video processing", Signal Processing, Vol. 80, November 2000, pp. 323-336.

[4] L. Marcenaro, F. Oberti, G.L. Foresti, C.S. Regazzoni, "Emerging Requirements for Multimedia Surveillance Systems", Proceedings of IEEE, in press, date of publication foreseen: July 2001.

[5] C. Sacchi, G. Gera, C. Regazzoni, "Intelligence distribution and radio link configurability in wireless video-based surveillance networks", in: Software Radio, E. Del Re (ed.), Springer, 2001, in press.

[6] C. Sacchi, G. Gera, L. Marcenaro, C. Regazzoni, "Advanced image processing tools for counting people in tourist site monitoring applications", Signal Processing, Vol. 81 (2001), pp. 1017-1040.

[7] J. G. Proakis, "Digital Communications", 3rd Edition, McGraw-Hill International Editions, New York: 1995.

[8] M. Tarkiainen, I. Niva, P. Kemppainen, "Performance of WCDMA system with space diversity and power control in slowly fading channel models", Proc. of 1999 IEEE WCNC Conference, 21-24 September, 1999 New Orleans (USA), Vol. 2, pp. 674-678.

[9] C. Sacchi, F. Granelli, C. Regazzoni, F. Oberti, "A real-time algorithm for error recovery in remote video-based surveillance applications", Signal Processing: Image Communications, 2001, in press.

Chapter 22

# A Comparison between Continuous And Burst, Recognition Driven Transmission Policies in Distributed 3GSS

Franco Oberti, Giancarlo Ferrari and Carlo S. Regazzoni
*Department of Biophysical and Electronic Engineering (DIBE) - University of Genoa, Via all'Opera Pia 11A, 16145 Genova (Italy), e-mail: carlo@dibe.unige.it*

**Abstract**: This paper introduces an architecture of a third generation surveillance system (3GSS). In particular a method for choosing the optimal distribution of intelligence required by 3GSSs is presented. The effect of introducing recognition tasks, which can cause the interruption of flow in the system, is discussed. Experimental results over a simulated system illustrate the presented approach.

**Key words**: Distributed intelligence, Dynamic resource allocation, multimedia surveillance systems

## 1. Introduction

In this work an example architecture and optimal configuration of a third generation surveillance system (3GSS) [1] is studied.

First generation video-based surveillance systems are based on analog signal and image transmission and processing. In these systems, scenes remotely collected from a set of sites are proposed to human operators after analog communication of the video signal; a set of monitors placed in a remote control room are used to present continuous flow visual sequences to operators. Second generation systems (1980-2000), that constitute current state of the art, are mainly characterized by different intermediate degrees of digital video signal transmission and processing [2] trying to solve separately some of the drawbacks of 1GSSs. Third generation surveillance systems are coming in the next years which will take advantage of progress in low cost high performance processors and multimedia communications. A multimedia 3GSS can be defined as a distributed multimedia computer system [1] based on multisensory input data and provided by advanced scene understanding capabilities oriented at automatic alarm generation or attention focusing of human operators. Distribution of intelligence is a crucial step in configuration of a 3GSS (installation or dynamically reconfiguration). Distribution of intelligence can involve partitioning of main logical tasks used in a 3GSS among digital processing capable ("intelligent") cameras, hubs and central control rooms. The distribution of the intelligence in 3GSS is related with the optimal task allocation in the video-surveillance network. The optimization of the system should take into account the performances of the overall surveillance network in terms of processing time, bandwidth occupation and computational load of each single processing unit. An approach for choosing the optimal task allocation in a 3GSS is presented in this work. Main constraints in a distributed surveillance system are related with the features of the involved processing units and the complexity of each task in the video-surveillance system; in particular the available bandwidth of the links between each processing unit and the computational capabilities of each processor will be considered together with the number of operation required for each task to be completed. In Figure 1 a general architecture for a 3GSS is presented.

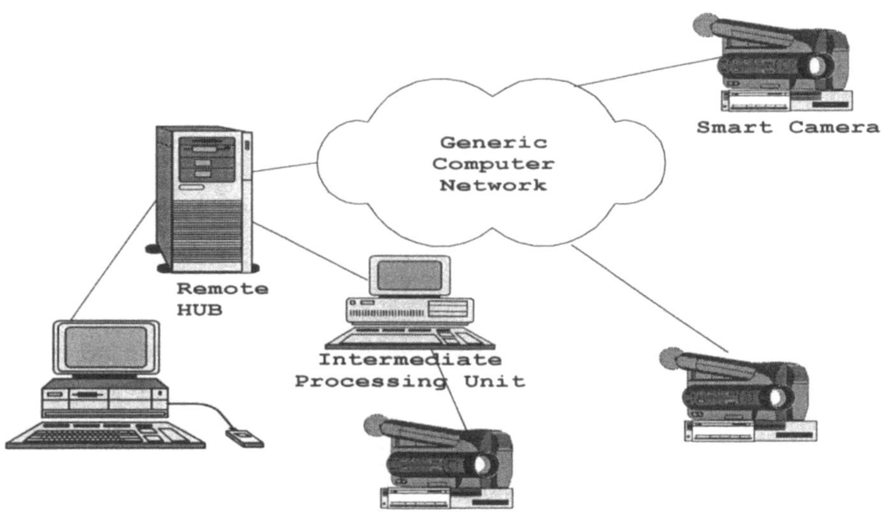


*Figure 1.* A 3GSS multimedia surveillance network

The paper is organized as follows: Section 2 presents the physical architecture of a 3GSS; Section 3 is related with the logical architecture and the distribution of intelligence. In Section 4 an approach is discussed for optimal dynamic task allocation, while Section 5 presents experimental results. Future work is discussed in section 6. Finally, conclusions are drawn in Section 7.

## 2. Physical Architecture

The general model of the physical architecture of a hierarchical 3GSSs can be defined in terms of a graph whose node are sensors, hubs, and control centers and whose links are associated with their communication channels, as shown in figure 1. Processing power at the hub and central control unit level can be associated with standard general-purpose processors. An intelligent camera should meet specific characteristic such as low-cost, low-power consumption, etc. Intelligent cameras for 3GSS will take advantage from improvements in DSP, general-purpose processors and memory capability. Examples of such cameras featuring processing capabilities in the order of 240/750Mhz [3,4] able to execute one or several parts of the surveillance algorithms, are coming on the market.

The communication link could be wired or wireless. A simple model of each link in a 3GSS system can be given by specifying: (a) channel bandwidth, (b) physical link length, (c) channel capacity, (d) channel noise, (e) number of users.

## 3. Logical Architecture and Distribution of Intelligence

The architecture of a 3GSS can be described also at a logical level, this level is related to:

- The definition of the domain-specific logical information process to obtain a given system functionality;
- A hierarchical decomposition of the whole logical process into a set of basic processing tasks.

A task can be defined as a well-defined set of operations of the video-surveillance process. Tasks are logical entities and they need a physical architecture to be executed. They can be organized into hierarchical chains, where lower level tasks provide output data sequences to higher-level tasks. The higher the level of a task, the closer its output should be to the functionality to be pursued, i.e. the lower must be the amount of processing necessary to serve the functionality of the 3GSS for which the task chain has been designed.

Distribution of intelligence in a 3GSS is described as the process of mapping the logical architecture onto the physical one, i.e., to allocate tasks to processors. The resulting mapping is the solution of the intelligence distribution problem.

We can identify two classes of tasks oriented to pursue the above functionalities:

1. *Representation tasks*: Representation tasks process information coming from the previous tasks in order to provide a higher-level representation of data more strictly related to events of interest in a scene. In general, the output of a representation task is more synthetic than its input.
2. *Recognition tasks:* Recognition tasks aim at comparing a signal representation at a given level with a pre-defined set of event models. Models are related to descriptions of events associated with possible situations of interest. In a 3GSS, recognition tasks can be coupled with a representation task along a hierarchy that can be defined as alarm-generation hierarchy.

The alarm-generation introduced by the recognition tasks can control the flow of information in the processing and transmission chain. This can be seen as a burst behavior in which the execution of tasks can be interrupted if a desired event is not present at some level (alarm not detected). For example if a recognition task allocated to a camera does not detect the presence of objects in a scene (alarm), the system can decide to interrupt the processing of consecutive tasks and to not transmit any processed information because revealed information is not helpful for consecutive tasks. These impulsive interruptions can free processing and transmission resources and should be considered in order to maximize optimum task allocation.

Let's consider for simplicity a single hub sub-system collecting data from related camera, this sub-system can be represented as a sub-tree in the physical architecture and we can extend results found in the following to the analysis of the whole 3GSS without loss of generality. Let's study the case in which the desired events are present and the processing and transmission chain is not interrupted. Let's suppose $N_c$ cameras compose the terminal nodes of the sub-tree and a fixed number $F$ of data fusion tasks is pre-assigned to the hub for fixing boundary conditions. Let $t_i$ be a generic task, $T$ the whole set of tasks and $N$ the total number of tasks to be performed by the surveillance system in the presence of desired events (alarm-generation). If $C_i$ indicates the i-th camera of the system, let $T_{C_i} = \{t_0, \ldots, t_{x_i}\}$ be the set of tasks associated to $C_i$: for example $t_0$ can be the acquisition task, $t_1$ can be the change detection task, etc. The total number of tasks $N$ performed by the systems can be considered as the sum between data fusion tasks $F$ and lower level processing tasks $I$: $N=I+F$.

In this case the distribution of intelligence problem correspond to assign $I$ tasks either to an intelligent cameras or to the hub.

On the above basis, a possible solution can be described by the set of tasks allocated to the hub, i.e.:

$$T_h = \left( \bigcup_{i=0}^{N_C} T_{h_{C_i}} \right) \cup T_F \text{ where } T_{h_{C_i}} = \{t_{x_i+1},\ldots,t_I\} \text{ and } T_F = \{t_{I+1},\ldots,t_N\}. \quad (1)$$

The overall processing time for a given frame can be considered as a significant parameter as it is directly related to reaction time of the surveillance system to an event, as well as to available bandwidth and computational resources of processing units and links, and to computational load of each task.

## 4. Dynamic Task Allocation

Dynamic task allocation can be useful for example when the capacity resources of a communication channel are time variant. Flexibility in terms of possibility of re-configuring the system when either computational power or bandwidth is locally changed can be an important technological characteristic to be efficiently exploited. To this end, a computational model is needed to relate optimal task allocation solution to 3GSS parameters.

Different tasks are characterized by different computational complexity. Let $Y=Y(t_h)$ be the processor instructions needed to complete the $h$-th task. Since the tasks are executed in an independent and sequential way, the total number of instructions for executing the task set $T_{c_i}$ can be defined as:

$$Y(T_{c_i}) = \sum_{h=1}^{x_{C_i}} Y(t_h) \quad (2)$$

The execution time $p(t_h)$ of the $h$-th task in the $k$-th intelligent camera, measured as instructions per second is given by:

$$p(t_h) = \frac{Y(t_h)}{CPU_k} \quad (3)$$

if $CPU_k$ represents the processing power of the camera. The total execution time in the $i$-th camera is given by:

$$P_{C_i} = \sum_{h=0}^{x_{C_i}} p(t_h) \tag{4}$$

If $CPU_H$ is the processing power of the hub, the processing time, neglecting hub idle times for synchronizing processing tasks, is given by:

$$P_h = P_F + \sum_{i=0}^{N_C} P_{h_{C_i}} = \frac{Y(T_F)}{CPU_H} + \sum_{i=0}^{N_C} \frac{T_{h_{C_i}}}{CPU_H} \tag{5}$$

If we neglect transmission time, the processing time of the system is given by:

$$J = \max_i \left\{ P_{C_i} \right\} + \sum_{i=0}^{N_C} P_{h_{C_i}} + P_F \tag{6}$$

Let's $P_{Tx_i}$ is the transmission time from the i-th camera

$$J = \max_i \left\{ P_{C_i} \right\} + \sum_{i=0}^{N_C} P_{h_{C_i}} + P_F + \max_i \left\{ P_{Tx_i} \right\} \tag{7}$$

The above functional represents the time to process a frame in each camera (synchronous 3GSS system). Functional is given by the sum of information processing and transmission times by considering the whole set of tasks of the surveillance chain.

The optimal solution to the intelligence distribution in a 3GSS correspond to the parameter setting minimizing the above functional. Such a minimization can be considered over tables of processing and information transmission time derived from available architectural solutions as well as on varying bandwidth availability conditions. Functionals for a large set of realistic more complex hierarchical physical multi-sensor architectures e.g. multiple hubs collecting separate sets of multiple cameras converging to single control center, can be derived from the above one.

# 5. Experimental Results

## 5.1 Test Architecture

The considered 3GSS is composed by a set of non-overlapped static cameras and one hub. The goal of the following analysis will be to find the parameters that can drive to the optimum allocation of tasks and to study the impulsive behaviors introduced by recognition tasks. This is done in order to minimize system reaction time. Different configurations of available bandwidth and image resolution are considered. Figure 2 shows the block diagram of the studied system (described in [5]) represented by its representation tasks.

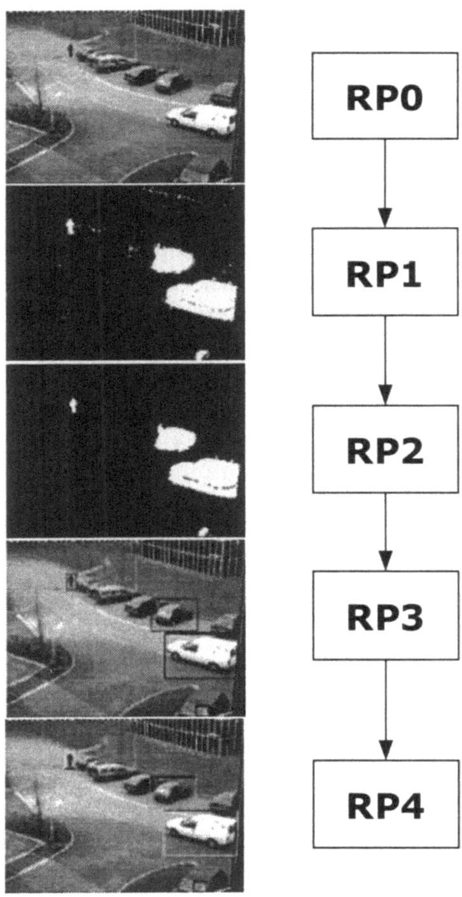


*Figure 2.* Representation tasks processing chain

*RP0: Image acquisition.* Task devoted to the quantized digital representation of the video sequence. The acquisition time depends on several hardware characteristics and on the desired digital image format (image size, black and white or full colour.

*RP1: Change detection.* This task aims at comparing the current image with a reference background one to establish areas of change [5].

*RP2: Non-linear filtering.* Morphology algorithms such as erosion and dilatation are necessary [5] to filter out noise in the change detection image.

*RP3: Focus of attention.* The aim of this task is to identify as separate entities (i.e. blobs) all connected regions starting from a filtered change detection image [5]. To this end, a blob-coloring algorithm associates the same label with changed pixels in connected regions. Output representation is constituted by a list of connected regions, each characterized by its position and size.

*RP4: Tracking.* This task provides a representation of behaviours of objects in a sequence in terms of a temporally tagged list of attributes of object instances separately detected in subsequent frames [5].

In this analysis recognition tasks are introduced into the previous described system. In particular, after each representation task (excluding *RP0*), a recognition task is included, which validates the output of the recognition task and produces a binary output (alarm), which decides if the processing and/or transmission chain should be interrupted. For the purpose of this paper, simple recognition task are considered for studying the behavior of a burst transmission, but more complex and precise ones could be implemented [1]. In particular the considered recognition tasks are:

*RC1: Scene change detection.* This task evaluates the number of changed pixels obtained by *RP1*. If the number of changed pixels is greater than a threshold (*th1*) then an alarm is generated and the chain is not interrupted, otherwise is interrupted.

*RC2: Filtered change detection.* This task is similar to *RC1* but compares the result of *RP2* with a different threshold *th2* for alarm generation. *RC2* should be more precise than *RC1* since many changed pixels relative to noise were filtered by *RP2* and this involves that $th2 < th1$.

*RC3: Intrusion detection.* This task searches for the presence of objects in the scene with a considerable area. In particular this task evaluates the size of the blobs obtained by *RP3*. If at least the size of one blob is greater than a fixed threshold (*th3*) then an alarm is generated and the chain is not interrupted.

*RC4: Motion detection.* The motion detection basically evaluates the displacement of each blob in the scene using the information provided by *RP4*. If the displacement of at least one blob in the scene is greater than a threshold (*th4*) then an alarm is generated.

For obtaining an estimate of transmission and processing times needed for each task, experiments have been performed using a standard PC network architecture, simulating the intelligent camera as a standard PC with images stored locally on a hard disk. The network management has been implemented on Java and images processing modules have been implemented in C++ as Dynamic Link Libraries (DLLs).

The information collected after the test phase is used to set the optimal parameters in terms of task distribution between camera (client) and hub (server) in conformity with functional (7) and the available bandwidth.

## 5.2 Results

Two configurations of available bandwidth (100Kb/s, 50Kb/s) are considered at an image resolution (R2) of 320x240 pixels. The camera is simulated using a 200Mhz processor and the Hub using a 750MHz one. The used thresholds were: *th1*=1000, *th2*=700, *th3*=250 and *th4*=10.

In the following description, T1 indicates the processing up to module *RC1* (*RP0+RP1+RC1*), T2 the processing up to *RC2* (*RP0+RP1+RC1+RP2+RC2*) and so on.

The required bandwidth for each configuration (amount of data to be transmitted to the hub, related with the last task of the chain allocated to the camera) in the presence of alarm is presented in table 1.

| *T* | *Required BW without compression* | *Required BW with compression* |
|---|---|---|
| T1 | *Isize* | $1/20 \times Isize$ |
| T2 | *Isize* | $1/20 \times Isize$ |
| T3 | $\sum_{i=1}^{n}(blobsize_i + featsize_i)$ | $\sum_{i=1}^{n}(blobsize_i/20 + featsize_i)$ |
| T4 | $\sum_{i=1}^{n}(blobsize_i + featsize_i)$ | $\sum_{i=1}^{n}(blobsize_i/20 + featsize_i)$ |

*Table 1.* Bandwidth required for frame per second

The measured functional (7) for each configuration CPUClient - CPUServer, is shown in fig. 3 (a, b).

In this figures A*i* (*i=1,2,3,4*) represents the level (T*i*) at which the chain is interrupted for the absence of an alarm generation by *RCi*. Note that if the interruption happens at a task allocated to the camera, the information of the representation task (*RPi*) is not transmitted to the Hub. "None" means that the chain is not interrupted at any task (all alarms were generated).

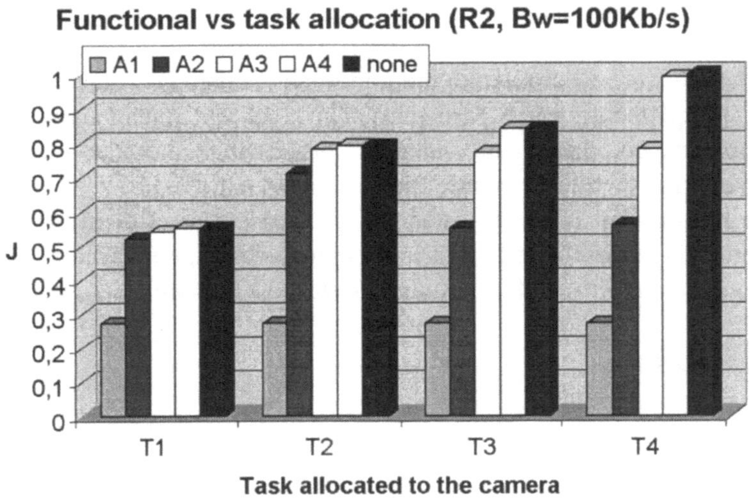


(a)

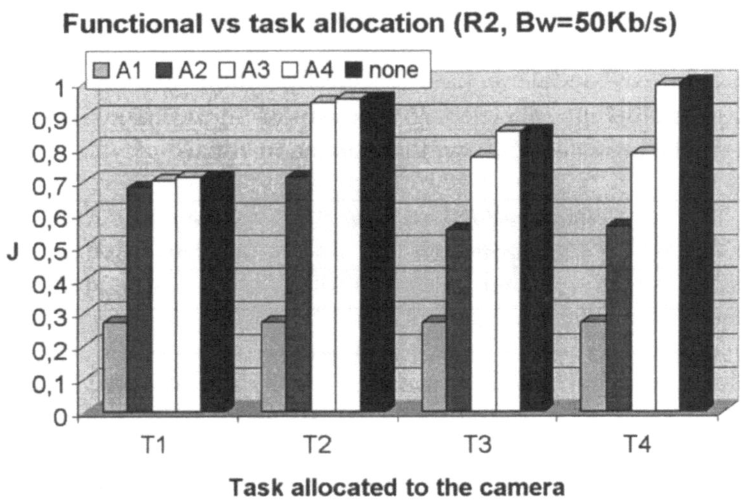


(b)

*Figure 3:* Evaluation of functional in the presence of interruption generated by recognition tasks. a) BW=100Kb/s b) BW=50Kb/s

## 5.3 Results Analysis

Let us consider first the situation in which all the alarms are detected, thus the processing and transmission change is not interrupted at any level.

Using results from figure 4 it is possible to select the optimal tasks allocation in case of different available bandwidth configurations. One way to do this, is to consider the maximum execution time per frame, that in this case is the one obtained when all the alarms were generated and there where no interruptions ("none" bar).

In both figures the best functional value (lower one) is obtained by allocating only T1 to the camera (0.5 and 0.7 respectively). This is reasonable since the processing power of the Hub is superior, but if the processing power of the camera and/or the number of cameras is increased then the allocation will change and possibly more tasks should be allocated to the camera [1].

The best functional for each considered bandwidth also gives us the processing frame rate which is 2 frames per second for the first configuration and 1.43 for the second one.

Let us now consider the case when a recognition task does not detect an alarm and interrupts the processing chain (A1, A2, A3 and A4). One can see the advantages of using recognition tasks in order to implement a burst system because a lower functional is obtained when an absence of alarm is revealed. The early the absence of an alarm is detected, the greater the decrease of the functional comparing to the "none" situation.

As expected the lowest functional is obtained when there is an absence of alarm in T1 (A1), because this is the case when the lowest number of tasks are processed and the transmitted information is almost none. (The current implementation transmits one byte for synchronization purposes, which informs the server of the absence of an alarm). Note that for A1, the functional value is almost independent of the number of task allocated to the camera and of the available Bandwidth.

The gains obtained by introducing recognition tasks *RC3* and *RC4* are not very significant since the amount of data to be transmitted after *RP3* and *RP4* is low.

It is also important to note that if the absence of an alarm is detected at the hub, the gain in functional is low since there is no gain in the transmission.

## 6. Future work

In the presented case the optimal task allocation was selected by evaluating the functional when all the alarms were present (no interruption). If the statistics of the percentage of alarms for each camera are known, the optimal task could be selected by considering situations A1-A4, thus obtaining a better allocation of resources, specially bandwidth. Note also that introducing recognition task, the performance of the VS system are modified. A more complete functional should consider also the performance (in terms of detection) of the system.

## 7. Conclusions

An architecture for a general third generation surveillance system have been discussed. In particular an approach for selecting the optimal task allocation have been presented.

The introduction of recognition tasks, which interrupts the processing, and transmission when an absence of alarm is detected was discussed. Experimental results illustrate the presented approach for optimal distribution of intelligence.

## Acknowledgements

This work was partially supported by the University and Scientific Research Ministry (MURST) of the Italian Government under the National Interest Scientific Research Program.

## References

[1] L.Marcenaro, F. Oberti, G.L. Foresti and C.S Regazzoni, "Emerging Requirements for Multimedia Surveillance Systems," submitted for publication to the *Proceedings of the IEEE*, Special Issue on Video Communications, Processing and Understanding for Third Generation Surveillance Systems.

[2] Special Section on Video Surveillance, IEEE Trans. On Pattern Analysis and Machine Intelligence, August 2000, Volume 22.

[3] Vision Components, http://www.vision-comp.com.

[4] Tattile SrL, http://www.tattile.com.

[5] L. Marcenaro, F. Oberti and C.S. Regazzoni, "Short-memory Shape Models for Ground-plane Predictive Object Tracking", *Proc. First IEEE Int. Workshop on Performance Evaluation of Tracking and Surveillance*, PETS'2000, Grenoble, France, pp.50-56, 2000.

# Index

MIX
Papier aus verantwortungsvollen Quellen
Paper from responsible sources
FSC® C105338

If you have any concerns about our products,
you can contact us on
**ProductSafety@springernature.com**

In case Publisher is established outside the EU,
the EU authorized representative is:
**Springer Nature Customer Service Center GmbH**
**Europaplatz 3, 69115 Heidelberg, Germany**

Printed by Libri Plureos GmbH
in Hamburg, Germany

# SpringerBriefs in Psychology



For further volumes:
http://www.springer.com/series/10143

Andrea J. Romero · Lisa M. Edwards
Sheri Bauman · Marissa K. Ritter

# Preventing Adolescent Depression and Suicide Among Latinas

## Resilience Research and Theory

Springer

Andrea J. Romero
Marissa K. Ritter
Family Studies and Human Development
University of Arizona
Tucson
AZ, USA

Sheri Bauman
Department of Disability and
Psychoeducational Studies
University of Arizona
Tucson
AZ, USA

Lisa M. Edwards
Department of Counselor Education and
Counseling Psychology
Marquette University
Milwaukee
WI, USA

ISSN 2192-8363 ISSN 2192-8371 (electronic)
ISBN 978-3-319-01380-0 ISBN 978-3-319-01381-7 (eBook)
DOI 10.1007/978-3-319-01381-7
Springer Cham Heidelberg New York Dordrecht London

Library of Congress Control Number: 2013945792



Printed on acid-free paper

Springer is part of Springer Science+Business Media (www.springer.com)

# Contents

# Chapter 1
# *La Llorona* Calling: The Epidemic of Latina Adolescent Depressive Symptoms and Suicidality

*La Llorona* (the weeping woman) is a powerful image in the mind of many Latinos throughout the Americas who have been told the story as young children. A beautiful young woman who becomes caught in a love affair gone wrong chooses to drown her children and ultimately herself in the river. She becomes an unsettled soul, doomed to roam rivers, and arroyos wailing for her lost children crying, "mis hijos (my children)." Embedded within the story are messages to young women, and men, about the potentially fatal outcomes of unhappy love affairs, and the possibility of profound depression and suicidal inclinations. The story has many variations depending on who the speaker is and who their audience is, but the final ending is always the same: tragic and long-lasting torment for the young woman.

Scholars have reclaimed this story of *La Llorona* by re-envisioning the woman as one who is angry, crying over the patriarchal oppression, and has fear of the future for her children (Garcia 1997). In this version, the *La Llorona* story is retold as a powerful and resilient woman who is fighting to determine her own future and that of her children (Anaya 2011). We begin the discussion of the 30-year epidemic of Latina adolescent depressive symptoms and suicide attempts by also reclaiming the interpretation of the epidemiological findings from a perspective that is focused on the strength and power of young women—rather than as victims of their class, culture, or families. Additionally, we include a review of recent literature of protective factors as well as risk factors, in order to focus on prevention efforts that may be consistent with Latina adolescents' gendered and cultural roots. Our ultimate goal is to raise awareness among researchers, scholars, counselors, and prevention specialists about the public health issue of Latina adolescent depression and suicidality as a means to inspire new research and interventions that may save the lives of more young women in the future.

## Latina Adolescents in the United States

Latina adolescents are among the fastest growing ethnic and gender subgroups in the United States today, and in several states, they are the largest ethnic minority under the age of 18 years (U.S. Census Bureau 2010). Currently, Latinos of

A. J. Romero et al., *Preventing Adolescent Depression and Suicide Among Latinas*, SpringerBriefs in Psychology, DOI: 10.1007/978-3-319-01381-7_1,

all ages make up 16.7 % of the population and account for 43 % of the population growth in the past 10 years, which far surpassed the 39 % growth projected in 2000 and is primarily attributed to births rather than immigration (U.S. Census Bureau 2011). Mexican descent individuals comprise the majority of the U.S. Latino population at 64 %, with the next largest ethnic groups being Puerto Rican at 9.0 %, Central American at 7.6 %, and Cuban at 3.4 %. Most Latinos in the United States are citizens who speak English very well, although many are also bilingual and speak Spanish at home. In fact, 92 % of Latino children are U.S. citizens, although a large proportion (52 %) has at least one parent who is an immigrant; 36 % of the entire Latino population is foreign born (Fry and Passel 2009). The majority of Latino children have two parents living at home (U.S. Census Bureau 2012). Among children under the age of 18 years old, Latinos make up 24 % of the U.S. children's population (Annie E. Casey Foundation 2013). Latino children are overrepresented among U.S. children living in poverty at 34 %, which has increased by 5 % over the past 11 years (Annie E. Casey Foundation 2013). Estimates range from 20 to 37 % for Latino children who do not have health insurance (Annie E. Casey Foundation 2013; Fry and Passel 2009).

Latinas are quickly becoming a significant population in the United States, with current demographic data indicating that there are 24.2 million Latinas (5.96 million between ages 10 and 24) (U.S. Census Bureau 2011). Many Latinas are young and are more likely than Latino men to graduate from high school and enroll in college (Center for American Progress Action Fund 2008). Furthermore, many are choosing careers in politics and starting their own businesses (Center for American Progress Action Fund 2008). Yet Latina adolescents have long been more likely to suffer from depression (see Fig. 1.1) and engage in suicidal behavior (Rasmussen et al. 1997; Roberts and Chen 1995; Roberts et al. 1997) than their male counterparts. We refer to these statistics as an epidemic because depressive symptoms affect a disproportionately large percentage of Latina adolescents, and this health disparity has been problematic for over 30 years (CDC 2008). Moreover, it is associated with disproportionately high rates of suicide attempts among Latina teens. A persistent disparity has been documented for Latina girls since the 1960s; they have among the highest rates of depressive symptoms and suicidal ideation

**Fig. 1.1** Percentage of Adolescent Female Depressive Symptoms Over 12 years: *Source* Centers for Disease Control Youth Risk Behavior Surveillance System (YRBSS)

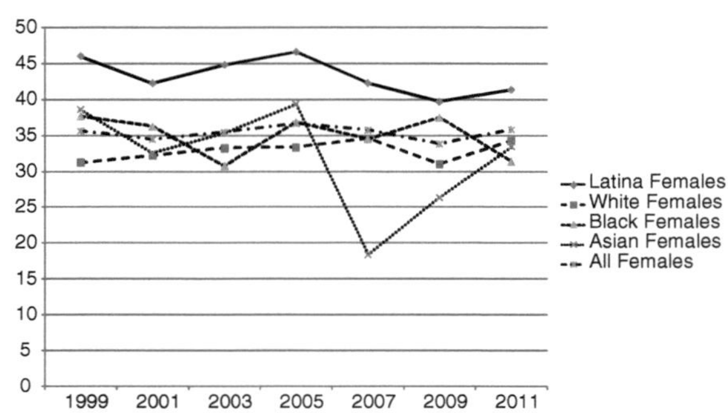

compared to other ethnic and gender groups (Rasmussen et al. 1997; Roberts and Chen 1995; Roberts et al. 1997; Trautman 1961). In 2007, an estimated 42.3 % of Latina girls reported feeling sad or hopeless almost every day for two or more weeks in a row, a rate higher than females of other ethnic groups (White: 34.6 %; Black: 34.5 %) and males (White: 17.8 %; Black: 24 %; and Latino: 30.4 %) (Eaton et al. 2008). Our goal in this book is not to pathologize the development of Latina adolescents by suggesting that all Latinas suffer from depression or attempt suicide. In fact, most Latinas develop in positive and successful ways. We hope, instead, to provide a better understanding of the myriad factors that may help us to understand when Latina teens do get depressed and suicidal and to learn from those that are resilient so that we can develop prevention and intervention treatments using evidence-based strategies.

Eaton et al. (2011) noted that Latina adolescents are twice as likely to engage in suicidal behavior than their White or Black peers (Fig. 1.2). The CDC indicates that nationally, Latino female youth report higher levels of suicide-related thoughts and behaviors compared to males, with 11.1 % reporting having made a suicide attempt, 15.4 % reporting having made a suicide plan, and 20.2 % reported having seriously considered attempting suicide during the past year (compared to 5.1, 9, and 10.7 % of Hispanic males, respectively) (Fig. 1.3) (Suicide Prevention Resource Center 2012). Research has demonstrated that

**Fig. 1.2** Percentage of Adolescent Female Suicide Attempts Over 20 years: *Source* Centers for Disease Control Youth Risk Behavior Surveillance System (YRBSS)

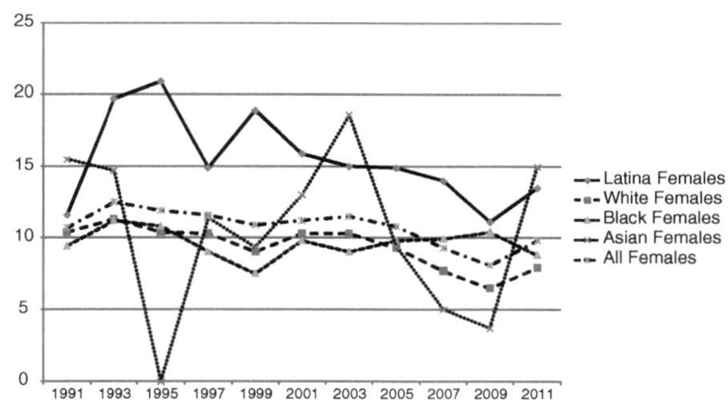


**Fig. 1.3** Percentage of Latino Adolescent Suicide Attempts Over 20 years: *Source* Centers for Disease Control Youth Risk Behavior Surveillance System (YRBSS)

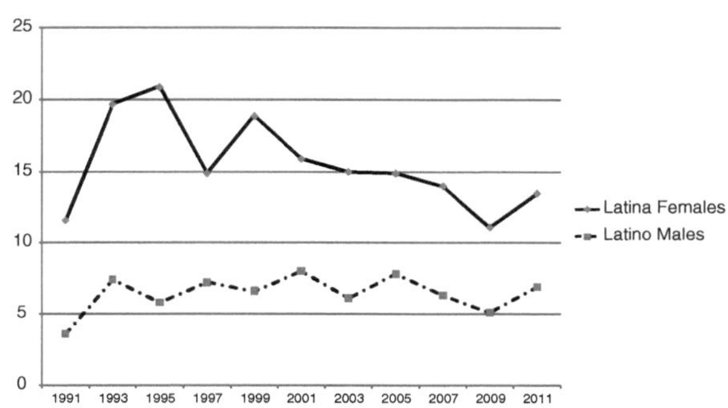

suicide attempts among Latina adolescents are distributed among all socioeconomic classes in the United States, among all ethnic subgroups and among all regions of the United States (Zayas 2011)

## Definitions of Suicide, Suicide-Related Behaviors, and Depressive Symptoms

*Suicide* refers to a deliberate act to end one's life. The term *suicidal behaviors* refers to thoughts of suicide (suicidal ideation), plans to commit suicide, and attempting suicide. Suicide plans refer to more than just thoughts; it is preparing oneself in a more active fashion that can include both mentally and materially for the act (Zayas 2011). *Suicidal behavior* can be considered an umbrella term for thoughts of killing oneself (suicidal ideation), suicide attempts, and deaths by suicide (Goldston et al. 2008; O'Carroll et al. 1996).

Depression, also known as *clinical* or *major* depression, is the term used to describe a mood disorder. According to the Diagnostic and Statistical Manual of Mental Disorders (DSM-IV-TR; American Psychiatric Association 2000)[1], a diagnosis of major depressive disorder requires a minimum of five symptoms to be present for at least 2 weeks. At least one of the symptoms must be either a depressed mood or a loss of interest or pleasure in activities previously found to be pleasurable. The possible symptoms are depressed mood (sadness) most of the time, although in children and adolescents it can be irritable mood, weight loss or weight gain when not trying to do so, sleep disturbances (inability to sleep or sleeping for very long duration), restlessness or feeling of slowness, feeling worthless, difficulty concentrating, or recurrent thinking about death or suicide. The term depression is frequently used rather casually to describe feelings of sadness, but clinical depression is a constellation of symptoms that persist and prevent the person from engaging in their normal activities.

## Warning Signs of Depression and Suicide

According to the American Psychological Association (2013), depressive symptoms may include lack of interest and pleasure in daily activities, significant weight gain or loss, insomnia or excessive sleeping, lack of energy, inability to concentrate, feeling worthless, excessive guilt, recurrent thoughts of death or suicide. Suicide warning signs according to the American Psychological Association

[1] Please note that at the time this manuscript was developed the publication of the DSM-V was imminent. Readers are encouraged to consult the DSM-V to better understand any recent changes in cultural formulation of illness.

(2013) include many similar symptoms to depression, but also include talking about dying, significant recent loss (death, divorce, separation, broken relationship), change in personality, fear of losing control, suicide plan, suicidal thoughts, and no hope for the future. There are many available free screening tools that are of high quality.

## Adolescent Suicide in the United States

In the United States, suicide is the third leading cause of death among young people (Cash and Bridge 2009; CDC 2012; Eaton et al. 2012; Steele and Doey 2007). Even though data trends from the 2011 Youth Risk Behavior Survey (YRBS) (CDC 2011) indicate that prevalence of youth suicide-related behaviors has mostly decreased over the last decade, current data indicate that 15.79 % of ninth- through twelfth-grade students in public and private schools throughout the United States reported having seriously considered attempting suicide and 7.76 % of students attempted suicide one or more times during the 12 months prior to the survey (CDC 2011). While suicide is uncommon in childhood and early adolescence, it increases markedly in the late teens and continues to rise until the early twenties (Gould et al. 2003). Of note is the fact that most Latinos who report at least one lifetime suicide attempt made the attempt before the age of 18 years (Fortuna et al. 2007). It is also notable that while Hispanic adolescents in the United States experience suicidal ideation and attempt suicide at higher rates than other ethnic groups, the rates of suicide completion are not disproportionately high (Cash and Bridge 2009). Suicide rates in the general population tend to be higher among boys than girls, but typically girls have higher rates of suicidal ideation and attempted suicide (Cash and Bridge 2009; Gould et al. 2003).

## Interpreting Ethnic Health Disparities: Avoid the Gloss

Ethnic/racial mental health disparity data based solely on ethnic labels do not provide information about risk or protective factors that could be amenable to change through prevention strategies or mental health services (Bradby 2003; Nazroo 2001; Walsh et al. 2000). In order to truly reduce existing ethnic/racial health disparities, we need to go beyond ethnic label-based differences and examine the impact of ecological contexts on risk and protective factors (Bradby 2003; Nazroo 2001). The high rates of Latina girls' suicide attempts were initially concentrated among Puerto Rican teens in large urban areas, but since the 1960s, suicide and depression rates have also increased among girls of other Latino ethnic subgroups (see Zayas et al. 2005 for a review). Pan-ethnic labels tend to gloss over important differences between Latina subgroups; within the United States, Latinos comprise such diverse cultural and national backgrounds as Mexican (64 %),

Cuban (3.4 %), Puerto Rican (9.0 %), Dominican (2.8 %), and Central American (7.6 %) (Trimble 1990–1991; U.S. Census Bureau 2010). Each of these distinct Latin American nations has their own unique culture, history, economic status, politics, and immigration contexts that need to be considered in order to make research and treatment culturally relevant for each Latino ethnic subgroup (Dana 1993; Phinney 1996; Umaña-Taylor and Fine 2001). To understand the complexities of Latino mental health, we must consider the deeper structure of culture and move beyond generalizations across Latino groups in order to embrace the implications for mental health research and treatment.

One study compared a number of ethnic groups on suicidal behavior (thoughts, plans, attempts) and found that Mexican Americans had the highest rates of all group, (Duarté-Vélez and Bernal 2008). Puerto Ricans had higher representation than expected, whereas Central American and Cuban Americans had lower representations than expected (Duarté-Vélez and Bernal 2008). That is, the percentage of the suicides that were from those groups were greater or less than their proportion of Miami's Latino population at the time data were collected. These examples illustrate within-group differences among the Latino pan-ethnic label and demonstrate how important it is to consider subgroups (Duarté-Vélez and Bernal 2008). What Latino groups have in common are the Spanish language, certain cultural stories (such as *La Llorona*), being of mixed races, and certain values such as *familismo, respeto,* and *personalismo* (Duarté-Vélez and Bernal 2007).

## No Easy Answers: Transcending Cultural Deficit Models

Perhaps it is too easy to interpret epidemiological health disparities, such as higher rates of depressive symptoms, as a result of the culture; this is an easy answer because there are no other social factors that are included in most large nationwide studies. However, this simplistic answer is not only incorrect, it actually impedes research and clinical advances because culture and gender are not modifiable within prevention or intervention programs. Moreover, this easy answer avoids more profound analysis and in fact reinforces stereotypes of Latina adolescents as trapped within dysfunctional cultural and gender roles. Unfortunately, cultural deficit models have hindered the advancement of research and interventions to reverse the epidemic of Latina adolescent depression and suicide attempts. It is apparent to many scholars that it is necessary for the health of Latina girls and for the development of effective prevention/intervention programs to avoid what might appear to be "easy answers" and delve deeper into the complex and multilayered experience of life of Latina adolescents in the United States (Garcia-Coll et al. 1996).

Psychology has a history of using cultural deficit models to explain psychological health disparities. It is likely that the prevalent cultural assumption within the United States about the "myth of the melting pot" may have influenced psychologists to assume that mental well-being is positively influenced by assimilation into U.S. culture and that the retention of traditional culture may lead to more mental conflict and dysfunction (Rosaldo 1993). Psychologists assumed that individuals who did not assimilate would be more susceptible to mental illness (Born 1970). These interpretations have since been debunked and are referred to as the cultural deficit model, which attributes the cause of mental illness to "minority" culture, which is viewed as lacking or as a deficit (Ramirez 1998). Peña (2003) describes how the history of psychoanalysis included instances where psychologists chose not to treat Latinos, or blamed ineffective treatment on the patients themselves, because they were too "unsophisticated" to psychoanalyze. The myth of the melting pot and the cultural deficit model both "blame the victim" for mental health problems and create barriers to identifying and implementing effective treatment. In fact, Latino patients were described at times as "pathological, underdeveloped" and thus "untreatable" (Peña 2003). In fact, existing stereotypes of Latina adolescents further diminish their inherent strengths, agency, and internal power because they too often reiterate overt generalizations of Latino culture as patriarchal with strictly divided gender role socialization. However, a significant amount of current research has more than debunked these myths of Latino families and Latina women and teens. Thus, Peña (2003) challenges psychologists and psychiatrists to not only discard the cultural deficit model, but to re-envision the Latina client as one who is multi-voiced with cultural strengths and as a full and equal human being.

In order to avoid the cultural deficit model interpretation, clinicians and researchers must first move beyond ethnic labels because they gloss over important differences (Pedrotti and Edwards 2009; Pedrotti et al. 2009; Peña 2003; Ramirez 1998). In order to begin this process, we begin with definitions of culture as complex and multi-dimensional, and we focus on cultural strengths and protective factors within this book. Cultural psychologists argue that ethnicity is complex and involves the *lifeways* and *thoughtways* of individuals in a manner that is far more complex and profound than ethnic labels (Helms et al. 2005; Phinney 1996; Shweder 1993; Trimble 1990–1991). Culture is in many ways like the water that fish swim in; it permeates how we breathe, live, how we interact, and our mental well-being. It is fluid, dynamic, changeable, and it permeates our surroundings, our interactions, and our interpretation of our experiences. Culture can be defined as the shared set of learned experiences that encompasses a broad range of beliefs, values, language, norms, and shared symbols (Marsella and Yamada 2000). As such, it has been argued that culture has an impact on numerous aspects of mental illness (American Psychological Association 1994; Guarnaccia and Rogler 1999; U.S. Department of Health and Human Services 2001) and its treatment (Angel and Thoits 1987; Cuéllar 1982; Donlan and Lee 2010; Draguns 1973; Lopez 1989; Marsella and Kameoka 1989). Clearly, there is a need to debunk myths and stereotypes of Latina women if we are determined to effectively prevent and treat Latina adolescent depression and suicide in order to reduce mental health disparities. Peña (2003) argues that we need to reclaim the stereotyped images of Latinos, particularly of Latinas, in order to be able to approach mental illness prevention and treatment from a strength perspective that will focus on empowerment, current, and future.

## Culture-Bound Definitions of Suicide-Related Illness

Culture has been purported to influence the perception and definition of mental illness (Kleinman and Good 1985) as well as the manifestation and expression of the symptoms (Draguns 1973; Guarnaccia and Rogler 1999; Kleinman 1988). There are several identified culture-specific disorders in Latino cultures, or "lay idioms of physical and mental distress" (Bayles and Katerndahl 2009, p. 16). Some of these *culture-bound syndromes* (e.g., *mal de ojo*, or evil eye, *nervios*, or nerves, and *susto* or soul loss) are included in a glossary of the DSM-IV-TR (American Psychiatric Association 2000), and clinicians are encouraged to be aware of these behaviors or experiences. One of the most common, *nervios* or *ataque de nervios*, is a culturally appropriate demonstration of strong emotions due to stressful life events, and an outlet for anger, grief, and family disruptions. This is often characterized by shouting, swearing, and striking out at others and falling to the ground. Often an *ataque de nervios* proceeds to a worsened condition of *susto*, which is a more prolonged and chronic condition, which might be manifested in a manner similar to depression. *Susto* is described as having lost one's soul due to a traumatic life event; treatments by *curanderos* may include *limpia*s or sweeping ceremonies (Gonzalez 2012). Suicide attempts among Latina adolescents have been likened to an *ataque de nervios*, which suggests that suicide attempts are not premeditated or based on long-standing depression, but rather are a reaction to an immediate traumatic event (Zayas and Pilat 2008; Zayas et al. 2012). This approach to understanding suicide attempts may provide insight into Latina adolescent distress as well as family responses to their treatment adherence and resilience after the event. Ramirez (1998) and Gonzales (2012) argue that indigenous perspectives about the world are infused with spirituality and this may influence the expression of mental illness and the manner in which individuals understand the causes and cures. It is clear that symptoms could be understood and manifested in different forms in different cultures.

## In Summary

This book provides an overview of research on the mental health of Latina adolescents in the United States with emphasis on those cultural constructs and theories with relevance for assessment, prevention, and interventions for depression and suicide. We include both protective factors and risk factors in order to provide a balanced understanding of the experience of Latinas. This approach may also help researchers and service providers to find avenues to prevent depression and suicide by building on girls' internal and cultural strengths. This chapter provides a framing of the historical and current situation of Latina adolescent depressive symptoms and suicide attempts at national levels. Chapter 2 summarizes the most common theories of depression and suicidality. We use previous theories to help provide insight into the factors that may lead to depression and suicide. Chapter 3 then introduces a multiracial feminist framework that provides an ecodevelopmental resilience view of young Latinas. Chapter 3 also describes the current research on protective factors within ecological contexts. Chapter 4 extends our discussion on how positive development can be derailed for Latina girls by risk factors at micro-, meso-, and macroecological levels. Finally, Chap. 5 provides recommendations for researchers, clinicians, and schools about how they can help prevent and intervene among Latina adolescent populations with issues of depression and suicide. Our ultimate goal is to critically think through the epidemic of Latina adolescent suicide attempts and to motivate individuals to action that can help improve and save the lives of Latina adolescents in the United States.

# Chapter 2
# What Drove Her to Do It? Theories of Depression and Suicide

Like many versions of *La Llorona*, there are different explanations for what leads a young woman to choose to end her life. It is for this reason that we turn now to theoretical perspectives on depression and suicide in order to consider the primary factors that may contribute to Latina adolescents' high rates of depression and suicide attempts. We will focus on those psychosocial theories that seem to provide the most promise for understanding and treating young Latinas. Much of suicide research focuses on the epidemiological level of interactions between risk factors and outcome variables (e.g., identifying the statistical associations between substance use and suicidal ideation). These studies are useful in helping identify possible red flags for those who are suicidal; however, they often utilize self-report surveys or psychological autopsies, and they rarely examine *why* someone attempts suicide. The theories explained in this chapter include theories that focus on depression (Cognitive Behavioral Therapy and Acculturative Stress Models), as well as theories that are specific to suicidal behavior, including the interpersonal theory of suicide, and Zayas's ecodevelopmental model for Latina adolescents.

## Depression and Suicide

Depression is one of the leading risk indicators for suicide. Over the life course, people with untreated depression have a nearly 20 % risk of suicide (Gotlib and Hammen 2002); about two-thirds (2/3) of people who commit suicide are believed to be depressed at the time of their deaths (American Association of Suicidology 2012). Unsurprisingly, depression was the most robust predictor of suicidality in a sample of urban Latino and Black adolescents (O'Donnell et al. 2004). Adolescent Latinas have among the highest rates of depressive symptoms and suicidal ideation and attempts compared to other ethnic and gender groups, a persistent disparity that has been documented since the 1960s (Rasmussen et al. 1997; Roberts and Chen 1995; Roberts et al. 1997; Trautman 1961). New information regarding biological influences on

A. J. Romero et al., *Preventing Adolescent Depression and Suicide Among Latinas*,
SpringerBriefs in Psychology, DOI: 10.1007/978-3-319-01381-7_2,

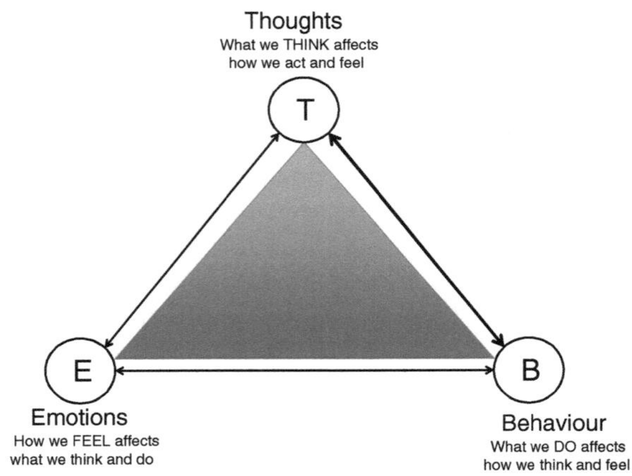


**Fig. 2.1** Cognitive Behavioral Theory

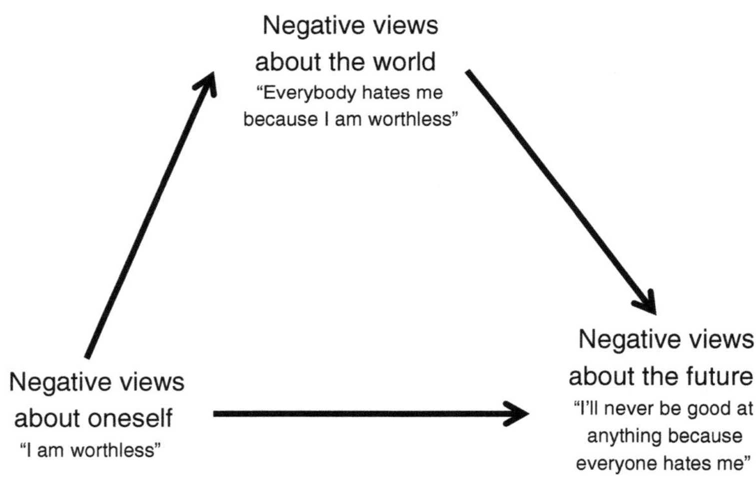


**Fig. 2.2** Cognitive Behavioral Theory Negative Thoughts Triad

depression is emerging. Among these are genetic markers that suggest a genetic component to suicidal behaviors (Brent and Mann 2005). As Beck and Alford (2009) noted, although much has been learned about the biological factors in depression, the lines of inquiry have focused on identifying involved brain systems, understanding how the brain responds to various medications, and identifying genetic markers that might suggest more targeted pharmacological interventions. Although we will not spend time discussing genetic factors in this book, it is important to remember that individuals in all groups vary in their predisposition to develop depressive disorders.

Any theory that increases our understanding of suicidal thoughts and behaviors among Latina adolescents should take into account depression as a precursor (Duarté-Vélez and Bernal 2007). In one longitudinal study, Latino adolescents who were depressed at the start of the study were also more likely to be depressed in the fifth year compared to other ethnic/gender groups (Dawson et al. 2012). Similarly, Latina youth who reported experiencing depression in the previous year were 5.1 times more likely to consider suicide than those who were not depressed (Romero et al. 2012). Clearly, depression as a risk factor is indicated among Latina adolescents and is a critical variable to consider in developing prevention and intervention programs.

## Cognitive Behavioral Theory

A well-known theory of depression is the cognitive behavioral theory (CBT) which underlies the therapy also known as CBT. CBT draws from a variety of sources, most notably the work of Aaron Beck and Albert Ellis. This theory is based on the premise that thoughts are primary in depression, and those (maladaptive or irrational) thoughts are the cause of depressive feelings and behaviors (Fig. 2.1). It is assumed that as cognitive distortions are identified and modified, the depressive feelings will be ameliorated. A behavioral component of this approach was added to the original cognitive schema. Because thoughts are inextricably linked to behaviors, this theory includes the notion that changing behavior will also have a positive effect on both feelings and thoughts. For example, depressed adolescents may lose interest in previously enjoyable activities and lack motivation to pursue those or new activities. When they can understand the importance of engaging in positive behaviors (e.g., exercise, music, volunteering), they are advised to push themselves to take steps to include more uplifting activities. The notion of the *cognitive triad* as a critical symptom of depression is also a central feature of this framework. There are three beliefs that comprise the triad: (1) I am worthless, a failure, hopelessly inadequate; (2) the world is a terrible, cruel, inhospitable place; and (3) the future is likely to be worse (Fig. 2.2). While the CBT is helpful to explain the effect of depression in general, there is limited research with Latino populations. The majority of research with Latinos and depression has focused on the cultural context. Because this theory and treatment approach has been so successful with a variety of populations, it should be tested with Latinas, with culturally appropriate adaptations as needed.

## Acculturative Stress Model

Acculturation helps researchers begin to conceptualize the complexity of the cultural context of mental health. According to a recent definition, "acculturation is proposed as a multi-dimensional process consisting of the confluence among heritage-cultural and receiving-cultural practices, values, and identifications" (Schwartz et al. 2010, p. 237). As such, the concept of acculturation has played a central role in understanding within-group differences among Latinos as well as factors that may influence assessment and treatment of Latinos (Cuéllar et al. 1995; Cuéllar 2000), including the areas of depression and suicidality. A significant amount of research in this area has focused on the linkage between acculturation and depression among Latinos, proposing that depression may be attributed to stress that results from cultural changes when different cultures interact (Redfield et al. 1936). It was originally assumed that the acculturative process was stressful and that assimilation to the U.S. dominant culture would result in better mental health; however, empirical findings did not fully support that traditional culture was associated with more mental illness or that assimilation leads to better mental health (Cervantes and Castro 1985; Rogler et al. 1991).

Theories of stress and coping improved the understanding of the impact of cultural change on mental health, by including measures with individual-level subjective perception of stress within experiences of intercultural interaction and individual-level cultural changes (Aldwin 2007; Lazarus 1997). *Acculturative stress* is stress that results from the acculturation process as individuals come into contact with different cultures and the resultant changes that they may experience in their own attitudes, behaviors, or cognitions. *Bicultural stress* is everyday stress that may result from pressure both to adopt the majority culture's values and to preserve one's original culture (Romero and Roberts 2003a). Acculturative and bicultural stress include not only major life events' stressors (immigration), but also daily hassles (prejudice) that may accumulate over time to lead to chronic high stress with negative physical and mental health outcomes, such as depression (Lazarus 1997; Lewis et al. 2009). The loss of home, nation, language, friends, and family through experiences of immigration and acculturation may be perceived as stressful and challenging at any age and thus may negatively impact stress levels and mental health. Among Latino adolescents, bicultural stress has been linked to more depressive symptoms and more substance use, as well as lower optimism among Latina girls and lower self-esteem for Latino adolescents (Romero and Roberts 2003a, b; Romero et al. 2007a, b). Although being bicultural and having a bicultural identity have been associated with better mental well-being for Latino youth, there can also be stress within bicultural contexts for some adolescents. For example, Latino youth may feel pressure from the school environment to adopt U.S. values; at the same time, their Latino peers may pressure them to maintain their Latino identity (Romero and Roberts 2003a).

Another example of bicultural stress is experienced within the family, such that adolescents may experience conflict with family members who are adapting at a different rate to the U.S. culture in terms of language and/or values. Different rates of acculturation within families may lead to acculturation gaps between parents and children that may contribute to family conflict and stress (Baptiste 1993; Arnett 1999; Szapocznik et al. 1978). Peña et al. (2011) describe a theory of *segmented acculturation* which may be quite relevant to Latino family dynamics. This theory proposes that there are three acculturation patterns: selective, dissonant, and consonant. In *selective acculturation,* children retain the values of their culture of origin, but incorporate selected aspects of the dominant culture in a slow-paced manner. *Dissonant acculturation* occurs when the youth discard the cultural values of their cultural of origin and acculturate rapidly to the dominant culture at a pace that is much faster than that of their parents. *Consonant acculturation* is a pattern exhibited by families in which both parents and children acculturate quickly to the dominant culture.

Several national and regional studies have found that U.S.-born Hispanics have higher rates of suicide than first-generation Hispanics (Peña 2008). These findings have been interpreted as perhaps due to acculturative stress within families due to an acculturation gap (American Association of Suicidology 2012b). It has also been found that acculturation gaps between parents, their children, and their children's peers heighten peer–parent conflicts, which may create additional stressors for Latina girls (Baptiste 1993; Coatsworth et al. 2000; Duarté-Vélez and Bernal 2007). Research in this area argues that tension arises when traditional family values of strong parental authority, with concomitant restrictions on the Latina adolescent's autonomy, contrast with the dominant cultural norms that encourage adolescent autonomy and exploration. Zayas and colleagues (2008, 2011, 2012) have demonstrated in multiple studies that internal family stress is a precursor to Latina adolescent suicide.

## Interpersonal Theory of Suicide

While cognitive behavioral therapy and acculturative stress models help to understand the processes of depression and the unique cultural context of depression among Latinos, they do not fully explain what leads some individuals to take a more serious step toward ending their life. Clinical depression is described in Joiner's theory as a mental disorder (i.e., an illness), which means it is not due to some deficit in the individual that must be corrected (Klerman et al. 1984). This perspective views symptoms as markers of the disorder and adds that the symptom presentation can create difficulties in relationships, which then increase the symptoms. The most important aspect of depression is social and interpersonal relationships. The assumption is that when more satisfying relationships are attained, the depressive symptoms will diminish. This theory proposes that the primary problem area will be one of four general themes: (1) grief, (2) interpersonal disputes, (3) role transitions (such as changes in the family, recent life events such as changes in school), or (4) interpersonal deficits (the lack of lasting close relationships) (Klerman et al. 1984; Rosselló and Bernal 1999). Future research may consider how immigration, acculturation, and gender may also factor into these general themes. This certainly appears consistent with research findings on suicidal Latina adolescents (Zayas 2011). Thus, the focus of treatment is on the person's closest social relationships—within the family, in romantic relationships, and in friendships (Fig. 2.3).

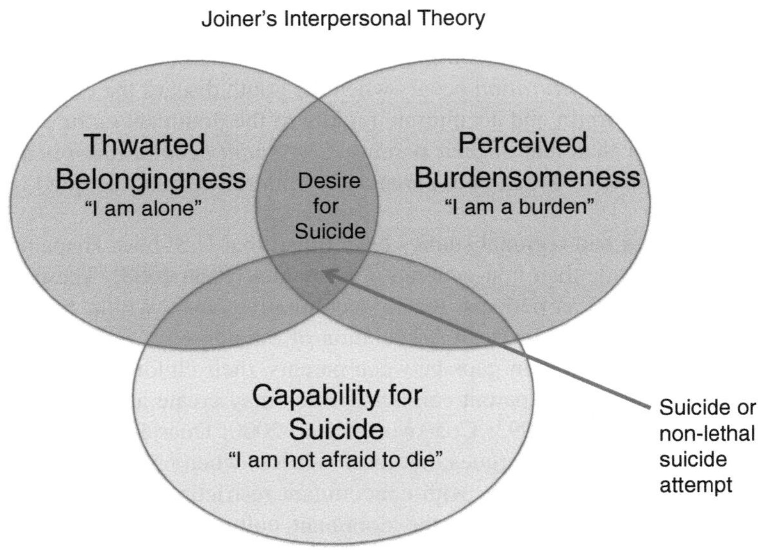


**Fig. 2.3** Joiner's Interpersonal Theory of Suicide (Van Orden et al. 2010)

Joiner's interpersonal theory of suicide (Joiner 2005; Van Orden et al. 2010) posits that serious suicide attempts result when a person feels both suicidal ideation and also has the capacity to carry out an attempt. The interpersonal theory of suicide posits that there are three conditions that promote suicidal thoughts and behaviors: *thwarted belongingness*, *perceived burdensomeness,* and *capability for suicide* (Joiner 2005; Van Orden et al. 2010). Thwarted belongingness and perceived burdensomeness must be accompanied by an acquired capability to commit suicide (often acquired through progressively more lethal attempts, engaging in self-harm, or doing and surviving high-risk behaviors) in order to lead to suicidality. Thwarted belongingness refers to the idea that the individual feels that they do not belong to any particular group or maybe even society in general. Perceived burdensomeness refers to feelings of being a burden on those around them and can lead to ideas about how the world might be "better off" without the individual in it. Collectivistic cultures place high priority on the family unit, often to a degree that it comes before individual well-being; future studies may consider how strong collectivistic values and gender values of self-sacrifice for family (such as *marianismo)* may contribute to perceived burdensomeness among Latina adolescents (Castillo et al. 2010; Gaines et al. 1997).

Thwarted belongingness is consistent with the observation that the suicide attempts by Latina adolescents were often preceded by conflicts with either parents, a boyfriend, or parents about a boyfriend (Zayas et al. 2005). Damage to the most valued and central relationships of belonging in a Latina's young life may result in powerful negative emotions. It is plausible that the Latina who is in protracted conflict with family and/or has serious conflicts with other significant people in her life may not only feel that she does not belong, but she may also feel that she is a burden to her family (or friends, or boyfriend) (Zayas 2011). Van Orden and her colleagues (2010) observed that social isolation is one of the most robust predictors of suicidality. Given recent research on bullying, peer discrimination, and school discrimination, it is also likely that these factors may also contribute to a sense of thwarted belongingness (Cammarota 2007; Romero et al. 2013).

Joiner's theory was created with a focus on adult populations, and there is relatively little research with adolescent populations or with Latinos. One qualitative study evaluated the utility of the interpersonal theory of suicide with 33 adolescent females who are described as "mostly Puerto Rican" (p. 101) and who were in the hospital due to suicidal behavior. They were compared with similar-aged females who were also hospitalized, but were not suicidal (Razin et al. 1991). In the group who had attempted suicide, some patterns were notably more prevalent, including recent conflicts with a mother or a boyfriend, poor school performance, sexual activity, suicidal history in the family, and poor self-esteem. These patterns are consistent with the primacy of interpersonal disputes in Joiner's theory.

## Zayas's Ecodevelopmental Model of Latina Suicide

There is one model that is very specific to the epidemic of Latina adolescent suicide and is based on several years of clinical research with suicidal Latina teens. Zayas et al. (2000) proposed a model for understanding suicide that we include here because it links the risk factors in a meaningful way that reflects the cultural context. This model makes it clear that a single risk factor is unlikely to lead directly to suicide (although there may be a precipitating event), but the combined influence of multiple factors is what creates the conditions that lead to suicidal behaviors (see Fig. 2.4). The reader may note that many of the factors in the model would be found in other groups of adolescents; however, Zayas argues that it is the unique prioritization of family conflict that makes the model specific to Latinas.

Zayas and colleagues (2005) further revised their conceptual model to guide research on Latina suicide rates that moves beyond documentation of ethnic disparities and leads to a richer understanding of the complexity of Latina adolescents' mental health context (Fig. 2.5). Zayas and colleagues' (2005) conceptual model is based on ecodevelopmental theory (Szapocznik and Coatsworth 1999), which argues that social interactions between family, society, and individuals are critical ecological factors to consider for the healthy development of children and adolescents. The young person's developmental context is described as the center of concentric circles that are comprised of microsystem (e.g., family, school, peers, individual cultural values), mesosystem (e.g., parent–school, parent–peer), and macrosystem (e.g., culture, politics) elements based on the social ecology of human development (Bronfenbrenner 1979, 1986). Given the importance of the family context among Latinos (Gloria et al. 2004), Zayas and colleague's (2005) model emphasizes the central nature of familial context for Latina adolescent

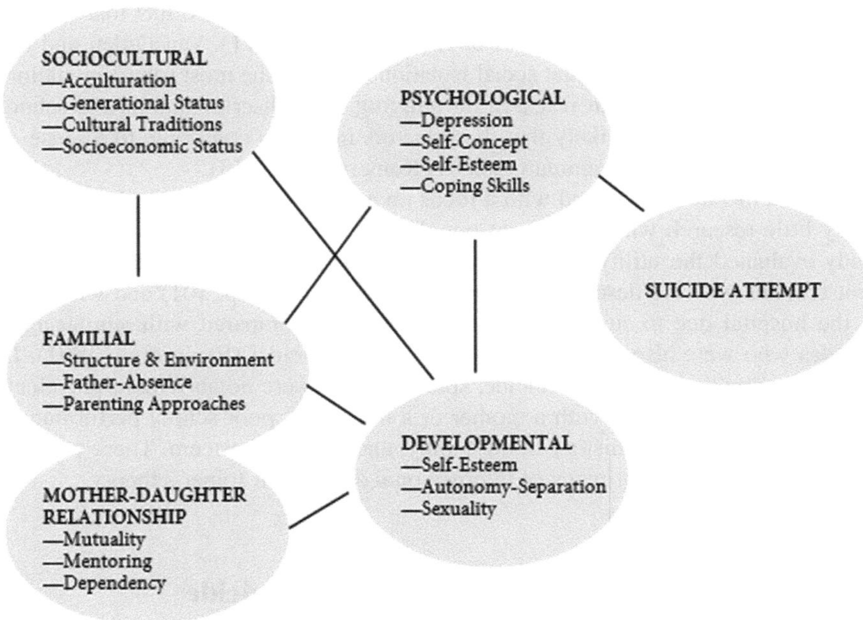


**Fig. 2.4** Zayas et al. (2000) Integrative model of suicide attempts by adolescent Hispanic females

development in terms of cultural values, socialization, family functioning, and interactions with family members. The revised model specifically identifies the interactions between family functioning, adolescent development, and cultural traditions that may lead to emotional vulnerability and psychosocial functioning of young women. They argue that the suicide attempt then is influenced by this combination of factors and is mediated by the adolescent's subjective experience of adolescent-family crisis. This model is an important step forward in understanding the nuances of family and adolescent mutual influence during development and the impact on mental health. Certainly, this model goes far more in depth to explain the central role of family for Latina adolescents, which may be more unique than other ethnic/gender groups during the teen years. Additionally, Zayas's (2000, 2005) research explains the distinctive mother–daughter relationship that plays a central role in the daughter's behavior. Given the bulk of previous research on mother–daughter connection and relationship, it is not surprising that there is a focus on this dyadic relationship within the family system; however, future research will need to also make more efforts to include fathers. Recent research has further uncovered the important role that Latino fathers may play within the family and specifically on the development of their daughters that can be positive (Ojeda et al., in press). This version of this model does not fully explain the role of the ecodevelopmental context, because it does not include the influence of peers, school, and United States/Mexico society. Furthermore, lack of clear inclusion

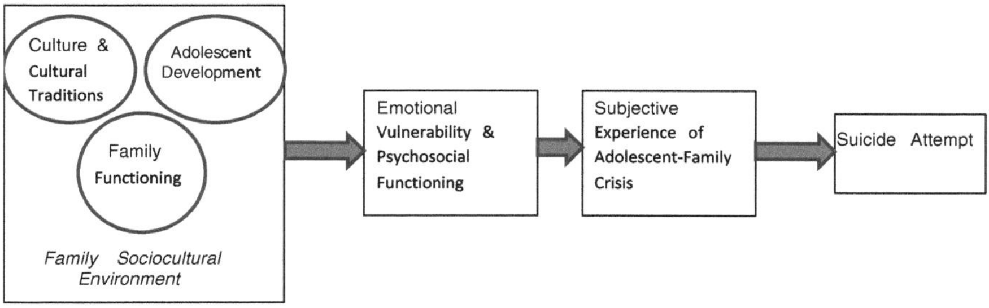


**Fig. 2.5** Zaya's et al. (2005) Conceptual model of Latina suicide attempts

of cultural strengths and protective factors may lead individuals unfamiliar with Latino culture to assume that culture may only be a risk factor for adolescent girls. In the next chapter, we describe the multiracial feminist framework that may help advance our understanding of resilience among Latina adolescents from a feminist perspective that can also help identify key modifiable factors to consider in prevention efforts (Fig. 2.5).

Although adolescence is typically portrayed as a period during which peers have increasing importance for development of values and behaviors, research with Latinos demonstrates that family continues to play a significant role even during this developmental period (Zayas et al. 2005). Empirical research with Latina middle school–aged adolescents in Florida based on an ecodevelopmental framework found that the dual aspects of support and conflict within families accounted for the majority of differences in internalizing and externalizing behaviors (Coatsworth et al. 2000). Given the importance of the family context among Latinos (Gloria et al. 2004), Zayas and colleague's (2005) model emphasizes the central nature of familial context for Latina adolescent development in terms of cultural values, socialization, family functioning, and interactions with family members.

## In Summary

The intention of this chapter was to provide a basic background about the theories that help explain factors that lead to suicidal behavior and depression among all populations, among Latinos, and specifically among Latina adolescents. This background may lead to greater understanding by mental health researchers and service providers of their realistic ecological contexts and resilience. CBT explains how depressive symptomatology may lead to cognitive distortion of repeated negative thoughts that may lead to negative behaviors. Acculturative stress models extend the research on depression to incorporate the cultural context of Latinos in the United States that may lead to major life event stress and daily chronic stressors that are also linked not only to depressive symptoms, but also lower self-esteem and lower optimism.

Specific theories of suicidality improve our understanding of the specific factors that may lead to up to suicidal thoughts, plans, and attempts. Joiner's interpersonal theory of suicide juxtaposes thwarted belongingness along with perceived burdensomeness that can become exacerbated into suicidal behavior when the individual also has acquired the capacity for self-injury (through behaviors such as a cutting or other sources of non-lethal self-injury or repeated suicide attempts). Although Joiner's theory was not developed specifically based on Latina adolescent research and there is limited empirical work with this population, we feel that the basic concepts have significant relevance for understanding the impact of social relationships on suicidal behavior of Latinas. Lastly, we described the conceptual models of Zayas and colleagues based on their clinical work with Latina suicide attempters. Zayas argues that the root cause for most attempters lies in the subjective experience of adolescent-family crisis and is influenced by the teen's psychological functioning and the combined interaction of their preexisting family functioning, cultural traditions, and their own development. Latina adolescent health disparities cannot be properly understood without considering the risk and protective factors in their environment (Szapocznik and Coatsworth 1999).

What is missing all in of these theories is a feminist gender-rooted approach to understanding the experience of young Latina women. As such, we hope to contribute to the future prevention efforts with Latina adolescents by integrating a cultural strength approach that is firmly centered in a balanced view of the Latino culture that not only acknowledges, but also celebrates sources of resilience and strength within the cultural and gendered lives of Latina girls and portrays a view of their lives and experiences as positive and agentic despite the challenges and adversity that some Latina youth encounter in the United States.

# Chapter 3
# Latina Adolescent Resilience Rooted Within Cultural Strengths

Many Latina adolescents cope effectively with stress and experience positive physical and mental health in spite of high levels of poverty, discrimination, bicultural stress, and lowered expectations. Understanding the resilience inherent in how these young Latinas "make it" is a critical component of prevention and intervention. Indeed, in the past 10 years, researchers have attempted to better understand within-group variability among Latinas and identify protective factors that might mediate the relationship between multiple risks and negative adolescent outcomes (Gutman et al. 2002; Klein and Forehand 2000). In this chapter, we review the Multiracial Feminist Framework (MFF); (Arellano and Ayala-Alcantar 2004) to identify macrosystem, mesosystem, and microsystem factors within a resilience perspective as a way to frame our discussion of protective factors and cultural strengths. Drawing from current research about the role of these variables in the lives of Latino/a youth, we provide an overview of the most relevant resources that can positively influence the mental health of Latina youth in the U.S. Specifically, we focus on protective factors at multiple ecodevelopmental levels including, macrolevel, mesolevel, and microlevel (Bronfenbrenner 1979, 1986). In the following chapter, we will include more discussion of the risks within the ecodevelopmental context of Latina girls in the United States.

## Multiracial Feminist Framework

Previous theories of depression and suicide, discussed in Chap. 2, have not integrated ecodevelopmental contexts at macrolevels (societal) and also levels that include school and peer influences. U.S. society views of Latina women contribute to inequalities based on negative stereotypes, discrimination, sexism, and homophobia. The MFF (Arellano and Ayala-Alcantar 2004) is not specific to suicidal behaviors, but it does provide us with a gendered perspective that is focused on the resilience, equality, and empowerment of young women. There are indications that gender-specific issues of sexism, gender identity, and gender roles all play a critical

A. J. Romero et al., *Preventing Adolescent Depression and Suicide Among Latinas*, SpringerBriefs in Psychology, DOI: 10.1007/978-3-319-01381-7_3,

role in young Latinas' lives, development, and mental health (Arellano and Ayala-Alcantar 2004; Arredondo 2004; Céspedes and Huey 2008). The MFF advances existing models for the mental health of Latina adolescents that still too often portray them as victims of their culture and socioeconomic status (McLean et al. 1995; Pedrotti and Edwards 2009). Furthermore, we feel that this framework takes us one step closer to prevention of depression and suicidal behavior by including positive and protective factors that are modifiable in intervention programs and therapy.

To begin with, a multiracial perspective is particularly relevant for Latina girls given that there are a multitude of races represented by the ethnic group "Latino," including indigenous, African, Spanish, French, and Chinese (Fernández 1992). Racial phenotypes among Latinos range across a continuous spectrum of extremely fair skinned, blond, and blue-eyed to dark skin with indigenous or African facial features to a mixture of each of these elements (Fernández 1992). Studies demonstrate that more African or indigenous racial phenotypes are associated with more discrimination among Latinos (Espino and Franz 2002; Vasquez et al. 1997). The term *multiracial feminism* reflects an inclusive view of gender that reflects the socially constructed nature of gender and the duality of both male and female spectrums of experience and gender roles. Moreover, this term indicates that gender is experienced simultaneously with privilege and power associated with race, class, and sexual orientation; in other words, Latina girls do not experience gender, race, and class separately, but they are all integrated and interact to influence discrimination at the same time (Arellano and Ayala-Alcantar 2004). This complex and continuous view of gender and race is a central and critical component to MFF, and in many ways, it defies outdated portrayals of Latino culture as perpetuating strictly divided gender roles (Castillo et al. 2009; Hurtado and Cervantes 2010). Thus, we take a continuous and multi-dimensional approach to understanding the diversity of the expression of gender among Latina adolescents. This perspective also places the teen adolescent as agentic in a manner that she is also acting upon the systems and contexts around her in a manner that is active and creative.

The term "feminist" is commonly assumed to be unknown among Latinas and to be rarely used, yet there is a substantial history of Latina girls and women throughout the United States and Latin American countries who have stood up for equal rights for women and fought against patriarchy at all levels (Arellano and Ayala-Alcantar 2004). Contemporary research has documented that, among Latina adolescents, there is a strong sense of feminism and awareness of sexism, even if they do not use the specific term "feminist" to describe these concepts in their lives or their community (Manago et al. 2009). Some research describes cultural concepts of *marianismo*, based on the Virgin Mary as a role model, which is said to create a cultural script that describes women as pillars of their family, self-sacrificing and virginal (Castillo et al. 2009). In fact, some authors have gone as far to describe the "Maria paradox" where young Latinas feel trapped between traditional gender values for women and modern contemporary opportunities (Gil and Vasquez 1996). Yet, there are some who do not even agree with this basic representation of the Virgin de Guadalupe; rather, some portray her as subversive, powerful, and a voice of the indigenous, oppressed, and poverty stricken (e.g., see Castillo 1996). Furthermore, we argue that portrayals of any culture in such Black and White perspectives on gender gloss over the authentic complexity of gender and the diversity of gender socialization. Additionally, it is possible that while some women might agree with the *marianismo* values, it is likely that the values do not necessarily reflect the daily behaviors that women engage. Research has demonstrated that the behaviors of women in Mexico and the United States, as well as men in both countries, do not necessarily align with these values (Carrillo et al. 2010). Moreover, an essentialization of Latina women's experience to *marianismo* contributes to the silencing of the ways in which women subvert and transform their interpretations of societal gender values. Thus, in order to focus on resilience, or how young women overcome adversity, we will focus on the agentic ways in which Latinas navigate issues of sexism and rigid gender roles.

The MFF (Arellano and Ayala-Alcantar 2004) was created by two Chicana/Latina clinical psychologists in order to advance mental health research and services with women and men of color by providing a paradigmatic structure rooted in concepts of resiliency and strength in the face of subjugation. As we discussed in Chap. 1, cultural deficit ideology limits the understanding of internal knowledge and strengths that Latina adolescents have learned through overcoming adversity (Pedrotti and Edwards 2009; Pedrotti et al. 2009). MFF is consistent with cutting-edge models of ecodevelopmental frameworks (see Fig. 3.1) in that it includes several levels of ecological contexts (Zayas et al. 2005). As such, micro- (individual), meso- (community) and macrosystems (societal and policy-level) need to be considered in a way that addresses the historical and modern contexts of power and subordination to fully comprehend risk factors as well as cultural strengths of Latinos (de la Torre and Estrada 2001; García-Coll et al. 1996; Ramirez 1998; Szapocznik and Coatsworth 1999).

While power differentials cannot be overlooked, we must treat individuals holistically and acknowledge the agency of each person to navigate through all of these spaces in a manner that is strength-based. The model acknowledges the differences

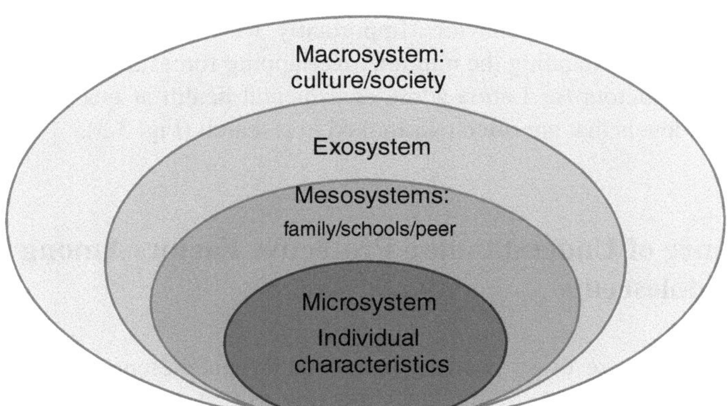


**Fig. 3.1** Bronfenbrenner's Ecological Development Model

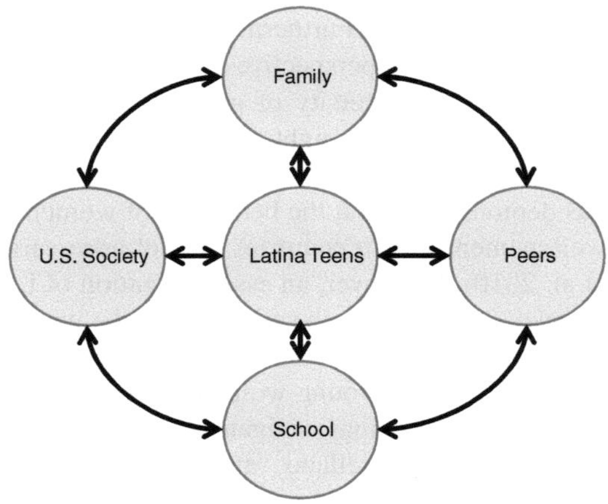


**Fig. 3.2** Latina teens agentic influences on contexts

in power and privilege in the United States that are based on race, gender, language, socioeconomic status, and sexual orientation. Latina women are likely to face multiple layers of subordination and prejudice within the United States and within their families and communities. Inequalities that result from power and privilege in the United States will be interlocking via the interactions between gender, race, class, and sexuality. Furthermore, MFF demonstrates how *identities and images of women of color* are essential to consider and critique because Latina women are often portrayed in stereotyped and negative ways in society and media; however, women's identities are not only centered within their gender and messages of society; rather, they are derived from multiple facets of their life and experiences that are also positive and empowering.

Despite these challenges, MFF states that the center of the entire model is based on the *resilience and strengths of women of color* that are derived from generations of transforming and subverting social structures within society and their families (Delgado-Bernal 2001) (see Fig. 3.2). This final model helps us to situate Latina girls as active agents in their world. Moreover, it helps us identify protective factors and sources of resilience that can help Latinas overcome depression and suicidal thoughts and behaviors. Importantly, we will apply the core elements of the MFF to understanding the multiple overlapping forces that influence the risk and protective factors for Latina adolescent mental health at macro-, meso-, and microsystem levels that are often overlooked in research (Fig. 3.3).

## Importance of Understanding Protective Factors Among Latina Adolescents

Protective factors are those characteristics that, when present, serve as a buffer between a stressor and the individual's reaction to that stressor. They reduce the impact of the stressor, or chronic adversity, and thus act as a shield that protects the person from the potential negative outcomes and allows for a more adaptive

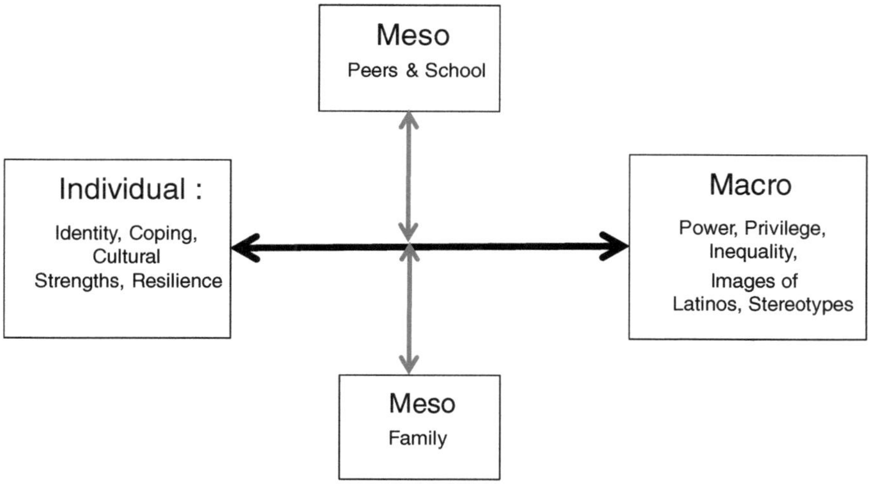


**Fig. 3.3** Multiracial Feminist Framework

response. According to Dekovic (1999), protective factors might be better conceptualized as "compensatory factors," as they appear to be beneficial for all adolescents, regardless of the presence of risk factors.

The concept of protective factors grew out of research on resiliency. This research sought to understand how some individuals who experience negative life events become depressed and suicidal, while others are able to continue to function or even gain strength from managing adversity. Years ago, Rutter (1987) questioned whether protective factors were simply risk factors with the opposite valence. For example, *family conflict* could be a risk factor, while *family harmony* would be a protective factor. We argue that while this position has merit, there are protective factors that are more robust and more than simply the opposite of the corresponding risk factor. A cumulative risk perspective (Forehand et al. 1998; Rutter 1979; Sameroff et al. 1997) would suggest that as the number of risk factors increases, the level of maladjustment for youth will also increase (Loukas and Prelow 2004). For Latina youth, whose cultural contexts include daily experiences of discrimination, lowered expectations, and bicultural challenges, in addition to poverty and poorer health status, the number of risks that they experience can be quite high. Importantly, it has been noted that protective factors can lessen the impact of a risk irrespective of the level of risk experienced (Klein and Forehand 2000). Again, research suggests that Latina youth may be faced with numerous challenges, but that protective resources are effective in buffering the negative effects of these risks on their mental health.

*The Role of Family.* The centrality of family has been noted as one of the most prominent cultural values in the lives of Latinas (Gloria et al. 2004). *Familismo*, or familism, has been described as the strong feelings of commitment, loyalty, and obligation to family members that extends beyond the nuclear family (Arredondo and Perez 2003; Gloria et al. 2004; Lugo Steidel and Contreras 2003). The interdependent nature of family (including extended family) involves the sharing of parenting tasks and financial responsibility and includes family-based problem solving and the priority of family needs over individual needs. Peña et al. (2011) describe four attitudinal components of familism: (1) the needs of the family are more important than the needs of individual members; (2) members of a family should remain emotionally close and involved in each other's lives; (3) members can rely on one another and can depend on family members when needed; and (4) family members are expected to protect the family name and family honor. The family context is of unique relevance during the normative development of identity for Latina girls. Familism's cultural values that prioritize close family relationships and unity among extended family members also translate into behaviors, such as parental and family closeness and monitoring, that are associated with fewer risk behaviors (Romero and Ruiz 2007). Given the strong cultural values of familism within all Latino cultures, perhaps it is not surprising that family context can have the strongest impact on adolescent girls' mental health in both risk and protective ways (Arellano and Ayala-Alcantar 2004; Zayas et al. 2005). Although *familismo* may increase resiliency at some levels, it may increase vulnerability to stressors at other levels (Luthar et al. 2000).

This attachment to family has generally been considered a protective factor in the lives of Latinas youth and adults, buffering them against the negative effects of stress and mental health problems. The role of families has been the most frequently researched proximal variable with respect to protecting against depression and suicidality. For example, among adults, large-scale prevalence studies have shown that familism is associated with decreased lifetime suicidality among Latino adults (Fortuna et al. 2007). Studies have also found that low levels of family involvement are related to higher rates of psychological distress among Mexican immigrants (Rodriguez et al. 2007) and Latino/a college students (Castillo et al. 2004). Cultural values of familism are pervasive in most Latino cultures and influence child psychopathology and help-seeking behavior (Abad 1987; Badillo-Ghali 1977; Canino and Canino 1980; Comas-Diaz 1988; Cuéllar and Glazer 1995; Gloria et al. 2004). Since family members often live close by and have a sense of obligation toward each other (Molina and Aguirre-Molina 1994), it is not surprising that health and sickness both become family affairs. Often family members are the ones that teens turn to first for help seeking (Caldwell et al. 2008); however, immigration may disrupt family patterns of social support and help seeking.

For Latino/a adolescents in particular, families play a role in monitoring and involvement in youth behavior and providing social support, among other things. Additionally, researchers have suggested that keeping family ties and having a strong sense of familism allows Latino/a adolescents the ability to sustain their cultural values (Dawson et al. 2012). The role of family can be complex and may influence outcomes in different ways, suggesting that understanding the particular kind of family-related behavior is important in interpreting its effects. For example, feeling a sense of support from one's family is clearly different than having a parent who monitors one's behavior and provides structure and routine in the home. In recent research, authors have attempted to better understand the behaviors that family members do and how they influence mental health outcomes of Latino/a youth.

With respect to social support, a study of 254 first-generation, immigrant Latino/a adolescents demonstrated that the risk of depression was most strongly associated with having experienced discrimination in the United States and that social support at home ameliorated the negative effects of discrimination and other stressors (Potochnick and Perreira 2011). Similarly, studies with young Latinas showed that perceived mutuality (bidirectional exchange of feelings and thoughts) between mother and daughter was associated with decreased internalizing behaviors and suicidal behaviors (Turner et al. 2002; Zayas et al. 2011).

In a study using the Reach for Health dataset (O'Donnell et al. 2004), 879 Latino and African American youth were surveyed on prevalence and predictors of suicidality. Across all youth, family closeness was a strong predictor of both ideation and suicide attempts, above and beyond having an intact (two-parent) family. In other words, feeling that family was available and supportive in times of need was protective for youth, regardless of the actual composition of their families (O'Donnell et al. 2004). In contrast, adolescent Latinas who made suicide attempts had lived fewer years with their fathers than non-attempters (Turner et al. 2002). While the majority of research has focused on mother–daughter dynamics, it is apparent that the fathers also play an influential role. However, significantly less research has investigated the positive sources of resiliency and socialization that fathers can provide; in fact, some empirical research is demonstrating that traditional views of Latino fathers are stereotyped and misrepresent their parenting behaviors (Gonzalez-Lopez 2004).

In another study about poverty-related risks, family routines, mother–child relationship quality, and maternal monitoring were explored among 520 low-income Latino early adolescents (Loukas and Prelow 2004). Providing support for the cumulative risk framework, girls exposed to four or more risk factors (e.g., neighborhood disadvantage, single-parent household) had scores on internalizing problems (e.g., depression, anxiety) 2.5 times higher than their female peers exposed to zero risk factors. Findings also revealed that among girls, family routines (consistent routines for mealtimes, homework, and bedtimes) served as a protective factor, whereas among boys, the mother–son relationship quality was protective.

These specific studies provide support for the protective nature of a sense of *familismo* and family-related behaviors on Latinas adolescent depression; however, it is important to note that recent research has shown that familism can have a "double-edged effect" (Kuhlberg et al. 2010), contributing to positive outcomes but also symptoms of distress in youth. In early research about Latina middle school–aged adolescents in Florida, the dual aspects of support and conflict within families were found to account for the majority of differences in internalizing and externalizing behaviors (Coatsworth et al. 2000).

In a recent study of 226 adolescent Latinas (Kuhlberg et al. 2010), higher levels of internalizing behaviors predicted suicide attempts, and self-esteem protected against these attempts. Familism was found to be a protective factor against parent–adolescent conflict, and parent–adolescent conflict was in-turn related to higher levels of internalizing behaviors and lower self-esteem. However, familism was also found to predict higher levels of internalizing behaviors, which was unexpected. In other words, Latinas who perceived more conflict with their parents also reported higher levels of internalizing behaviors and lower self-esteem, which led to suicide attempts through indirect paths. These findings are explained by the authors as a possible case of stress from family being *turned inward*. Specifically, as conflict with parents escalates, Latinas may feel guilty for not conforming to the deferential posture expected of them, including traditional gender norms that urge them to sacrifice autonomy for their parents and family. This guilt may lead to the higher reported levels of internalizing behaviors and lower self-esteem. These interesting findings suggest that familism can be protective and enhance self-esteem, but can also reduce the conflict between adolescents and parents.

As research has progressed in this area, authors have begun to make more nuanced explorations of within-group variables that might influence the relationship between stress, protective factors, and depressive symptoms. In one of the few studies that has probed within-group differences among Latino adolescents, Dawson and colleagues (2012) studied the association between family involvement and depressive symptoms over 5 years among 187 Mexican, Dominican, and Central American immigrant youths. Results for gender were similar to other studies, with Latinas more likely to experience depressive symptoms than males across all ethnic groups. Findings also revealed that among Dominican adolescents, lower family involvement was more strongly associated with depression than for Mexican or Central American groups. Though one of the only studies that has examined group differences, this research suggests that important differences may be muted when we group all Latino adolescents together (Villarruel et al. 2009; Umana-Taylor and Fine 2001), and family involvement may be more a more important protective factor to accentuate for Dominican adolescents than for Mexican or Central American groups.

*Biculturalism.* As had been discussed elsewhere in this volume (Chap. 2), understanding the role of acculturation and acculturative stress among Latino/a adolescents is fundamental to adequately address the mental health needs of this population. Researchers have speculated that differences in acculturation between Latino parents and their daughters may lead to depressive symptoms and/or suicidal behaviors (Zayas 1987; Zayas and Pilat 2008). Specifically, studies with Latino/a adolescents have indicated that experiencing acculturative or bicultural stress has been associated with higher levels of depression among immigrant Mexican–American adolescents (Hovey 1998; Hovey and King 1996), and Latina adolescents (Romero et al. 2007). The bicultural orthogonal model of cultural orientation argues that adaptation to the new culture and maintenance of the original culture includes more than one continuum of culture changing at the same time (Keefe and Padilla 1987; Oetting and Beauvais 1990–1991; Rivera 2008; Schwartz et al. 2010). Thus, bicultural individuals can maintain their culture of origin while adapting to a new culture at the same time. A significant component of being bicultural for many Latinos is the ability to speak both Spanish and English fluently. Thus, we see in many cases that being bilingual is a protective factor for Latinos and is now linked with many other physical and developmental sources of positive development (Craik et al. 2010; Barac and Bialystok 2011; Brice et al. 2008; Buriel et al. 1998).

With respect to acculturation, research has shown that certain orientations are more adaptive than others. In a study of 705 Latino/a adolescents, youth who were bicultural (those with high levels of connection to Mexican culture and mainstream, U.S. culture), as compared to those who were considered marginalized (not connected to their culture of origin or mainstream, U.S. culture), had less depressive symptomatology and were more likely to be optimistic (Carvajal et al. 2002). Biculturalism appears to be healthy, and it is perhaps the most authentic representation of the manner in which an individual navigates more than one culture on a regular basis (Berry 1980; Delgado-Gaitan 1994). Research demonstrates that bicultural individuals seem to have the most adaptive mental health; in contrast, the worst health outcomes are noted among marginalized people, who reject, or are rejected by, both cultures (Kim 2007; Negy and Woods 1992; Schwartz et al. 2010). Being bilingual and bicultural may be highly adaptive for Latinos in the United States because they may continue to have regular interactions with both the U.S. dominant culture and their culture of origin. This may be due to the proximity of Latin American nations, and via multi-generational families, family, or friends, or within ethnic enclaves in the United States (Carvajal et al. 2002). Taken together, these findings suggest that a bicultural orientation, which is reflected in a balance of heritage and host cultural orientation, is protective for Latino/a youth.

*School Connectedness.* School connectedness, or a sense of belonging to one's school, has been found to be negatively related to emotional distress and suicidality in several studies (Resnick et al. 1997; Borowsky et al. 2001). Other studies, however, have not resulted in similar findings about the importance of school connectedness as a protective factor (Loukas et al. 2009; O'Donnell et al. 2004), and one study found that self-esteem accounted for most of the variance in the relationship between school connectedness and depression (Millings et al. 2012). Lower teacher support was associated with higher risk of suicide attempt among Latina adolescents, which may suggest that teacher support can be protective (DeLuca et al. 2012). The difficulty in determining whether attachment to school serves as a protective factor is that many studies that include this variable either did not include Latinas or did not disaggregate the data by gender/ethnic groups; clearly, more research is necessary in this area.

There have been dramatic increases in Latina education achievement in the past 40 years. In 1970, one-third of Latinas were enrolling in college, whereas by 2007, about half were enrolling in college, a sevenfold increase (Fry 2009). In 2012, a record seven out of ten Latinos (69 %) enrolled in college putting them 2 % points higher than White non-Hispanics in terms of higher education enrollment (Fry and Taylor 2013). The rates of enrollment have been higher among Latina women than Latino men (Fry 2009). While achievement of higher education is associated with higher wages and better employment opportunities (stability and health insurance), this rapid sociological change may also impact the internal family dynamics of Latinos. In 2008, 42 % of Latino children were living with mothers who did not have a high school diploma, the increase in maternal education has been happening rapidly for Latinos with a drop of 10 % points for children living with mothers who did not have a high school diploma between the years 2000 and 2008 (Mather and Foxen 2010). More research will need to be conducted to ascertain the influence of educational on Latina mental health and suicidality.

## Individual Variables

Several factors at the level of the individual have been investigated as potential protective factors against depression in the lives of ethnic minority (and Latina) youth, including coping strategies, self-esteem, socioemotional competence, and ethnic identity. Below, we describe each of these constructs and relevant research in the area.

*Coping strategies.* Young Latinas often develop coping strategies to deal with stressors (e.g., discrimination, stereotypes, family conflict) in their lives, and researchers have attempted to better understand the specific strategies that are effective in helping youth cope. The ways in which the youth manage those stresses can influence their vulnerability or resilience to suicide and depression (O'Donnell et al. 2004). Coping strategies refer to actions taken to manage stressful situations. Coping strategies can be positive (i.e., effective at reducing stress), or negative (ineffective). The following strategies are considered to be positive: distraction, cognitive restructuring, problem solving, emotional regulation, emotional expression, social support, and acceptance. Negative strategies include social withdrawal, self-criticism, wishful thinking, and blaming others. In one study of Latino/a youth, results indicated that primary engagement coping strategies (e.g., make a plan, talk to the person) buffered the effect of discrimination stress on depressive symptoms (Edwards and Romero 2008). In a recent qualitative study, Latina adolescents utilized primary engagement coping strategies (Compas et al. 2001) to deal with discrimination stress (Edwards et al. under review). Specifically, they used acceptance/letting go for discrimination stressors and compensation (e.g., proving others wrong) for others' lowered expectations.

In a qualitative study about Latina girls and stereotypes (Denner and Dunbar 2004), participants reported being conscious of the negative effect of stereotypes and sexism in their family and societal contexts. These young Latinas report that they are attuned to these stereotyped and derogatory images of women through their mothers' descriptions of their own experiences of sexism in the workplace. The authors suggest that this process of socialization and consciousness-raising on the parts of these mothers may help young Latinas to actively resist the sexism and discrimination that is around them. Similarly, authors (Delgado-Bernal 2001) have suggested that the family environment teaches girls to negotiate struggles inherent in their social positions of gender, ethnicity, and class through the communication of narratives about resistance. These narratives are shared in legends, storytelling, *corridos*, prayer, and the wisdom of older women (mothers, *tias*/aunts, *abuelas*/grandmothers, *madrinas*/godmothers) in the family.

Researchers have proposed that minority individuals, by virtue of the need to survive in a less-than-hospitable climate, develop coping mechanisms that actively reduce stress (O'Donnell et al. 2004). Some evidence suggests that a passive coping style is associated with depression and suicide (Canino and Roberts 2001). Turner et al. (2002) measured coping strategies in a sample of 31 Latinas aged 12–20 who were receiving mental health services. Subjects responded to a scenario in which they want to do something they know their parents will not permit. Those who had made a suicide attempt chose significantly more negative strategies than those who had not made an attempt. This finding is particularly important because we know that suicidal behaviors often follow conflict in the family; although the scenario in the study did not describe a conflict, it was clear that if the girl decided to join her friends, a conflict would ensue. Learning how to cope with these situations in an effective way could be an additional strategy for reducing suicidal behaviors in Latina adolescents.

*Socioemotional competence.* In a study of 510 Latino/a early adolescents, Loukas and Prelow (2004) found that socioemotional competence, or a prosocial/sociable orientation, served as a resource for girls by protecting against the negative effects of risk factors on internalizing symptoms such as depression. Specifically, socioemotional competence served as a resource for young Latinas, but not necessarily for Latino boys. These findings highlight the importance of investigating gender differences and within-group variation among young Latinos to better understand the complexity of influences on depressive symptoms. Clearly, socioemotional competence has been identified in both the interpersonal theory of suicide and Zaya's ecodevelopmental model. Thwarted belongingness and perceived burdensomeness may be reduced through socioemotional competence, or both may be exacerbated if the youth lacks it.

*Self-esteem.* In recent research with a large ethnically diverse sample of adolescents, Costello et al. (2008) found that self-esteem at baseline was associated with a higher probability of membership in a non-depressed mood group compared to a depressed mood group. Low self-esteem has also been found to predict future depression at follow-up of up to 5 years in adolescents and young adults (Orth et al. 2008). Studies that focus more closely on Latinas would be important contributions to the literature.

*Ethnic identity.* Ethnic identity and its protective role on the self-esteem of Latino adolescents have been investigated in several studies. Phinney and Kohatsu (1997) describe ethnic identity as, "a complex, multi-dimensional concept including, at a minimum, self-identification, a sense of belonging and commitment to one's ethnic group, and the cognitive and affective meanings of one's group membership" (p. 422). In a review of 21 studies done with Latino/a adolescents, Umaña-Taylor et al. (2002) found an overall positive relationship between ethnic identity and self-esteem. More recently, Umaña-Taylor and Updegraff (2007) found that aspects of ethnic identity positively predicted self-esteem among Latino/a adolescents and that self-esteem partially mediated the relationship between perceived discrimination and depressive symptoms. It seems, therefore, that ethnic identity is a valuable resource for Latinas, similar to self-esteem. In one of the few studies to investigate suicidality (O'Donnell et al. 2004), however, ethnic identity was not found to be a resiliency factor. The authors note that because their sample resided in a neighborhood comprised of participants' own ethnic group, ethnic identity levels were high so perhaps this variable was less influential.

While ethnic identity is experienced at the individual level, a collective sense of cultural identity may also serve as a protective factor for young Latinas, particularly if they engage in civic and collective action. Recent qualitative data suggest that educational success among Latina girls is associated with involvement in activism, which builds on cultural strengths and resistance to traditional gender roles (Cammarotta 2007). Additionally, Latino youth activism and civic engagement have been linked to fewer depressive symptoms and higher self-esteem (Romero and O'Leary 2011). Linking a positive ethnic identity with larger action against negative societal stereotypes or larger issues of discrimination and sexism may lead to greater resilience and empowerment of young women. These concepts are worthy of future quantitative research.

*Religiosity.* Religion and spirituality have been examined in relation to numerous indicators of health, well-being, and adaptive behavior in adolescents. A literature review by Milevsky and Levitt (2004) showed a negative relationship between religiosity and rates of adolescent drug use, suicide attempts, depression, psychological distress, and engagement in sexual intercourse, and positive correlations were noted among religiosity, self-esteem, and self-concept. Overall, the literature has supported the idea that religious and spiritual practices and affiliations in the lives of adolescents may serve as important resources during this period of development, though more research is needed with ethnically diverse samples.

Only a handful of studies have investigated the role of religiosity as a protective factor against suicidality among Latino youth. In one study of influences on suicidal ideation in 114 Mexican immigrants to the United States (Hovey 2000), findings suggested that low levels of religiosity and infrequent church attendance were correlated with elevated levels of depression, although the perceived influence of religion on participants' lives was not. However, all three variables were associated with high levels of suicidal ideation, suggesting that religion is protective against suicidal behaviors. In a study of risk and resiliency against suicidality among Latino and African American youth (O'Donnell et al. 2004), the influence of religiosity was only marginally significant but still somewhat protective against suicidal thoughts and behaviors.

## In Summary

As can be seen from this review of protective factors, researchers are only beginning to understand the possible resources that may help young Latinas prevent depression and suicide; there are promising variables that have been identified.

The majority of research in this area has focused on the role of family, which is logical given its prominence in Latino culture. Indeed, the conflicts that arise because of the convergence of cultural expectations and developmental, and social and individual factors have been posited to play a role in the development and maintenance of depressive symptoms and suicidality (Zayas and Pilat 2008). Research has also focused predominantly on depressive symptoms, while more recent studies are beginning to explore the role of protective variables on suicidality. Finally, more research is warranted that explores the experience of young Latinas and their cultured and gendered contexts in particular.

# Chapter 4
# Risk Factors for Latina Adolescents' Mental Health and Well-Being

Understanding how various risk factors influence the decision to take one's own life is a critical aspect of being able to plan and implement effective prevention programs, interventions, and recovery treatments. According to the Center for Disease Control (2012), known risk factors for suicide include the following: a history of prior attempts, abuse or mental illness (particularly depression), exposure to suicidal behaviors in others (especially if there is a family history of suicide), exposure to a relatively recent life crisis or loss, feelings of isolation or hopelessness, alcohol or substance abuse, access to lethal methods, impulsive or aggressive tendencies, cultural beliefs that promote suicide as a resolution to certain problems, physical illness, and barriers to or unwillingness to access mental health treatment (CDC 2012). For Latinos in general (not solely limited to Latina adolescents), risk factors for suicide include the following: alcohol use (CDC 2009; Cannon and Levy 2008), alienation (Olson 2011), barriers to mental health services (Larkin 2011; Cabassa 2006), hopelessness and fatalism (Jamieson 2008), ineffective coping strategies (Olvera 2001), acculturation stress (Peña 2008; Zayas 2008, 2010), firearm access (Spicer 2000), affluence and mobility (Wadsworth and Kubrin 2007), divorce (Wadsworth and Kubrin 2007), and immigrant status (Wadsworth and Kubrin 2007). Recognizing risk and protective factors for depression and suicide in Latina adolescents reveals opportunities for prevention and intervention (see Chap. 3, this volume, for a discussion of protective factors). In this section, we review what is known about factors that place these youth at greater risk for developing depression and engaging in suicidal behaviors.

Both the Multiracial Feminist Framework (MFF; Arrellano and Ayala-Alcantar 2004) and Zaya's (2005) Ecodevelopment Model of Suicide Attempts among Latina adolescents use an ecological framework. Based on ecological perspectives, we envision that Latina adolescents utilize their individual characteristics to navigate microsystems, which include family, peers, and school. There are also unique interactions in the mesosystem between the young Latina and her family and school, family and peers, and peers and school. Importantly, the MFF reminds us of the U.S. societal norms and perspectives that also shape the acculturation processes,

A. J. Romero et al., *Preventing Adolescent Depression and Suicide Among Latinas*, SpringerBriefs in Psychology, DOI: 10.1007/978-3-319-01381-7_4, 

stereotypes of Latina girls, and policy-level factors of immigration. Thus, we will begin first with macrosystem factors that have been found to increase stress and risk for depression and suicide among Latina adolescents. While we already discussed in Chap. 3 the cultural norms of Latino families, family-based strengths, and protective factors, in this chapter, we will highlight research that introduces dysfunctional family typologies and family systems' (mother–daughter dyad) approaches to understanding internal family conflict that is associated with depression and suicide attempts. Lastly, we will include research that describes individual-level behaviors that may contribute to depression and suicide, such as alcohol and substance use.

## Macrosystem-Level Risks and Influences

Latinas face prejudice, sexism, homophobia, discrimination, negative stereotypes, and the lasting intergenerational effects of historical traumas. Additionally, the political climate in the U.S. has become increasingly anti-immigrant, and even those born and raised in America can find themselves as targets of discriminatory political agendas. Developmental scholars argue that in order to understand the normative experience of underrepresented ethnic minority adolescents, we must consider the effects of prejudice and discrimination because they are so pervasive (Eccles et al. 2006; García-Coll et al. 1996; Spears-Brown and Bigler 2005). Negative images and stereotypes not only essentialize Latina teens, but may also lead to prejudice or discrimination related to race, language/accent, cultural and gendered stereotypes, or immigration status (García-Coll et al. 1996; Neimann 2004). *Prejudice* is defined as the negative attitudes toward ethnic groups; *discrimination* is characterized by unfair actions toward people/groups based on ethnicity; *sexism* is unfair treatment based on gender. Discrimination and prejudice effects can range from major trauma (e.g., abuse from police/border patrol) to daily hassles and microaggressions[1] (Meyers 2004; Romero and Roberts 2003a; Sellers and Shelton 2003). Prejudice and discrimination negatively affect mental health of U.S. minority ethnic groups across the lifespan, beginning as young as elementary age children through older adults (Eccles et al. 2006; García-Coll et al. 1996; Spears-Brown and Bigler 2005). It can lead to feelings of devaluation that can negatively affect self-esteem and mental health (Flores 2013).

Additionally, the overlapping nature of social positions uniquely impacts Latina girls; in other words, prejudice and discrimination are influenced not only by outward

[1] "Racial microaggressions are brief and commonplace daily verbal, behavioral, or environmental indignities, whether intentional or unintentional, that communicate hostile, derogatory, or negative racial slights and insults towards people of color." Those who inflict racial microaggressions are often unaware that they have done anything to harm another person. From: Sue D.W., Capodilupo, C., Torino, G, Bucceri, J., Holder, A., Nadal, K., & Equin, M. (2007). Racial Microaggressions in Everyday Life: Implications for Clinical Practice. The American Psychologist, 62 (4) 271–286.

racial phenotype, but also by gender, accent, language fluency, socioeconomic status, immigrant status, and sexual orientation (Edwards et al. under review; Romero and Roberts 2003). Latina girls may experience discrimination, sexism, and prejudice in multiple contexts ranging from family members, peers, teachers, store clerks, and police to border patrol agents (Gonzalez and Kim 1997; O'Leary 2009). There are indications that gender-specific issues of sexism, gender identity, and gender roles all play a critical role in young Latinas' lives, development, and mental health (Arellano and Ayala-Alcantar 2004; Arredondo 2004; Céspedes and Huey 2008). Espín (1997) describes how homophobia within the context of racism and sexism may further contribute to pain, anger, and frustration among Latina girls that can lead to negative mental health outcomes. But is it invisibility rather than discrimination that many Latina lesbians may navigate. Espín (1987) writes that while most Latinos do not reject homosexual children, there is an invisibility rooted within a "silent tolerance" that can influence family dynamics. However, there are very few studies on Latina queer youth, and one national study reports that they do not have any greater likelihood of mental health problems than heterosexual Latina adolescents (Consolacion et al. 2004).

Discrimination and prejudice against Latinos have become more noticeable in recent years in media (Santa Ana and de Bustamante 2012). Since September 11, there have been several systemic policies enacted in response to rapidly increasing Latino populations. These policies deal with language accessibility in schools and government institutions, racial profiling, and anti-immigrant policies (O'Leary 2009). There have also been increases in deportation raids targeting undocumented immigrants in the U.S., many of whom are from Latin-American countries (predominantly Mexico and Central America; Capps et al. 2007; Shields and Berhman 2009; Thompson 2008). These raids often result in traumatic separation of parents from children and leave children in vulnerable situations without parents or without documentation (Thompson 2008). Children and families can be caught in a tangled web of legal systems and policies that span across agencies, jurisdictions, countries, and languages. Given the high rates of Latino children (52 %) that live in families with at least one immigrant parent, these are grave considerations for their well-being. The impact of the immigration enforcement policies, such as Arizona's SB 1070, can lead to chronic daily stress to not reveal immigration status. Families often discuss with children what do if a parent is unexpectedly deported, which may increase children's fear and stress. Moreover, the acute traumatic stress of a parent or parents' deportation or related incarceration may severely negatively impact depression and increase hopelessness. Furthermore, it is this situation that may also lead to long periods of time where family members are not living in the same place and are limited in their visitation opportunities. It is not only the absence, but also the reunification of families during teen years that can set in motion internal family turmoil. Emerging empirical research has been documenting how anti-immigrant political climates and the associated rhetoric increase negative stereotypes, discrimination, racial profiling, fear, stress, and barriers to seeking help for Latino populations (McNeill et al. 2001; O'Leary 2009).

## Stereotypes

There are very few images of Latinas in U.S. media, and as they are often negative, they can have a negative influence on Latino mental well-being (Mastro et al. 2008). Research has demonstrated that women of color are even more attuned to media images of women of color and may be even more likely to be attracted to those images (Stephens and Few 2007). Some of the stereotyped images of Latina women in the U.S. society that continue to be represented in media reflect the overlapping nature of social positions and may include the image of the self-sacrificing passive mother based on the Virgin Mary, a maid/domestic worker, a pregnant adolescent, or the hot sexy Latina image (Arellano and Ayala-Alcantar 2004; Greenberg and Mastro 2008; Merskin 2007; Niemann 2004). Additionally, transnational and bilingual media, such as Spanish language telenovelas, bilingual hip hop, or Spanish language radio/TV stations, (which are targeted to Latino populations), often represent Latina women in a stereotypical manner and with an overemphasis on idealized body images and sexuality that may leave Latina girls without many other healthy images (Mayer 2003). Although Latina teens (15–19 years old) used to have disproportionately high pregnancy rates, those have decreased significantly over 11 years (from 1999 to 2010), from 87 per 1,000 to 56 per 1,000 (compared to 51/1,000 for African-American teens and 23/1,000 for White non-Hispanic teens) (Fig. 4.1). This decrease may be in part due to the rise in pregnancy prevention programs targeted to Latina teens and the rise in educational achievement of Latinas. Other impacts from stereotypes of women and Latinos are revealed in experimental study designs, which show negative impacts on academic performance and decrease in working memory capacity (Schmader and Johns 2003; Schmader et al. 2001). All of these factors may contribute to Latina adolescents' mental health inequalities through increasing adolescent risk factors, such as prejudice and discrimination, and decreasing self-esteem.

**Fig. 4.1** Reduction in Latina teen birth rates over 11 year period. Rates are per 1,000 teens 15–19 years old

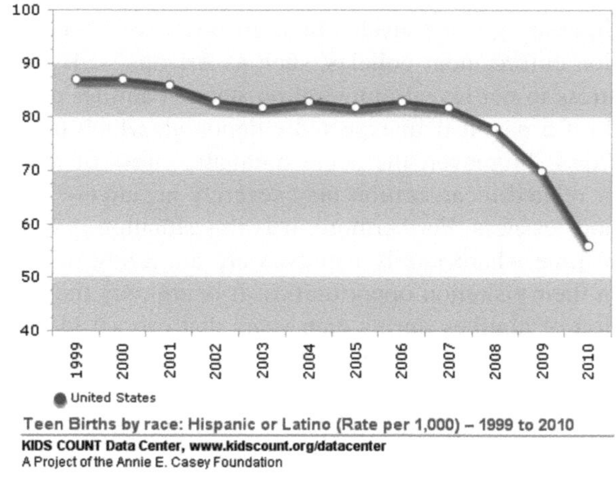

Research has found that stereotypes can create unconscious bias, which if remaining unconscious, can lead to discrimination or bias against individuals within the targeted group (Greenwald et al. 2009). Unfortunately, these biases can and do extend into professional settings. For example, among clinicians and in mental health counseling settings, these unconscious biases may lead to stereotypes, prejudice, or discrimination that may perpetuate documented inequalities in mental health help-seeking, treatment quality, and adherence to maintaining treatment (Arredondo 2004; Arellano and Ayala-Alcantar 2004). Qualitative research has documented that Latina girls report being conscious of the negative effect of stereotypes to an extent that, in addition to noticing stereotypes at a societal level, they also notice sexism and unfair treatment based on gender within their own family contexts (Denner and Dunbar 2004; Edwards et al. under review). They report that their mothers describe their own experiences of sexism in the workplace to them, perhaps as a means to socialize them and raise their consciousness (Denner and Dunbar 2004).

Stereotyping is not something unique to Latina adolescents, but it does impact their mental well-being. Latina teens face stereotyped images about both their gender and their ethnicity; as noted by Multiracial Feminist Framework, these images influence the identity and well-being of young women. While being aware of stereotypes can be a starting point for consciously resisting their effects, it is also the case that negative stereotypes of Latinas perpetuated by individuals and media can have lasting negative effects on the development of Latina children and adolescent behavior, social interactions, and internalized identity (Greenberg and Mastro 2008). The internalization of stereotypes occurs when individuals believe the negative generalizations about their group, which may lead to feelings of shame about one's ethnic group, difficulty developing a healthy/positive ethnic identity, and lower self-esteem in general (Niemann 2004). One aspect of prejudice that can be particularly painful is rejection felt within one's own ethnic, gender, or family groups, such as when individuals internalize stereotypes and perpetuate discrimination or sexism against individuals who share other social groups with them (Romero and Roberts 2003a). For example, Latina adolescents may feel sexism within their own families or homophobia within their own ethnic or religious groups (Comas-Diaz 2000; Espín 1997). Research suggests that identity integration where one's multiple identities (racial, gender, and sexual orientation) can be successfully resolved, and thus, contribute to better mental well-being (Rosario et al. 2004).

## Stereotypes and Body Image for Young Women

Existing literature has not found many significant ethnic differences for internalization of media images or body satisfaction between White and Latina women (Grabe and Hyde 2006; Kornblau et al. 2007; Warren et al. 2005). One study did demonstrate that Mexican-American women showed less awareness of the thin ideal than both European-American and Spanish women (Poran 2002). However, they fell in between European-American and Spanish women on the internalization of media images of women's bodies. Latina women have been found to be very aware of U.S. consumerism to the extent that they mention it extensively in describing U.S. beauty ideals (Poran 2002). Frederick et al. (2007) found that self-objectification and appearance surveillance were stronger in women who most deviated from the slender White ideal: heavier women and women of color. Media images that include a greater representation of women of color, such as hip–hop, also portray idealized images of the body that young women of color may be more likely to may internalize since the women on the screen look more like them. Women of color are not only aware of how they are represented in the media, but also may internalize those images. In a qualitative study of how first-generation Mexican-American girls interpret *telenovelas*, Mayer (2003) found that the girls felt that not only the title character's body was unattainable, but also it was a sign of her success in the mainstream. The findings on body image may be contributing to lowered self-esteem and depression among Latina youth.

## Stress of Discrimination and Prejudice

Discrimination experiences may be in the form of daily hassles, or microaggressions, that may take the form of derogatory jokes or being ignored, any of which may seem minor but has an additive quality of stress over time that can lead to significant negative influences on depressive symptoms (Folkman et al. 1986; Sellers and Shelton 2003). Discrimination experiences may also be more acute and severe, such as serious academic setbacks in school settings (discipline, drop outs, or failing grades) or severe forms of abuse and trauma at the hands of police or border patrol agents (Capps et al. 2007; Romero and Roberts 2003; Shields and Behrman 2009; Thompson 2009). In fact, in recent years, changes in immigration policy have increased Latino's interactions with border patrol and police across the nation, especially in states that border Mexico (O'Leary 2009; Thompson 2008). It is important to note that these more acute and traumatic forms of discrimination, especially those involving state and federal agencies, take place against a background of historical traumas and abuse, which may also contribute to impulsive responses such as suicidal behaviors. The U.S., as well as many Latin-American counties, has historic and current legacies of collective action of oppressed groups who fight societal change at broader policy levels (e.g., civil rights movement; Martín-Baro 1994; Munoz 2007; O'Leary 2009). Psychologists argue that in order to change the negative effects of modern prejudices on an individual's mental well-being, ethnic group identity development needs to be based on an understanding of the historical context of discrimination and positive cultural values (Martín-Baro 1994; Padilla and Perez 2003; Ramirez 1998). Research indicates that Mexican-American Studies courses that educate students with critical analysis of the history of ethnic minorities in the U.S. and that are based on positive contributions of Mexican Americans are associated with better educational outcomes (Cabrera et al. 2012). Empirical studies have shown that self-reports of discrimination and prejudice are associated with more depressive symptoms, less optimism, and less self-esteem among Latino girls and boys (Hovey et al. 1996; Hovey 2000; Romero and Roberts 2003; Romero et al. 2007a, b).

## Family Stress and Conflicts

Many aspects of the family environment are relevant to the discussion of risk for suicide attempts in Latina adolescents. As noted previously, acculturation discrepancies between parents and daughters are frequently mentioned as a risk factor, but economic status may affect this dynamic. For example, several studies cited by Zayas et al. (2000) note that fewer suicide attempts are reported by middle class compared to lower class Latinas. They suggest that as parents' income and education increase, their parenting style often becomes less authoritarian, which will diminish the level of intrafamilial conflict. In addition, as parents' education and income increase, they may have more access, awareness, and familiarity with outside resource help. Interestingly, Guiao and Esparza (1995) found in their study of suicide in Mexican-American teens that family adaptability was not a significant predictor of suicidal behavior, but the level of family cohesion was highly significant and second only to depression in predictive power (see also Peña et al. 2011). In Chap. 3, we described the cultural values of familism that may make certain cultural norms of Latinos unique, such as daily family interaction and extended family member involvement in raising children. Zayas and colleagues' (2000, 2005) ecodevelopmental models focus on the family as the most central factor in Latina adolescent mental well-being, in part because family conflict is perceived as very serious and troublesome in the context of Latino cultural values and norms of family closeness and unity. In the next section, we will focus on studies that describe what happens when some Latino families develop unhealthy patterns of interactions that may lead to depression and suicide among adolescent girls.

Family history for suicidal behaviors may also impact Latina teens' mental health. In the Rew et al. (2001) study, Hispanic girls in grades 7–12 had the highest rates of all ethnic/gender groups for suicide attempt by family member and also for completed suicide by family member. They had the highest rates of suicide attempts by friends as well, but not for completed suicide by friends. Additionally, Latina girls are more likely than Latino boys or other ethnic groups to report family members have committed suicide, which may be a risk factor for their own depression or suicide attempt (Rew et al. 2001). Family conflict is often found to be the precipitant for suicidal behavior in adolescent Latinas (Gulbas et al. 2011). An interesting study categorized the types of family relationships found in a sample of Latina adolescents (122 of whom had attempted suicide and 110 who had not). From their extensive interviews with the adolescents and their mothers (and fathers when available), they described three patterns: *reciprocal* (in which everyone in the family contributes, and there is a strong sense of togetherness and *respeto*, which is respect from children to parents and also from parents to children); *asymmetrical* (relationships are one directional, with a member giving all to the family without receiving anything in return); and *detached* (lack of commitment to family, *respeto* is absent, authority expressed via physical abuse, or rules not enforced). The reciprocal families were almost all non-attempters (7 of 8), while all the detached families' adolescent daughters had made a suicide attempt (5). The asymmetrical families (11) were almost evenly divided between non-attempters (5) and attempters (6). The researchers believe this model can be used in clinical work with Latino families. The clinician could determine which pattern predominates in the family and help them increase the mutual expression of affection, understanding, and support. An important recommendation is to assist the family in developing realistic expectations for their daughters about dating, sexual behavior, and responsibilities at home, which are frequent sources of family conflict.

In a sophisticated analysis of data from 109 Latinas who had attempted suicide and 107 who had not (Peña et al. 2011), attempters reported more family conflict and less family cohesion than the non-attempters. Using latent class analysis, the researchers identified three family types: tight-knit (high cohesion, low conflict), intermediate-knit (moderate to low cohesion and moderate to high conflict), and loose-knit (moderate to low cohesion and moderate to high conflict). The attempters were significantly less likely to come from tight-knit families than from the other two types. Peña et al. (2011) speculate that in tight-knit families, Latinas are likely to have internalized the concepts of familism, and thus respond in culturally acceptable ways to parental expectations, leading to less conflict and more cohesion. This study is important because the findings include the perspective of parents as well as the adolescent girls. Much of the scant research relies on the girl's report only, which may differ significantly from a parent's report (Baumann et al. 2010).

Baumann et al. (2010) investigated 169 mother–daughter pairs; in half of the pairs, the daughter had attempted suicide. The two groups (attempters and non-attempters) were not significantly different on measures of acculturation and familism. Significant differences between mothers and daughters were found on measures of mutuality and internalizing (withdrawn, somatic complaints, anxious/depressed) and externalizing (delinquent and aggressive behavior). Mutuality with the mother, as rated by the daughter, was related to both internalizing and externalizing behaviors and suicide attempts. The mothers of girls who attempted suicide also rated their daughters high on internalizing and externalizing behaviors, while mothers of those who had not attempted suicide did not rate their daughters as high on those scales. Further analyses showed that low levels of mutuality predicted higher scores on internalizing and externalizing behaviors.

One specific source of intergenerational acculturation conflicts has been observed among Latino families in mental health counseling sessions and is termed *cultural freezing*. This phenomenon happens when older family members become culturally "frozen" when they immigrated, and they idealize the culture and gender patterns from this time period (Flores-Ortiz 1993). Fear of losing one's culture may accompany this experience that may result in more rigidity in enforcing cultural norms, gender roles, values, languages, and behaviors (Flores-Ortiz 1993, 2004). Cultural freezing may be a unique stressor for Latina girls who are learning to negotiate modern gender images and opportunities while elder family members may be resisting modern images based on a cultural argument (Céspedes and Huey 2008). These types of gendered and cultural conflicts may be particularly salient for Latina girls during the adolescent period, when U.S. teens are typically trying to increase their sense of autonomy and independence (Jensen-Arnett 1999). The development and expression of autonomy of Latina girls in the U.S. may be further exacerbated by cultural values that prioritize collective and family obligations over individual desires or goals (Gaines et al. 1997; Zayas et al. 2005). Some studies have indicated that Latinas perceive more family conflict and more cultural, gender-specific stressors than Latino males (Niemann et al. 2000). Family conflict has been identified as a major source of stress (Romero and Roberts 2003) and a precursor of depression and suicide attempts among Latina girls (Zayas et al. 2005). Given the cultural values of familism, it is likely that when faced with family dysfunction, Latina adolescents may feel that their safe base is crumbling. Moreover, if they feel that the family is falling apart because of their own behaviors, they are likely to have high rates of both thwarted belongingness and perceived burdensomeness as identified in Joiner's Interpersonal Theory of Suicide, which may lead to suicidal behavior. Latina adolescents may feel that there is no other option to resolve this level of internal family conflict other than to remove themselves from the family system.

## Peers and Bullying

Although the family exerts a strong influence upon Latinas' development and mental health, the influence of peer interactions cannot be overlooked. During adolescence, peer relationships begin to take on a more intimate and meaningful role in an individual's life compared to when they were children. In general, it can be very difficult when peer interpersonal conflict takes place, but it is made all the more complicated when the individual is also dealing with societal, cultural, and family stressors. Additionally, among youth in general, bullying-related behaviors and victimization also appear to have an impact on suicide ideation and attempts (Klomek et al. 2011; Kim and Leventhal 2008). Surprisingly, the influence of peers on suicidal behaviors in Latina adolescents does not appear much in the literature, unless one is referring to a break-up of a romantic relationship. Peers enter the equation at the point where friendships and romantic relationships create conflict with parents. However, one line of inquiry that has gotten only scant attention is the relationship between involvement in bullying and suicidal behaviors. Despite the fact that the relationship between bullying involvement and suicidal behaviors has been frequently observed among White adolescents (Patchin and Hinduja 2010; Schneider et al. 2011). Although the sample in the Patchin and Hindjua study included 23 % Hispanic/Latino participants, they did not examine gender differences by race. They concluded that suicidal ideation was significantly more common among non-Whites than Whites, but did not disaggregate the data any further. Schneider et al. did not detect any differences in bullying or victimization rates by race.

There is a complicated relationship between bullying/victimization, gender, and suicide risk. A recent review of over 37 studies that examined bullying and suicide among children and adolescents from the United States, Canada, several European countries, South Korea, Japan, and South Africa found that in almost all of the studies, a connection existed between being bullied and suicidal thoughts (Kim and Leventhal 2008). Research shows that even infrequent involvement in bullying behavior is associated with higher risk of depression and suicidality, particularly among girls (Klomek et al. 2007). There is evidence that relational victimization (where the victim's social relationships are the target of the bullying) has a greater impact on mental health status compared to overt victimization (e.g., name calling, physical bullying), and studies have generally shown that females tend to be more likely to experience relational victimization (Klomek et al. 2010). Students who are involved in frequent bullying—regardless of their role as the victim, perpetrator, or both—without concurrent depression or suicidal ideations at the time of bullying were found to later be at risk for depression and suicidality (Klomek et al. 2011; Klomek et al. 2007; Mills et al. 2004; van der Wal et al. 2003). Kaltiala-Heino et al. (1999) and Klomek et al. (2007) found that suicidal ideation was most common among bullies and victims, with no significant gender differences. Yet, one study suggests that girls who are victims of bullying are more likely to exhibit not only suicidal behaviors but also completed suicides compared to those who are neither bullies nor victims (Klomek et al. 2009).

One study suggested that being a victim of bullying is prevalent among Latino youth and comparable to other populations, and rates do not appear to vary by gender (Peskin et al. 2006). A study by Romero et al. (2013) of 650 self-identified female Latina Arizona high school students found that bullying rates are higher than previous studies have found for the following: victimization at school (23 %), cyber victimization (26 %), school bully (18 %), and cyberbully (18 %). Additionally, rates for depressive symptoms (49 %), suicide ideation (23 %), suicide planning (17 %), and suicide attempt(s) (13 %) are higher than national averages (see Chap. 1). After controlling for depressive symptoms, girls in the study who had been victimized by bullying were 1.5 times more likely to have attempted suicide than their peers who had not attempted. Those who bullied others were 1.5 times more likely to report suicidal ideation and 1.4 times to report making a suicide plan than those who had not made a suicide attempt. The link between bullying involvement and suicide merits more attention.

## Teen Dating Violence

Teen dating violence (TDV) refers to physical, emotional, psychological, or sexual abuse by one partner toward another in a dating relationship (Davis 2008). Similar to intimate partner violence (IPV), TDV tends to be the term utilized when describing adolescent relationships. Though prevalence rates among Latinas have varied, often due to issues of measurement and definitions, data suggest they experience IPV and TDV at equivalent or higher rates to women of other ethnic groups (Black et al. 2011) and that they experience the negative consequences of such violence. For example, in a study of domestic violence among ethnically diverse women in a shelter, a greater percentage of Latina survivors, as compared to non-Latinas, reported suicidal ideation or suicide attempts (Krishnan et al. 2001). In a recent study of IPV among Latina and non-Latina women (Bonomi et al. 2009), those who had survived partner violence scored lower on mental health and vitality measures, had higher physical symptoms, and were twice as likely to have depression than their non-abused Latina counterparts. Importantly, the adverse effects of partner violence on overall mental health were worse for Latinas than non-Latinas, suggesting a critical health disparity issue.

Research about TDV in adolescents (with all ethnic groups combined) has suggested that survivors are at an increased risk for negative outcomes such as substance use, risky sexual behaviors, pregnancy, sexually transmitted diseases, and depression and suicidal behaviors (Ackard and Neumark-Sztainer 2002; Coker et al. 2000; Roberts et al. 2005; Silverman et al. 2001, 2004). Among Latina adolescents, the consequences of TDV may even negatively influence the lives of Latina youth years after the actual experience of the dating violence (Teten et al. 2009).

Interestingly, in a study of acculturation and dating violence among Latino adolescents, young Latinas were less likely to hold pro-violence attitudes and were more knowledgeable about TDV and outcomes associated with dating abuse in comparison with their male Latino counterparts (Ulloa et al. 2004), highlighting the importance of understanding the intersection of gender role expectations and culture.

It should be noted that many of the correlates of TDV have been described as consequences or outcomes; however, they may also be considered risk factors. Clearly, future longitudinal research that focuses on Latinas is needed to establish causal links between TDV and outcomes such as depression and suicidality, as this topic is understudied among Latino adolescents (Ocampo et al. 2007).

## Alcohol and Substance Use: Individual and Peer Microsystems

Although substance abuse is often implicated in suicide and in suicide attempts among adolescents, there are mixed findings for Latina adolescents. Zayas and colleagues (2000) report that substance use is not prevalent in suicide attempts among Latinas. However, data from a large nationally representative sample revealed that Hispanic girls in grades 7–12 were 15.1 times more likely to attempt suicide if they used alcohol, as compared to 6.6 times for Black girls and 4.1 times for White girls, and 4.5 times more likely if they used marijuana (compared to 10.3 times for Black and 3.4 times for White girls) (Borowsky et al. 2001). The use of other illegal drugs also increased the likelihood of a suicide attempt by 2.3 times for Hispanic girls (compared to 6.2 times for Black girls and 2.9 times for White girls) (Borowsky et al. 2001). Substance use was also correlated with suicide attempts in a sample of 141 Latinas who were sentenced to probation for a variety of offenses. Results indicated that drug abuse (specifically marijuana and inhalants and total drug frequency) was strongly associated with suicide attempts (Cuellar and Curry 2007). It is notable that several large studies that examined suicidal risk among youth including Latinas did not include substance use in their analyses, although that is mentioned as a potential risk factor. For example, the authors of one study noted in the introduction that substance use has been associated with suicidal behaviors in African-American and Latino adolescents, but did not include substance use as one of the variables in their study (O'Donnell et al. 2004). Rew and colleagues (2001) mentioned "risky behaviors," which we assume would include substance use, but again these were not included in the analysis. Given the findings by Borowsky et al., it will be important for future research to include a measure of substance use to understand how this may function as a risk factor for suicide attempts.

## In Summary

Despite the multiple risks associated with being a Latino adolescent in the United States, not all Latino youths develop mental health problems (Loukas and Prelow 2004). In this chapter, we summarize the existing research to further identify key factors that contribute to depressive symptoms and suicidal behavior of Latina adolescents. We take an ecodevelopmental approach to understand multiple levels of ecological levels, so that we can really understand the Latina teen within a broader context of the U.S. environment. To begin with, there are many negative stereotypes and anti-immigrant policies that may shape the general climate that Latinas are experiencing as they move into larger school systems and as they engage with the U.S. society as emerging adults. As a result, they may encounter stressors that are atypical compared to other U.S. adolescents. In particular, stereotypes about Latina women may shape their exposure to sexism and body image.

As many studies have identified, family is so often central to the perceptions of Latina teens; however, when families face unusual distress or conflict, it can lead to extreme actions by Latinas, including suicide ideation, plans, and attempts. Several authors offer family typologies to help explain the combination of family context factors that are most likely to contribute to suicidal behaviors. However, recent research also indicates that peer context can contribute to bullying behaviors which are associated with depression and suicidality. A handful of studies suggest that being a victim and bully contributes to suicidal ideation among Latina adolescents. Additionally, abusive peer dating relationships have been shown to be particularly influential in Latina suicidality. Lastly, alcohol and substance use have often been overlooked in research, but it appears from national studies that there may be a strong link between substance use and Latina adolescent suicidal behaviors, which deserves further investigation. In the next chapter, we turn to researchers, clinicians, and service providers for recommendations to support their efforts to erase this deadly health disparity for Latina adolescents.

# Chapter 5
# Reclaiming *Malintzin*: Healing and Preventing Latina Depression and Suicide

*Hold on to what is good,*
*even if it's a handful of earth.*
*Hold on to what you believe,*
*Even if it's a tree that stands by itself.*
*Hold on to what you must do,*
*Even if it's a long way from here.*
*Hold on to your life,*
*Even if it's easier to let go.*
*Hold on to my hand,*
*Even if I've gone away from you. (Wood 1992).*

Latina young women are active and engaged in, reacting to, transforming, and subverting the world around them, as they have been doing for many generations (Arellano and Ayala-Alcantar 2004; Walters and Simoni 2002). Clearly, the mental health of Latina young women is complex and influenced by many factors including those that might increase risks and those that are protective. However, more than anything else, we have aimed to provide a multi-faceted perspective of the experience of Latina adolescents in the United States. We hope that rather than interpreting the health disparity findings as another sad case of *La Llorona* that our readers instead understand how Latinas actively negotiate and transform challenging economic, cultural, and gendered contexts. Some scholars argue that the origin of the *La Llorona* story is rooted in the story of the female indigenous slave who was given to Hernán Cortez when he arrived in the New World from Spain (Garcia 1997). Some versions of history have labeled this young woman "*Malinche*," which was used to indicate traitor, because she aided Cortez through her translations. Many felt that she led to the colonization of indigenous communities in the Americas. As such, they argue that she is the original *La Llorona,* crying for the children who she killed through her help of Cortez. To reclaim this "Eve" of the Americas, feminists argue that the true Nahuatl name of the young woman was *Malintzin*, and that it was because of her talent at languages that her

A. J. Romero et al., *Preventing Adolescent Depression and Suicide Among Latinas*, SpringerBriefs in Psychology, DOI: 10.1007/978-3-319-01381-7_5,

family gave her up as a slave to Cortez, who then raped her and forced her to assist with his plans (Garcia 1997; Hurtado and Cervantez 2009). In this interpretation of *Maltinzin*, we understand her ecological context and the forces that shaped her decision; moreover, in this version, she is actively negotiating the situation to minimize damage to her homeland and her people. In fact, it is argued that she is fierce and resilient in the face of great adversity, in so many ways parallel to the young Latina women of the United States who are fighting for their culture, families, identity, and mental well-being (Hurtado and Cervantez 2009). This is the image of Latina adolescents and their families that we hope to leave in the mind of our readers at the end of this book, along with inspiration to engage in prevention and intervention action research to eradicate health disparities of Latina adolescent depression and suicidality.

## Summary of Theory and Current Research Findings

Theory and research suggest that Latina girls may already have some of the resources that they need to thrive within themselves and their families. Perhaps, through acknowledging societal inequalities while also honoring existing cultural strengths, we, as researchers, educators, clinicians, and policy makers, can further support the healthy development of Latina adolescents. While Latina mental health disparities continue to exist, our theoretical and empirical understanding of contributing individual and environmental factors has significantly grown beyond cultural deficit model explanations. It is far too easy to blame a culture or family for complex mental health processes and suicidal behaviors; these *easy answers* only gloss over the real problem and distract us from determining the authentic aspects that we can modify to help prevent depression and suicide. The Multiracial Feminist Framework (Arrellano and Ayala-Alcantar 2004) and Zayas and colleagues' (2000, 2005) ecodevelopmental models both urge us to consider contextual influences, such as family and cultural contexts. Depression can take a serious toll on the lives of Latina adolescents, and the risk of suicide necessitates prevention and intervention on multiple levels. In this chapter, we provide suggestions for prevention and intervention efforts for a variety of stakeholders, including researchers, mental health practitioners, schools, and more. These suggestions have been compiled from current research and theory about best practices with Latino/a youth in the area of depression and suicide. The key points about Latina adolescent depression and suicide are as follows:

1. *There is a 30-year epidemic of Latina adolescent depressive symptoms and suicide attempts.* Rates do not appear to vary by Latina subgroup, economic level, or region.
2. *Latina girls are active agents in negotiating their identities, families, peers, and schools.* Sources of resilience and cultural strength are demonstrated in Latina's close family connectedness and communication, positive ethnic

identity development, religiosity/spirituality, biculturalism/bilingualism, and their engaged coping strategies.

3. *Despite adversity, young Latinas often emerge as resilient, even after suicide attempts.* Latina adolescents are negatively impacted by stereotypes, discrimination, and family factors, such as family dysfunction, family separation, and family suicides. Latinas also report higher than average levels of peer bullying that impacts their suicidality. Substance use and dating violence may be an emerging factors that are linked with suicidal behaviors of Latina adolescents.

## How We Can Use and Apply Research Findings to End the Epidemic

*Role of Family*. As has been highlighted throughout this volume, the role of family is prominent in Latino/a culture. Adolescent Latinas are likely to turn to peers and family when requesting help for depression or family problems, more so than health practitioners (Rew et al. 1997), suggesting that families may be the first and continued point of communication for these youth. However, the lack of health insurance among Latino children may hinder families and peers from obtaining health services, because many as 37 % of Latino children do not have health insurance. Several researchers and theorists have noted the importance of including family members in therapy (Celano and Kaslow 2000; Goldston et al. 2008; Zayas and Pilat 2008). Providing a welcoming environment that supports the inclusion of family, particularly extended family, is encouraged for mental health professionals. For young Latinas in particular, enhancing communication through developing a stronger emotional connection between girls and their parents has been shown to lead to positive outcomes (Kuhlberg et al. 2010).

*Acculturation Stress and Biculturalism.* The degree to which an individual adheres to mainstream and heritage culture and their respective values shapes their worldview, their interactions with others, and likely their mental health. Mental health professionals should integrate questions about acculturation and cultural values in their assessments as well as intervention approaches (See Edwards and Cardemil, in press, for an *Interview Guide for Assessing Latino/a Cultural Values*). In this process of understanding a particular client's level of acculturation or adherence to Latino cultural values, a clinician may discover that there are differences between the client and her parents, or the client and mainstream culture (or a combination of each). Supporting the client in addressing these areas of divergence and convergence in worldviews will likely enhance therapeutic engagement and also possibly target symptoms of distress.

Professionals are also encouraged to allow young Latinas to understand the value of both a strong cultural identity and skills in being bicultural and bilingual (Carvajal et al. 2002). The ability to navigate multiple cultural contexts and languages is a strength among Latino/a youth, and practitioners can highlight this skill set and help youth utilize these resources to interact effectively with their environments.

*Gender role expectations*. Differences in acculturation level, as mentioned above, may lead to conflict between parents and their children specifically in the area of gender role expectations (Cespedes and Huey 2008). Among some families, this acculturation conflict may emerge as young Latinas begin to demonstrate adolescent autonomy which may conflict with family members' resistance to acculturation to U.S. societal norms; as such, preexisting negative family patterns may result in even more rigid restrictiveness and control of girls in a manner that is different from the response to male children (Flores 2013; Kuhlberg et al. 2010). Additionally, gendered and cultural collectivistic expectations to place family as priority may also place young Latinas in a difficult situation negotiating family and cultural values within the U.S. environment (Goldston et al. 2008). Professionals working with young Latinas and their families should be attuned to the role of gendered conflicts and may choose to bring them to the forefront of the therapeutic process. Validating a Latina adolescents' struggle to maintain balance between her family's expectations and those of mainstream U.S. culture can be powerful. She can be assisted in developing coping strategies to deal with this type of conflict, and she can be supported in her journey toward clarification of values and personal and cultural expectations. Providing psychoeducation to parents of these girls can also serve to respectfully inform families about this developmental process and potentially allay fears they may have. Additionally, making sure that families are included in school and educational programs may be helpful to ensure that families are aware and supportive as Latinas become adults in the United States.

## Recommendations for Future Research

We have summarized a great deal of the existing research on Latino adolescents, with a particular focus on Latina adolescents. However, there are still many gaps in the existing empirical literature that can be addressed by future research. To begin with, using strong theoretical models is critical to identify modifiable factors that can help improve prevention and intervention. Additionally, longitudinal studies that include multiple key factors of depression, family conflict, peer bullying, cultural strengths, and protective factors [family (father and mother), school, peers, and individual] need to be carried out in order to determine how each factor contributes to Latinas' outcomes over time. Multiracial Feminist Frameworks (Arrellano and Ayala-Alcantar 2004) recommend using innovative and cutting-edge methodologies such as community-based action research approaches to create community partnerships to investigate and address this health disparity. They also highlight the importance of qualitative and mixed methods research because Latina adolescents have been understudied, and often even existing standardized measures are not appropriate or relevant.

Bernal et al. (2009) examined extant research on cultural adaptations of treatment and concluded that the research is still seriously lacking. Some models that have been tested were based on very small samples, and replication is needed. Further, it is important that such research incorporates fidelity checks so that researchers and clinicians can be certain that the described approach was actually implemented. We would add that the dissemination of culturally appropriate treatment manuals would be very valuable; while we are interested in those for Latina adolescents, we heartily support the development of similar documents for other cultural groups.

*Currently, there is an urgent need to develop evidence-based depression or suicide prevention programs targeted directly at Latina adolescents, because there are none in existence.* Current research has only modified mainstream therapies for Latino populations or provided guidelines for cultural awareness among clinicians. However, there are few studies that really investigate and compare different strategies of treatment that integrate cultural strengths and resilience of Latina girls. Once empirical evidence provides clarity on the most effective strategies to prevent and intervene with mental health issues of Latina girls, nationwide campaigns may be considered to truly eradicate this long-lasting health disparity.

## Mental Health Care Access and Treatment

Given that depression is the most significant risk factor for suicide and that Latina adolescents have the highest rates of depressive symptoms for more than 30 years in the U.S., this is a critical point of prevention and intervention. As we seek to address this persistent pattern, we must recognize that there are societal characteristics that are to be considered when making recommendations for treatment. For example, Hispanics/Latinos in the U.S. continue to face challenges of low socioeconomic status, lack of health insurance, and in some cases undocumented immigrant status. Low socioeconomic status, when it leads to basic needs being unmet, was a significant predictor of suicidality in a sample of Black and Latino urban adolescents (O'Donnell et al. 2004). This results in difficulty in accessing available assistance. For example, in a recent survey, Hispanics were less likely than any other racial/ethnic groups to call a suicide crisis line during a suicidal crisis (Larkin et al. 2011). Approximately, 31 % of Latinos do not have health insurance; these rates are even higher for foreign-born adults (34 %) and for those who are not citizens (45 %) (U.S. Bureau of the Census 2011). Compared to non-Hispanic Whites, Hispanics are less likely to receive care that follows recommended guidelines and rely more often on the primary care sector than on mental health specialists for mental health services (Cabassa et al. 2006). Clearly, access to mental health care services that are available to all individuals regardless of immigration status and economic status is important to Latina adolescent populations.

Researchers have noted that cultural factors still impact resource seeking and access to care even beyond socioeconomic factors that may limit access to health insurance and quality care providers (Alegria et al. 2007). Among immigrants, language can become a strong barrier to accessing services. Recent immigrants may not be familiar with the intricacies of the health care system, and bilingual services may not be available. Although most suicidal Hispanic youth are born in the United States and know English, many times, their parents are limited from participating in their treatment due to language barriers that call for bilingual service professionals (American Association of Suicidology [AAS] 2012). While family social support has been found to be a source of resilience for Latinos, at times, it can also be a barrier to seeking outside mental health if the family believes that problems should be addressed within the family (AAS 2012). Rotheram-Borus et al. (1999) found that these youths were less likely to attend follow-up counseling sessions if they reported high family values and unity. One interpretation of these findings was that the tendency to focus on family social support suppressed turning to outsiders for help (Rotheram-Borus et al. 1999). Additionally, service utilization may be impacted by how Latinos understand and attribute causes to mental illness, or how they treat it (e.g., the use of folk medicine such as herbs or visiting a *curandera*). The complexity of barriers to access are varied; however, there is evidence that treatment and therapy is worthwhile and can be effective for Latina adolescents.

## Recommendations for Mental Health Professionals

When considering the suggestions below, it is important to acknowledge the within-group heterogeneity of Latino groups and their families and how this might influence treatment adherence (Goldston et al. 2008). For example, differences in engagement in therapy were found between non-Cuban Latino families and Cuban families in an intervention in Miami with youth at risk of drug abuse (Santisteban et al. 1996). Few examples of evidence-based practice have been evaluated at this level (i.e., national origin), but these findings suggest that care should be taken to not assume that all Latino/as will respond in the same manner with interventions. While the extent of differences based on factors such as acculturation level of nationality are unknown, it is best to assess level of adherence to cultural values and acculturation before beginning intervention or programming to maximize the chances that any intervention be culturally relevant for the youth involved (Edwards and Cardemil, in press).

## Assessment and Evaluation

Assessing for suicidality is common clinical practice among mental health practitioners; however, it must be done in a thoughtful manner. This point is underscored by the fact that suicidality or distress may not express itself in same way for Latina adolescents as other youth. Indeed, authors have noted that screening for only depressive disorders without screening for dual or comorbid diagnoses may lead clinicians to miss important aspects of suicidality (Fortuna et al. 2007). For example, O'Donnell and colleagues (2004) found that not all inner city youth in their study who had made a suicidal attempt reported ideation or talked with someone about their distress. In addition, Goldston and colleagues (2008) noted that the behaviors that led to suicidal behavior among Latinas could be interpreted as the cultural syndrome *nervios*; in other words, sometimes it may be an impulsive response to a traumatic situation. However, these findings may also indicate that Latina girls are less likely to talk openly about thinking about suicide or making suicidal plans. These findings about cultural differences in the manifestation and interpretation of distress underscore the need for clinicians to move beyond relying on common warning signs (e.g., talking with others about distress and making suicide plans) to conducting complete biopsychosocial–cultural assessments in youth (Fortuna et al. 2007).

## Prevention and Treatment

Mental health professionals working with young Latinas who have already been identified as depressed or suicidal must consider treatments that are appropriate to utilize with this particular population. Meta-analyses indicate that cultural adaptation in psychotherapy is effective (e.g., Smith et al. 2011), and professionals are encouraged to understand and apply knowledge about culture to their work with diverse individuals (Edwards and Cardemil, in press).

Only a few studies have evaluated cultural adaptations of prevention or treatment interventions for youth who are depressed or suicidal. The Beardslee Preventive Intervention Program for Depression (PIP) is designed for use with low-income Latino families (D'Angelo et al. 2009). The authors describe their extensive process of utilizing experts and pilot testing to adapt their protocol by translating materials and also integrating cultural metaphors and other aspects of the Latino experience. The intervention was tested and compared to the original intervention, with results suggesting that it maintained fidelity and was feasible. In one study, Rossello and Bernal (2005) adapted cognitive behavioral and interpersonal therapies for Puerto Rican adolescents who were depressed by involving parents, among other methods. Similarly, Rotheram-Borus et al. (1996) found that engagement of Latino adolescent suicide attempters could be improved by using videotapes to modify treatment expectations and providing training for staff to enhance their sensitivity to parents' reactions to suicide attempts and the service process (Goldston et al. 2008). These studies provide examples of efforts being made to make existing interventions more effective and culturally relevant for young Latinos. In the sections below, we summarize prominent topics that would likely be included in any prevention or treatment with young Latinas.

Two frameworks, Cognitive Behavioral Theory and the Interpersonal Theory of Suicide, have served as the foundation for many psychotherapeutic interventions (Mufson et al. 1999). Both CBT and Interpersonal Theory have associated treatment protocols, and both systems have strong evidence of effectiveness, including that with Latina adolescents. The theoretical underpinnings of the theories are consistent with what we know about the dynamics of suicidal ideation and attempts in this group; culturally sensitive applications of these approaches to therapy should be considered preferred treatment for Latinas. Mufson and her colleagues assessed the efficacy of IPT for adolescents in a group of low-income Latinos (75 % female and 79 % Hispanic) and found that depressive symptoms decreased significantly in the treatment group compared to a control group, but no differences were found on suicide planning or attempts. Those in the control group had a much higher rate of non-completion than the treatment group. In another study, CBT was found to be effective with Latino adolescents in small, controlled trials when the concepts of *familismo* and *respeto* are incorporated (Miranda et al. 2005). In another study, Cognitive Behavioral Therapy and Interpersonal Therapy (delivered in group and individual formats) were compared in a randomized study of 112 Puerto Rican adolescents (55 % female). Both CBT and IPT resulted in a decrease in depressive symptoms; however, the decreases were larger in the CBT approach, and self-esteem and social functioning were increased in those who received IPT, so that the average participant in CBT made better progress than 67 % of those in IPT (Rosselló et al. 2008). These researchers were surprised by these results, as they considered IPT to be more consistent with Latino cultural values than CBT. They offer several possible explanations for these unexpected findings: (1) CBT may be more efficacious for treating depression in these adolescents, (2) there may be bias in favor of CBT in the outcome measures used, (3) IPT may not have been conducted as well as the CBT treatments; this is supported by the fact that treatment adherence was around 80 % for the IPT treatments and 90 % for CBT, (4) the treatment providers may have received training in incorporating culturally appropriate components in the CBT treatment, and (5) CBT is very structured and directive, consistent with the value of *respeto* (in which the authoritative position of the therapist is seen as positive).

## Cultural Adaptations of Treatments

One question that is important to consider is how to adapt an existing empirically validated treatment approach (such as CBT and IPT) to be more culturally sensitive. Bernal et al. (1995) proposed a model for doing so that was designed for Latino populations. There are eight dimensions that can be incorporated into a treatment protocol to increase the cultural relevance of the existing treatment approach that include language, cultural use of *cuentos* (*storytelling*)/*dichos* (*sayings*), and ecodevelopmental context. Ideally, treatment should be provided in the language of the client and their family, whether it is English, Spanish, or bilingual (code-switching). Language refers to dialect, local patterns, and the terminology of subgroups. In addition, the process of diagnosis needs to be careful not to pathologize practices or experiences that are normative in the culture. It is also important to not assume that all Latinos will have similar norms and values. The authors cite work (e.g., Bernal et al. 1995) that finds family therapy approaches to be well-suited to Latino clients because of the high value placed on family involvement. They stress the importance of actively incorporating such contextual elements as acculturative stress, migration history, developmental stages of individuals, the extent of social support, and the family's relationship with the country of origin.

It is clear that treatment considerations for young Latinas must consider individual needs and differences among youth, including issues of acculturation, historical context, current political context, economic, linguistic, and racial/ethnic factors, among others. In order for mental health service providers to be effective with Latino populations, it is important to understand the complex nature of culture, so that clients are not viewed in a stereotypical manner but in a holistic manner that is inclusive and respectful of their cultural context. Evidence-based techniques (such as storytelling or integration of spiritual beliefs) to intervene and treat mental illness among Latinos are available and can improve the quality of treatment. These can be integrated into more common, evidence-based therapeutic approaches (CBT and Interpersonal Therapy) that are based on theories that are consistent with what is known about the dynamics of depression and suicidal behavior in this population.

## Recommendations for School Professionals

Schools have an important role to play in prevention and treatment of depression and suicidal behaviors in Latina adolescents. First, given the increasing size of this population, it is imperative that schools ensure that staff are culturally competent and treat young women with respect. School counselors can be particularly helpful to combat depression and suicidal behaviors by forming relationships with Latinas. Demonstrating empathy for the possible challenges faced by Latina adolescents will establish a sound relationship for source of support. Thus, if depressive symptoms appear, Latinas may be more likely to seek assistance from the counselor because of the relationship that has formed. The counselor should also be alert to any changes in behavior or demeanor that might signal something is amiss. Certainly, schools and primary care physicians are often the first-line responders, particularly for depression.

A more politically charged issue related to schools is the way in which the use of Spanish is treated, although it can be very important within a clinical setting and inclusive of parents and family members. If the message is that one's language is unacceptable in school, it is logical to believe that this language and culture are less valuable than others. This devaluing of language, and by default, culture, may play a role in decreased self-esteem and vulnerability for depression. The English-only approach may also make it difficult for monolingual Spanish-speaking parents to collaborate with the school to support the teen who may be displaying depressive symptoms or suicidal thoughts. Schools that distribute information in Spanish as well as English are acknowledging the importance of communicating in the best way possible, but there is considerable room for improvement in valuing the language and culture that are part of the lives of Latina adolescent girls. In addition, the presence in the school of role models (teachers, counselors, administrators) who are Latina sends a hopeful message about future possibilities, and provides access to mentors who can share with young Latinas what they have learned about managing to live in two cultures.

## Conclusion and Recommendations

It was our aim to bring attention to the epidemic of Latina adolescent depression and suicide attempts, which has not abated in over 30 years in the United States. Clearly, there is a call to action that includes teens, parents, schools, service providers, clinicians, and policy makers to work together to address this issue through prevention and intervention strategies that are based on the most cutting-edge research findings, which we have summarized and presented here. There are many effective strategies that are proposed here to provide prevention and intervention programs for Latina adolescents, their families, and their schools. In summary, we offer the following recommendations:

1. Research: *There is an urgent need to develop evidence-based depression or suicide prevention programs targeted directly at Latina adolescents, because there are none in existence.* More empirical longitudinal research is needed from a resilience and cultural strengths perspective.
2. Policy: Mental health care access, within schools and communities, is critical to the prevention and treatment of depression and suicide attempts. Immigration reform policies need to consider the family context and mental health context of children. More funding for high-quality research and the development of prevention and intervention programs is needed.
3. Service Providers (e.g., psychologists, school counselors, social workers, or primary care physicians) need to be made aware of the epidemic of Latina adolescent depression and suicide. They also need to be trained to know the warning signs of depression and suicide in this population and to develop cultural competence to raise awareness of client diversity as a means to the most effective treatment.
4. Treatment: Latina-specific treatment options need to be developed and evaluated. There is also a need for more culturally inclusive adaptations of common therapies (e.g., including families in treatment contexts).
5. Schools: Schools can create school-wide practices that create safe and healthy climates that are inclusive of all gender, racial, ethnic, and sexual orientations as a means to minimize negative messages from teachers, curriculum, and other students. Additionally, anti-bullying campaigns have been very effective in school-wide settings to improve overall school climate. Additionally, school and family linkages can be critical to helping provide programs to strengthen families, identifying warning signs early, and to connect families with appropriate mental health resources.

6. Community: Early prevention of depression and depressive symptoms needs to happen as early as fifth and sixth grade. Awareness-raising campaigns in Latino communities can help improve awareness of depressive symptomatology, the fact that depression is treatable, and that suicide is linked to depressive symptoms.

Our discussion of Latina adolescent's depression and suicide has not portrayed them as victims of their culture, class, or families. Rather we have focused on young women's strengths in order to focus prevention efforts on resiliency in here in Latina's cultural and gendered roots in order to inspire new research, policy, and interventions to save the lives of more young women in the future. Yet, it is that balance between raising awareness about health disparities rooted in inequalities of discrimination and sexism and with the inclusion of cultural strengths that will lead to effective prevention and intervention efforts.

# References

Abad, V. (1987). Mental health delivery systems for Hispanics in the U.S.: Issues and dilemmas. In M. Gavira and J. Arana (Eds.), *Simon bolivar research institute monograph: Hispanic American psychiatric research program* (Serial no. 1, pp. 278–293). Chicago: University of Illinois.

Ackard, D. M., & Neumark-Sztainer, D. (2002). Date violence and date rape among adolescents: associations with disordered eating behaviors and psychological health. *Child Abuse and Neglect, 26*, 455–473.

Aldwin, C. M. (1994). *Stress, coping, and development: An integrative perspective*. NY: The Guilford Press.

Aldwin, C. M. (2007). *Stress, coping, and development: An integrative perspective* (2nd ed.). NY: The Guilford Press.

Alegria, M., Mulvaney-Day, N., Woo, M., Torres, M., Gao, S., & Oddo, V. (2007). Correlates of past-year mental health service use among Latinos: Results from the National Latino and Asian American study. *American Journal of Public Health, 97*, 1–8.

American Association of Suicidology. (2012a). *Facts about suicide and depression based on 2010 data.* Retrieved from http://www.suicidology.org/c/document_library/get_file?folderId=262&name=DLFE-622.pdf.

American Association of Suicidology. (2012b). *Hispanic suicide fact sheet based on 2010 data.* Retrieved from: http://www.suicidology.org/c/document_library/get_file?folderId=262&name=DLFE-623.pdf.

American Association of University Women Educational Foundation. (2001). *Beyond the "gender wars": A conversation about girls, boys, and education.* The American Association of University Women Educational Foundation.

American Psychiatric Association. (1994). *Diagnostic and statistical manual of mental disorders* (4th ed.). Washington, D.C.: American Psychiatric Association.

American Psychiatric Association. (2000). *Diagnostic and statistical manual of mental disorders, fourth edition, text revision*. Washington, D.C.: Author.

Anaya, R. (2011). *La llorona: The crying woman*. New Mexico: University of New Mexico Press.

Andrews, P. W., & Thomson, J. A, Jr. (2009). The bright side of being blue: Depression as an adaptation for analyzing complex problems. *Psychological Review, 116*(3), 620–654.

Angel, R., & Thoits, P. (1987). The impact of culture on the cognitive structure of illness. *Culture, Medicine and Psychiatry, 11*, 465–494.

Anzaldua, G. (1987). Ch. 7: La conciencia de la mestiza: Towards a new consciousness. *Borderlands: La Frontera*. (pp. 99–123). San Francisco: Aunt Lute Press.

Arellano, L. M., & Ayala-Alcantar, C. (2004). Multiracial feminism for Chicana/o psychology. In R. Velasquez, L. M. Arrellano, & B. McNeill (Eds.), *The handbook of chicana/o psychology and mental health* (pp. 215–230). NY: Routledge.

A. J. Romero et al., *Preventing Adolescent Depression and Suicide Among Latinas*, SpringerBriefs in Psychology, DOI: 10.1007/978-3-319-01381-7,

Arredondo, P. (2004). Psychotherapy with chicanas. In R. Velasquez, L. M. Arellano, & B. McNeill (Eds.), *The handbook of chicana/o psychology and mental health* (pp. 251–264). NY: Routledge.
Arredondo, P., Perez, P., Harper, F., & McFadden, J. (2003). Counseling paradigms and Latina/o Americans: Contemporary considerations. *Culture and Counseling: New Approaches*, 115–132
Arnett, J. J. (1999). Adolescent storm and stress, reconsidered. *American Psychologist, 54*(5), 317–326.
Badillo-Ghali, S. (1977). Culture sensitivity and the Puerto Rican client. *Social Casework, 58*(8), 459–468.
Balls Organista, P., Marin, G., & Chun, K. M. (2010). *The psychology of ethnic groups in the United States*. Thousand Oaks: Sage Publications.
Baptiste, D. A. (1993). Immigrant families, adolescents and acculturation: Insights for therapists. In B. H. Settles, D. E. Hanks, & M. B. Sussman (Eds.), *Families on the move: Migration, immigration, emigration and mobility*. London: Haworth Press.
Barac, R., & Bialystok, E. (2011). Cognitive development of bilingual children. *Language Teaching, 44*(1), 36–54.
Baumann, A. A., Kuhlberg, J. A., & Zayas, L. H. (2010). Familism, mother-daughter mutuality, and suicide attempts of adolescent Latinas. *Journal of Family Psychology, 24*(5), 616–624. doi:10.1037/a0020584.
Bayles, B. P., & Katerndahl, D. A. (2009). Culture-bound syndromes in hispanic primary care patients. *International Journal of Psychiatry in Medicine, 39*, 15–31.
Beck, A. T. (1967). *Depression: Clinical, experimental, and theoretical aspects*. Philadelphia: University of Pennsylvania Press.
Beck, A. T., & Alford, B. A. (2009). *Depression: Causes and treatments* (2nd ed.). Philadelphia: University of Pennsylvania Press.
Berry, J. W. (1980). Acculturation as varieties of adaptation. In A. M. Padilla (Ed.), *Acculturation: Theory, models, and some new findings* (pp. 9–25). Boulder: Westview.
Black, M. C., Basile, K. C., Breiding, M. J., Smith, S. G., Walters, M. L., & Merrick, M. T., et al. (2011). *The national intimate partner and sexual violence survey (NISVS): 2010 Summary Report*. Atlanta: National Center for Injury Prevention and Control, Centers for Disease Control and Prevention.
Bonomi, A. E., Anderson, M. L., Cannon, E. A., Slesnick, N., & Rodriguez, M. A. (2009). Intimate partner violence in Latina and non-Latina women. *American Journal of Preventive Medicine, 36*, 43–48.
Born, D. O. (1970). Psychological adaptation and development under acculturative stress: Toward a general model. *Social Science and Medicine, 3*, 529–547.
Borowsky, I. W., Ireland, M., & Resnick, M. D. (2001). Adolescent suicide attempts: Risks and protectors. *Pediatrics, 107*, 485–493.
Bradby, H. (2003). Describing ethnicity in health research. *Ethnicity and Health, 8*, 5–13.
Brent, D. A., & Mann, J. J. (2005). Family genetic studies, suicide, and suicidal behavior. *American Journal of Medical Genetics Part C: Seminars in Medical Genetics, 133C*, 13–24. doi:10.1002/ajmg.c.30042.
Brice, A. E., Shaunessy, E., Hughes, C., McHatton, P., & Ratliff, M. (2008). What language discourse tells us about bilingual adolescents: A study of students in gifted programs and students in general education programs. *Journal for the Education of the Gifted, 32*(1), 7–33.
Bronfenbrenner, U. (1979). *The ecology of human development*. Cambridge: Harvard University Press.
Bronfenbrenner, U. (1986). Ecology of the family as a context for human development. *American Psychologist, 32*, 513–531.
Buriel, R., Perez, W., De Ment, T. L., Chavez, D. V., & Moran, V. R. (1998). The relationship of language brokering to academic performance, biculturalism, and self-efficacy among Latino adolescents. *Hispanic Journal of Behavioral Sciences, 20*(3), 283–97.
Cabassa, L. J., Zayas, L. H., & Hansen, M. C. (2006). Latino adults' access to mental health care: A review of epidemiological studies. *Administration and Policy in Mental Health, 33*(3), 316–330.

Cabrera, N. L., Milem, J. F., & Marx, R. W. (2012). *An empirical analysis of the effects of Mexican American Studies participation on student achievement within Tucson Unified School District*. Tucson, AZ: Report to Special Master Dr. Willis D. Hawley on the Tucson Unified School District Desegregation Case.

Caldwell, A., Couture, A., & Nowotny, H. (2008). *Closing the mental health gap: Eliminating disparities in treatment for Latinos*. Retrieved from the Mattie Rhodes Center website: http://www.mattierhodes.org/web/english/Resources_and_Reports.html.

Cammarota, J. (2007). A social justice approach to achievement: Guiding Latina/o students toward educational attainment with a challenging socially relevant curriculum. *Equity and Excellence in Education, 40*(1), 87.

Canino, G., & Roberts, R. E. (2001). Suicidal behavior among Latino youth. *Suicide and Life Threatening Behavior, 31*, 122–131.

Canino, I. A., & Canino, G. (1980). Impact of stress on the Puerto Rican family: Treatment considerations. *American Journal of Orthopsychiatry, 50*(3), 535–541.

Cannon, E., & Levy, M. (2008). Substance-using Hispanic youth and their families: Review of engagement and treatment strategies. *The Family Journal, 16*, 199–204.

Capps, R., Castaneda, R. M., Chaudry, A., & Santos, R. (2007). *Paying the price: The impact of immigration raids on America's children*. Washington, D.C.: National Council of La Raza.

Carrillo, R., Moreno, M., & Zintsmaster, J. (2010). Cultural production of a decolonial imaginary for a young Chicana: Lessons from Mexican immigrant working-class woman's culture. *Educational Studies: Journal of the American Educational Studies Association, 46*(5), 478–502.

Carvajal, S. C., Hanson, C. E., Romero, A. J., & Coyle, K. K. (2002). Behavioural risk factors and protective factors in adolescents: A comparison of Latinos and non-Latino whites. *Ethnicity and Health, 7*, 181–193.

Casey Foundation, Annie E. (2013). *Kids Count Databook 2012*. Annie E. Casey Foundation. www.kidscount.com.

Cash, S. J., & Bridge, J. A. (2009). Epidemiology of youth suicide and suicidal behavior. *Current Opinion in Pediatrics, 21*(5), 613–619.

Castillo, A. (1994). *Massacre of the dreamers: Essays on xicanisma*. NY: Plume Publishers.

Castillo, L. G., Conoley, C. W., & Brossart, D. F. (2004). Acculturation, white marginalization, and family support as predictors of perceived distress in Mexican American female college students. *Journal of Curriculum Studies, 51*(2), 151–157.

Castillo, L. G., Perez, F. V., Castillo, R., & Ghosheh, M. R. (2010). Construction and initial validation of the marianismo beliefs scale. *Counseling Psychology Quarterly, 23*(2), 163–175.

Celano, M. P., & Kaslow, N. J. (2000). Culturally competent family interventions: Review and case illustrations. *American Journal of Family Therapy, 28*, 217–228.

Center for American Progress Action Fund. (2008). *U.S. Latinas by the numbers: A growing and ambitious population*. Retrieved October 10, 2009 from http://www.americanprogressaction.org/issues/2008/pdf/latinas_numbers.pdf.

Centers for Disease Control and Prevention (CDC). (2008). Youth risk behavior surveillance–United States, 2007. *MMWR Surveillance Summary, 57*(No. SS-4). Atlanta, GA: Author.

Centers for Disease Control and Prevention (CDC). (2009a). Alcohol and suicide among racial/ethnic populations – 17 states, 2005–2006. *MMWR, 58*(23), 637–641.

Centers for Disease Control and Prevention (CDC). (2009b). *Youth risk behavior surveillance*. United States.

Centers for disease control and prevention (CDC). (2011). *Youth risk behavior surveillance*. United States. Retrieved from Center for Disease Control website: http://www.cdc.gov/healthyyouth/yrbs/pdf/us_suicide_trend_yrbs.pdf.

Centers for Disease Control and Prevention (CDC). (2012, August). *Suicide prevention: Youth suicide*. Retrieved from Center for Disease Control website: http://www.cdc.gov/violenceprevention/pub/youth_suicide.html.

Centers for Disease Control and Prevention (CDC). (2012, Septemper). *Suicide: Consequences*. Retrieved from Center for Disease Control website: http://www.cdc.gov/ViolencePrevention/suicide/consequences.html.

Cervantes, R. C., & Castro, F. G. (1985). Stress, coping, and Mexican American mental health: A systematic review. *Hispanic Journal of Behavioral Sciences, 7*, 1–73.

Céspedes, Y. M., & Huey, S. J. (2008). Depression in Latino adolescents: A cultural discrepancy perspective. *Cultural Diversity and Ethnic Minority Psychology, 14*(2), 168–172.

Coatsworth, J. D., Pantin, H., McBride, C., Briones, E., Kurtines, W., & Szapocznik, J. (2000). Ecodevelopmental correlates of behavior problems in young Hispanic females. *Applied Developmental Science, 6*(3), 126–143.

Coker, A., McKeown, R. E., Sanderson, M., Davis, K. E., Valois, R. F., & Huebner, S. (2000). Severe dating violence and quality of life among South Carolina high school students. *American Journal of Preventive Medicine, 19*, 220–227.

Comas-Díaz, L. (1988). Hispanics. In L. Comas-Diaz & E. E. Griffith (Eds.), *Clinical guidelines in cross-cultural mental health* (pp. 183–268). NY: Wiley.

Comas-Diaz, L. (2000). An ethnopolitical approach to working with people of color. *American Psychologist, 55*(11), 1319–1325.

Comas-Díaz, L. (2006). Latino healing: the integration of ethnic psychology into psychotherapy. *Psychotherapy: Theory, Research, Practice, Training, 43*(4), 436–453.

Compas, B., Connor-Smith, J. K., Saltzman, H., Harding Thomsen, A., & Wadsworth, M. E. (2001). Coping with stress during childhood and adolescence: Problems, progress, and potential in theory and research. *Psychological Bulletin, 127*(1), 87–127.

Consolacion, T. B., Russell, S. T., & Sue, S. (2004). Sex, race/ethnicity, and romantic attractions: multiple minority status adolescents and mental health. *Cultural Diversity and Ethnic Minority Psychology, 10*(3), 200–214.

Coohey, C. (2001). The relationship between familism and child maltreatment in Latino and Anglo families. *Child Maltreatment: Journal of the American Professional Society on the Abuse of Children, 6*(2), 130–142.

Coonrod, D. V., Balcazar, H., Brady, J., Garcia, S., & Van Tine, M. (1999). Smoking, acculturation, and family cohesion in Mexican American women. *Ethnicity and Disease, 9*(3), 434–440.

Costantino, G., Malgady, R., & Rogler, L. (1988). Folk hero modeling therapy for Puerto Rican adolescents. *Journal of Adolescence, 11*(155–166), 166.

Costello, D. M., Swendsen, J., Rose, J. S., & Dierker, L. C. (2008). Risk and protective factors associated with trajectories of depressed mood from adolescence to early adulthood. *Journal of Consulting and Clinical Psychology, 76*(2), 173.

Craik, F. M., Bialystok, E., & Freedman, M. (2010). Delaying the onset of Alzheimer disease: Bilingualism as a form of cognitive reserve. *Neurology, 75*(19), 1726–1729.

Cuéllar, I. (1982). The diagnosis and evaluation of schizophrenic disorders among Mexican Americans. In R. Becerra, M. Karno, & L. Escobar (Eds.), *Mental health and Hispanic Americans: Clinical perspectives* (pp. 61–81). NY: Grune and Stratton.

Cuéllar, I. (2000). Acculturation and mental health: Ecological transactional relations of adjustment. In I. Cuéllar & F. A. Paniagua (Eds.), *Handbook of multicultural mental health* (pp. 45–60). San Diego, CA: Academic Press.

Cuéllar, I., & Glazer, M. (1995). The impact of culture on the family. In M. Harway (Ed.), *Treating the Changing Family* (pp. 17–36). NY: Wiley.

Cuellar, J., & Curry, T. R. (2007). The prevalence and comorbidity between delinquency, drug abuse, suicide attempts, physical and sexual abuse, and self-mutilation among delinquent Hispanic females. *Hispanic Journal of Behavioral Sciences, 29*(1), 68–82. doi:10.1177/0739986306295796.

D'Angelo, E. J., Llerena-Quinn, R., Shapiro, R., Colon, F., Rodriguez, P., & Gallagher, K., et al. (2009). Adaptation of the preventive intervention program for depression for use with predominantly low-income Latino families. *Family Process, 48*, 269–291.

Dana, R. H. (1993). *Multicultural assessment perspectives for professional psychology*. Boston: Allyn and Bacon.

Davis, A. (2008). *Interpersonal and physical dating violence among teens*. FOCUS: Views from the National Council on Crime and Delinquency.

Dawson, B. A., Perez, R. M., & Suárez-Orozco, C. (2012). Exploring differences in family involvement and depressive symptoms across Latino adolescent groups. *Journal of Human Behavior in the Social Environment, 22*(2), 153–171.

De la Torre, A., & Estrada, A. (2001). *Mexican Americans and Health: ¡Sana ! ¡Sana!*. Arizona: University of Arizona Press.

De Luca, S. M., Wyman, P., & Warren, K. (2012). Latina adolescent suicide ideations and attempts: Associations with connectedness to parents, peers, and teachers. *Suicide and Life-Threatening Behavior, 42*(6), 672–683.

Dekovic, M. (1999). Risk and protective factors in the development of problem behavior during adolescence. *Journal of Youth and Adolescence, 28*(6), 667–685.

Delgado-Bernal, D. (2001). Learning and living pedagogies of the home: The Mestiza consciousness of Chicana students. *International Journal of Qualitative Studies in Education, 14*(5), 623–639.

Delgado-Gaitan, C. (1994). Socializing young children in Mexican American families: An intergenerational perspective. In P. M. Greenfield & R. R. Cocking (Eds.), *Cross-cultural roots of minority child development*. Hillsdale: Lawrence Erlbaum Ass. Inc.

Denner, J., & Dunbar, N. (2004). Negotiating feminity: Power and strategies of Mexican American girls. *Sex Roles, 50*(5/6), 301–314.

Donlan, W., & Lee, J. (2010). Coraje, nervios, and susto: Culture-bound syndromes and mental health among Mexican migrants in the United States. *Advances in Mental Health, 9*, 290–303.

Draguns, J. G. (1973). Comparisons of psychopathology across cultures; Issues, findings, directions. *Journal of Cross-cultural Psychology, 4*(1), 10–47.

Duarté-Vélez, Y. M., & Bernal, G. (2007). Suicide risk in Latino and Latina adolescents. In F. T. Leong & M. M. Leach (Eds.), *Suicide among racial and ethnic minority groups: Theory, research, and practice* (pp. 81–116). NY: Taylor and Francis Group.

Duarté-Vélez, Y. M., & Bernal, G. (2008). Suicide risk in Latino and Latina adolescents. *Suicide among Racial and Ethnic Minority Groups,* 81–116.

Eaton, D. K. et al. (2012, June). Youth risk behavior surveillance—United States, 2011. *Morbidity and mortality weekly report, 61*(SS-4), 1–162. Retrieved from http://www.cdc.gov/mmwr/pdf/ss/ss6104.pdf.

Eaton, D. K., Foti, K., Brener, N. D., Crosby, A. E., Flores, G., & Kann, L. (2011). Associations between risk behaviors and suicidal ideation and suicide attempts: Do racial/ethnic variations in associations account for increased risk of suicidal behaviors among Hispanic/Latina 9th-to-12th grade females students? *Archives of Suicide Research, 15*(2), 113–126.

Eaton, D. K., Kann, L., Kinchen, S., Shanklin, S., Ross, J., Hawkins, J., & Harris, W. A., et al. (2008). Youth risk behavior surveillance- United States, 2007. *Morbidity and Mortality Weekly Report, 57*(4), 1–131.

Eccles, J. S., Wong, C. A., & Peck, S. C. (2006). Ethnicity as a social context for the development of African-American adolescents. *Journal of School Psychology, 44*(5), 407–42.

Edwards, L., & Cardemil, E. (in press). Clinical approaches to assessing cultural values among Latinos. In K. Geisinger (Ed.) *Psychological testing of hispanics: Clinical and intellectual issues*.

Edwards, L. M., & Romero, A. J. (2008). Coping with discrimination among Mexican descent adolescents. *Hispanic Journal of Behavioral Sciences., 30*(1), 24–39.

Edwards, L. M., Mayor, R. C., Barber, B. N., Jarrett, K. M., Newcomb, S., Kowalski, A. E., Green, M. B., & Stubbs, L. (under review). Beyond expectations: The cultural context of Latina adolescents' coping with stress.

Espín, O. (1997). *Latina Realities*. Los Angeles: Westview Press.

Espin, O. M. (1987). Issues of identity in the psychology of Latina lesbians. In B. L. P. Collective (Ed.), *Lesbian psychologies: Explorations and challenges* (pp. 348–363). Champaign: University of Illinois Press.

Espino, R., & Franz, M. M. (2002). Latino phenotypic discrimination revisited: The impact of skin color on occupational status. *Social Science Quarterly, 83*(2), 612–623.

Estrada, A. (2009). Mexican Americans and historical trauma theory: A theoretical perspective. *Journal of Ethnicity in Substance Abuse, 8*(3), 330–340.

Fernández, C. A. (1992). La Raza and the melting pot: A comparative look at multiethnicity. In. M. P. P. Root (Ed.), *Racially mixed people in America* (pp. 126–143). Thousand Oaks, CA.

Ferres, I., Garlikov, L., Oppenheim, K., Spoerry, S., Keck, K., & Whitbeck, H. (2006, May). *Thousands march for immigrant rights*. Chicago, Illinois: CNN.

Flores, Y. G. (2013). *Chicana and Chicano mental health: Alma, mente y Corazon*. Arizona: The University of Arizona Press.

Flores-Ortiz, Y. (1993). La mujer y la violencia: A culturally-based model for the understanding and treatment of domestic violence in Chicana/Latina communities. In N. Alarcon (Ed.), *Chicana Critical Issues* (pp. 167–182). Berkeley: Third Woman Press.

Flores-Ortiz, Y. (2004). Domestic violence in Chicano families. In R. Velazquez, L. Arellano, & B. McNeill (Eds.), *The handbook of chicana/o psychology and mental health* (pp. 267–284). New Jersey: Lawrence Erlbaum Associates.

Folkman, S., Lazarus, R. S., Dunkel-Schetter, C., DeLongis, A., & Gruen, R. J. (1986). Dynamics of a stressful encounter: Cognitive appraisal, coping and encounter outcomes. *Journal of Personality and Social Psychology, 50*(5), 992–1003.

Forehand, R., Biggar, H., & Kotchick, B. A. (1998). Cumulative risk across family stressors: Short-and long-term effects for adolescents. *Journal of Abnormal Child Psychology, 26*(2), 119–128.

Fortuna, L. R., Perez, D. J., Canino, G., Sribney, W., & Alegria, M. (2007). Prevalence and correlates of lifetime suicidal ideation and attempts among Latino subgroups in the United States. *The Journal of Clinical Psychiatry, 68*(4), 572.

Frederick, D. A., Forbes, F. B., Grigorian, K. E., & Jarcho, J. M. (2007). The UCLA Body Project I: Gender and ethnic differences in self-objectification and body satisfaction among 2206 undergraduates. *Sex Roles, 57*, 317–327.

Fry, R., Passel, J. S. (2009). *Latino children: A majority are U.S. born offspring of immigrants*. Washington, D.C.: Pew Hispanic Center.

Fry, R., & Taylor, P. (2013). *Hispanic high school graduates pass whites in rate of college enrollment*. Washington, D.C.: Pew Research Hispanic Center.

Gaines, S. O, Jr, Marelich, W. D., Bledsoe, K. L., Steers, W. N., Henderson, M. C., & Granrose, C. S., et al. (1997). Links between race/ethnicity and cultural values as mediated by racial/ethnic identity and moderated by gender. *Journal of Personality and Social Psychology, 72*, 1460–1476.

Garcia, A. (1997). *Chicana feminist thought: The basic historical writings*. NY: Routledge Publishers.

García Coll, C., Lamberty, G., Jenkins, R., McAdoo, H. P., Crnic, K., & Wasik, B. H., et al. (1996). An integrative model for the study of developmental competencies in minority children. *Child Development, 67*, 1891–1914.

Gil, A. G., Vega, W. A., & Biafora, F. (1998). Temporal influences of family structure and family risk factors on drug use initiation in a multiethnic sample of adolescent boys. *Journal of Youth and Adolescence, 27*(3), 373–393.

Gil, A. G., Wagner, E. F., & Vega, W. A. (2000). Acculturation, familism and alcohol use among Latino adolescent males: Longitudinal relations. *Journal of Community Psychology, 28*(4), 443–458.

Gil, R. M., & Vazquez, C. I. (1996). *The Maria paradox*. NY: Perigee Books.

Gloria, A. M., Ruiz, E. L., & Castillo, E. M. (2004). Counseling and psychotherapy with Latino and Latina clients. In T. B. Smith (Ed.), *Practicing multiculturalism: Affirming diversity in counseling and psychology* (pp. 167–189). Boston: Pearson Education, Inc.

Goldstond, D. B., Molock, S. D., Whitbeck, L. B., Murakami, J. L., Zayas, L. H., & Hall, G. C. N. (2008). Cultural considerations in adolescent suicide prevention and psychosocial treatment. *American Psychologist, 63*, 14–31.

Gomez, J., Miranda, R., & Polanco, L. (2011). Acculturative stress, perceived discrimination, and vulnerability to suicide attempts among emerging adults. *Journal of Youth and Adolescence, 40*(11), 1465–1476. doi:10.1007/s10964-011-9688-9.

Gonzales, N., & Kim, L. S. (1997). Stress and coping in an ethnic minority context: Children's cultural ecologies. In S. A. Wolchik & I. N. Sandler (Eds.), *Handbook of children's coping: Linking theory and intervention*. NY: Plenum Press.

Gonzales, P. (2012). *Red medicine: Traditional indigenous rites of birthing and healing*. Arizona: The University of Arizona Press.

Gonzalez, A. (2002). *Mexican Americans and the U.S. economy: Quest for buenos dias*. AZ: University of Arizona Press.

González-López, G. (2004). Fathering Latina sexualities: Mexican men and the virginity of their daughters. *Journal of Marriage and Family, 66*(5), 1118–1130.

Gotlib, I., & Hammen, C. (Eds.). (2002). *Handbook of depression*. NY: Guilford Press.

Gould, M. S., Greenberg, T., Velting, D. M., & Shaffer, D. (2003). Youth suicide risk and preventive interventions: A review of the past 10 years. *Journal of the American Academy of Child and Adolescent Psychiatry, 42*(4), 386–405.

Grabe, S., & Hyde, J. (2006). Ethnicity and body dissatisfaction among women in the United States: A meta-analysis. *Psychological Bulletin, 132*, 622–640.

Greenberg, B. S., & Mastro, D. E. (2008). Children, race, ethnicity, and media. In S. L. Calvert & B. J. Wilson (Eds.), *The handbook of children, media, and development* (pp. 74–97). Malden: Blackwell Publishing.

Greenwald, A. G., Poehlman, T. A., Uhlmann, E. L., & Banaji, M. R. (2009). Understanding and using the implicit association test: III. Meta-analysis of predictive validity. *Journal of Personality and Social Psychology, 97*(1), 17–41.

Guarnaccia, P. J., & Rogler, L. H. (1999). Research on culture-bound syndromes: New directions. *American Journal of Psychiatry, 156*, 1322–1327.

Guiao, I. Z., & Esparza, D. (1995). Suicidality correlates in Mexican American teens. *Issues in Mental Health Nursing, 16*(5), 461–479.

Gulbas, L. E., Zayas, L. H., Nolle, A. P., Hausmann-Stabile, C., Kuhlberg, J. A., & Baumann, A. A., et al. (2011). Family relationships and Latina teen suicide attempts: Reciprocity, asymmetry, and detachment. *Families in Society: The Journal of Contemporary Social Services, 92*(3), 317–323. doi:10.1606/1044-3894.4131.

Gutman, L. M., Sameroff, A. J., & Eccles, J. S. (2002). The academic achievement of African American students during early adolescence: An examination of multiple risk, promotive, and protective factors. *American Journal of Community Psychology, 30*(3), 367–399.

Helms, J. E., Jernigan, M., & Mascher, J. (2005). The meaning of race in psychology and how to change it: A methodological perspective. *American Psychologist, 60*, 27–36.

Hong, J. S., Espelage, D. L., & Kral, M. J. (2011). Understanding suicide among sexual minority youth in America: An ecological systems analysis. *Journal of Adolescence, 34*(5), 885–94.

Hovey, J. D. (1998). Acculturative stress, depression, and suicidal ideation among Mexican American adolescents: Implications for the development of suicide prevention programs in schools. *Psychological Reports, 83*(1), 249–250.

Hovey, J. D. (2000). Acculturative stress, depression and suicidal ideation in Mexican immigrants. *Cultural Diversity and Ethnic Minority Psychology, 6*(2), 134–151. doi:10.1037//1099-9809.6.2.134.

Hovey, J. D., & King, C. A. (1996). Acculturative stress, depression, and suicide ideation among immigrant and second generation Latino adolescents. *Journal of the American Academy of Child and Adolescent Psychiatry, 35*(9), 1183–1192.

Howard, D. E., Beck, K., Kerr, M., & Shattuck, T. (2005). Psychosocial correlates of dating violence victimization among Latino youth. *Adolescence, 40*(158), 319–331.

Humes, K., & Hogan, H. (2009). Measurement of race and ethnicity in a changing, multicultural America. *Race and Social Problems, 1*, 111–131.

Hurtado, A. (2003). *Voicing Chicana feminism: Young women speak out on sexuality and identity*. NY: New York University Press.

Hurtado, A., & Cervantez, K. (2009). A view from within and from without: The development of Latina feminist psychology. In F. Villarruel, G. Carlo, J. Grau, M. Azmitia, N. Cabrera, and T. J. Chahin (Eds.), *Handbook of U.S. Latino Psychology: Development and Community-based Perspectives*. Los Angeles, CA: Sage Publications.

Jamieson, P. E., & Romer, D. (2008). Unrealistic fatalism in U.S. youth ages 14 to 22: Prevalence and characteristics. *Journal of Adolescent Health, 42*(2), 154–160.

Joiner, T. (2005). *Why people die by suicide*. Cambridge: Harvard University Press.

Kaltiala-Heino, R., Rimpela, M., Mattunen, M., Rimpela, A., & Rantanen, P. (1999). Bullying, depression, and suicidal ideation in Finnish adolescents: School survey. *British Medical Journal, 319*, 348–351.

Keefe, S., & Padilla, A. (1987). *Chicano ethnicity*. Albuquerque: University of New Mexico Press.

Kim, B. S. K. (2007). Acculturation and enculturation. In F. T. L. Leong, A. G. Inman, A. Ebreo, L. Yang, L. Kinoshita, & M. Fu (Eds.), *Handbook of Asian American psychology* (2nd ed., pp. 141–158). Thousand Oaks: Sage.

Kim, Y. S., & Leventhal, B. (2008). Bullying and suicide: A review. *International Journal of Adolescent Medical Health, 20*(2), 133–154.

Klein, K., & Forehand, R. (2000). Family processes as resources for African American children exposed to a constellation of sociodemographic risk factors. *Journal of Clinical Child Psychology, 29*(1), 53–65.

Kleinman, A. (1988). *Rethinking psychiatry*. NY: Free Press.

Kleinman, A. M., & Good, B. (1985). *Culture and depression*. Berkeley: University of California Press.

Klerman, G. L., Weissman, M. M., Rounsaville, B., & Chevron, E. S. (1984). *Interpersonal psychotherapy of depression*. NY: Basic Books.

Klomek, A. B., Marrocco, F., Kleinman, M., Schonfeld, I. S., & Gould, M. S. (2007). Bullying, depression, and suicidality in adolescents. *Journal of the American Academy of Child and Adolescent Psychiatry, 46*(1), 40–49.

Klomek, A. B., Sourander, A., & Gould, M. S. (2010). The association of suicide and bullying in childhood to young adulthood: A review of cross-sectional and longitudinal research. *Canadian Journal of Psychiatry, 55*(5), 282–288.

Klomek, A. B., Sourander, A., & Gould, M. S. (2011). Bullying and Suicide. *Psychiatric Times, 28*(2), 27–31.

Klomek, A. B., Sourander, A., Niemela, S., Kumupulainen, K., Piha, J., & Tamminen, T., et al. (2009). Childhood bullying behaviors as a risk for suicide attempts and completed suicides: A population-based birth cohort study. *Journal of the American Academy of Child and Adolescent Psychiatry, 48*(3), 254–261.

Knight, G. P., Bernal, M., Garza, C. A., & Cota, M. K. (1993). A social cognitive model of the development of ethnic identity and ethnically based behaviors. In M. E. Bernal & G. P. Knight (Eds.), *Ethnic identity: Formation and transmission among Hispanics and other minorities* (pp. 213–234). Albany: State University of New York Press.

Kornblau, I. S., Pearson, H. C., & Breitkopf, C. R. (2007). Demographic, behavioral, and physical correlates of body esteem among low-income female adolescents. *Journal of Adolescent Health, 41*(6), 566–70.

Krishnan, S. P., Hilbert, J. C., & VanLeeuwen, D. (2001). Domestic violence and help-seeking behaviors among rural women: Results from a shelter study. *Family and Community Health, 24*, 28–38.

Kuhlberg, J. A., Peña, J. B., & Zayas, L. H. (2010). Familism, parent-adolescent conflict, self-esteem, internalizing behaviors and suicide attempts among adolescent Latinas. *Child Psychiatry and Human Development, 41*(4), 425–440.

Larkin, G. L., Rivera, H., Xu, H., Rincon, E., & Beautrais, A. L. (2011). Community responses to a suicidal crisis: Implications for suicide prevention. *Suicide and Life-Threatening Behavior, 41*(1), 79–86.

Lazarus, R. S. (1997). Acculturation isn't everything. *Applied Psychology, 46*, 39–43.

Lewis, T. T., Barnes, L. L., Bienias, J. L., Lackland, D. T., Evans, D. A., & de Leon, C. (2009). Perceived discrimination and blood pressure in older African American and White adults.

*The Journals of Gerontology: Series A: Biological Sciences and Medical Sciences, 64A*(9), 1002–1008.
Lopez, S. R. (1989). Patient variable biases in clinical judgment: Conceptual overview and methodological considerations. *Psychological Bulletin, 106*(2), 184–203.
Loukas, A., & Prelow, H. M. (2004). Externalizing and internalizing problems in low-income Latino early adolescents risk, resource, and protective factors. *The Journal of Early Adolescence, 24*(3), 250–273.
Lugo Steidel, A. G., & Contreras, J. M. (2003). A new familism scale for use with Latino populations. *Hispanic Journal of Behavioral Science, 25*(3), 312–330.
Luthar, S. S., Cicchetti, D., & Becker, B. (2000). The construct of resilience: A critical evaluation and guidelines for future work. *Child Development, 71*(3), 543–562.
Manago, A. M., Brown, C. S., & Leaper, C. (2009). Feminist identity among Latina adolescents. *Journal of Adolescent Research, 24*(6), 750–776.
Marsella, A. J., & Kameoka, V. A. (1989). Ethnocultural issues in the assessment of psychopathology. In S. Wetzler (Ed.), *Measuring mental illness: Psychometric assessment for clinicians*. Washington, D.C.: American Psychiatric Press, Inc.
Marsella, A. J., & Yamada, A. M. (2000). Culture and mental health: An introduction and overview of foundations, concepts and issues. In I. Cuéllar & F. A. Paniagua (Eds.), *Handbook of multicultural mental health* (pp. 3–22). San Diego, CA: Academic Press.
Martín-Baro, I., Aron, A., & Corne, S. (1994). *Writings for a liberation psychology*. Cambridge: Harvard University Press.
Martinez, E. (1998). *De colores means all of us: Latina views for a multi-colored century*. Cambridge: South End Press.
Mastro, D. E., Behm-Morawitz, E., & Kopacz, M. A. (2008). Exposure to television portrayals of Latinos: The implications of aversive racism and social identity theory. *Human Communication Research, 34*(1), 1–27.
Mather, M., & Foxen, P. (2010). *America's Future: Latino child well-being in numbers and trends*. Washington: National Council of La Raza.
Mayer, V. (2003). Living telenovelas/telenovelizing life: Mexican American girl's identities and transnational telenovelas. *Journal of Communication, 53*(3), 479–495.
McLean, J. T., Gilligan, C., & Sullivan, A. M. (1995). Girls, risk and resilience. In J. T. McLean, C. Giligan, & A. M. Sullivan (Eds.), *Between voice and silence: Women and girls, race and relationship*. Cambridge: Harvard University Press.
McNeill, B., Niemann, Y. F., & Pizarro, M., et al. (2001). Current directions in Chicana/o Psychology. *The Counseling Psychologist, 29*(1), 5–17.
Merskin, D. (2007). Three faces of Eva: Perpetuation of the hot-Latina stereotype in desperate housewives. *Howard Journal of Communications, 18*(2), 133–151.
Meyers, I. H. (2004). Prejudice as stress: Conceptual and measurement problems. *American Journal of Public Health, 93*, 262–265.
Mills, C., Guerin, S., Lunch, F., Daly, I., & Fitzpatrick, C. (2004). The relationship between bullying, depression and suicidal thoughts/behavior in Irish adolescents. *Irish Journal of Psychiatric Medicine, 21*(4), 112–116.
Miranda, J., Bernal, G., Lau, A., Kohn, L., Hwang, W-C., & LaFromboise, T. (2005). State of the science on psychosocial interventions for ethnic minorities. *Annual Review of Clinical Psychology, 1*, 113–142.
Moane, G. (2003). Bridging the personal and the political: Practices for a liberation psychology. *American Journal of Community Psychology, 31*(1), 91–101.
Molina, C. W., & Aguirre-Molina, M. (1994). *Latino health in the U.S.: A growing challenge*. Washington, D.C.: American Public Health Association.
Mufson, L., Weissman, M. M., Moreau, D., & Garfinkel, R. (1999). Efficacy of interpersonal psychotherapy for depressed adolescents. *Archives of General Psychiatry, 56*(6), 573–9.
Muñoz, J. P. (2007). Culturally responsive caring in occupational therapy. *Occupational Therapy International, 14*, 256–280. doi:10.1002/oti.238.
Muñoz, C. (1989). *Youth, Identity, Power: The Chicano Movement*. London: Verso.

Nazroo, J. Y. (2001). *Ethnicity class and health*. London: PSI Press.

Negy, C., & Woods, D. J. (1992). The importance of acculturation in understanding research with Hispanic Americans. *Hispanic Journal of Behavioral Sciences, 14*, 224–247.

Niemann, Y. F. (2004). Stereotypes of Chicanas and Chicanos: Impact on family functioning, individual expectations, goals, and behaviors. In R. Velasquez, L. M. Arellano, & B. McNeill (Eds.), *The handbook of chicana/o psychology and mental health (pgs* (pp. 215–230). Mahwah: Lawrence Erlbaum Associates, Inc.

Niemann, Y. F., Romero, A. J., Arredondo, J., & Rodriguez, V. (1999). What does it mean to be "Mexican" or "Mexican American": Lay person's construction of their ethnic identity. *Hispanic Journal of Behavioral Sciences, 21*(1), 47–60.

Niemann, Y. F., Romero, A. J., & Arbona, C. (2000). Effects of cultural orientation on the perception of conflict between relationship and education goals for Mexican American college students. *Hispanic Journal of Behavioral Sciences., 22*(1), 46–63.

O'Leary, A. (2009). *Arizona's legislative-imposed injunctions: Implications for immigrant civic and political participation.* Research Paper Series on Latino Immigrant Civic and Political Participation, No. 1. Woodrow Wilson Center.

O'Leary, A., & Romero, A. J. (2011). Undergraduate student's engagement with Arizona Senate Bill 1108 "Anti-Ethnic Studies": Influences of civic engagement, ethnic identity on well-being. *Aztlan, 36*, 9–36.

O'Donnell, L., O'Donnell, C., Wardlaw, D. M., & Stueve, A. (2004). Risk and resiliency factors influencing suicidality among urban African American and Latino youth. *American Journal of Community Psychology, 33*(1–2), 37–49.

Ocampo, B., Shelley, G., & Jaycox, L. (2007). Latino teens talk about help seeking and help giving in relation to dating violence. *Violence against Women, 13*(2), 172–189.

Oetting, E. R., & Beauvais, F. (1990–1991). Orthogonal cultural identification theory: The cultural identification of minority adolescents. *The International Journal of the Addictions, 25*(5a–6a), 655–685.

Ojeda, L., & Liang, C. T. H. (in press). Ethnocultural and gendered determinants of coping among Mexican American adolescent men. *Psychology of Men and Masculinity*.

Olson, L. M., Wahab, S., Thompson, C. W., & Durrant, L. (2011). Suicide notes among Native Americans, Hispanics, and Anglos. *Qualitative Health Research, 21*(11), 1484–1494 (published online 17 June 2011).

Orth, U., Robins, R. W., & Roberts, B. W. (2008). Low self-esteem prospectively predicts depression in adolescence and young adulthood. *Journal of Personality and Social Psychology, 95*(3), 695–708. doi:10.1037/0022-3514.95.3.69.

Pabon, E. (1998). Hispanic adolescent delinquency and the family: A discussion of sociocultural influences. *Adolescence, 33*(132), 941–955.

Padilla, A. M., & Perez, W. (2003). Acculturation, social identity, and social cognition: A new perspective. *Hispanic Journal of Behavioral Sciences, 25*, 35–55.

Patchin, J. W., & Hinduja, S. (2010). Cyberbullying and self-esteem. *Journal of School Health, 80*(12), 614–621.

Pedrotti, J. P., & Edwards, L. M. (2009). *The intersection of positive psychology and multiculturalism in counseling. Handbook of multicultural counseling* (3rd ed.). Thousand Oaks: Sage Publications.

Pedrotti, J. P., Edwards, L. M., & Lopez, S. J. (2009). Positive psychology within a cultural context. In S. J. Lopez (Ed.), *Handbook of positive psychology* (2nd ed., pp. 49–58). NY: Oxford Press.

Peña, E. (2003). Reconfiguring epistemological pacts: Creating a dialogue between psychoanalysis and Chicano/a subjectivity, a cosmopolitan perspective. *Journal for the Psychoanalysis of Culture and Society, 8*(2), 308–319.

Peña, J. B., Kuhlberg, J. A., Zayas, L. H., Baumann, A. A., Gulbas, L., & Hausmann-Stable, C., et al. (2011). Familism, family environment, and suicide attempts among Latina youth. *Suicide Life Threat Behav, 41*(3), 330–341. doi:10.1111/j.1943-278X.2011.00032.x.

Peña, J. B., Wyman, P. A., Brown, C. H., Matthieu, M. M., Olivares, T. E., & Hartel, D., et al. (2008). Immigration generation status and its association with suicide attempts, substance use, and depressive symptoms among Latino adolescents in the USA. *Prevention Science, 9*(4), 299–310.
Peragallo, N. (1996). Latino women and AIDS risk. *Public Health Nursing, 13*(3), 217–222.
Phinney, J. S. (1991). Ethnic identity self-esteem: A review and integration. *Hispanic Journal of Behavioral Sciences, 13*, 193–208.
Phinney, J. S. (1996). When we talk about American ethnic groups, what do we mean? *American Psychologist, 51*(9), 918–927.
Phinney, J. S., & Alipuria, L. L. (1990). Ethnic identity in college students from four ethnic groups. *Journal of Adolescence, 13*, 171–183.
Poran, M. A. (2002). Denying diversity: Perceptions of beauty and social comparison processes among Latina, black, and white women. *Sex Roles, 47*(1/2), 65–81.
Potochnick, S. R., & Perreira, K. M. (2010). Depression and anxiety among first-generation immigrant Latino youth: Key correlates and implications for future research. *The Journal of Nervous and Mental Disease, 198*(7), 470.
Queralt, M. (1993). Risk factors associated with completed suicide in Latino adolescents. *Adolescence, 28*(112), 831–850.
Quintana, S. M., & Vera, E. M. (1999). Mexican American children's ethnic identity, understanding of ethnic prejudice, and parental ethnic socialization. *Hispanic Journal of Behavioral Sciences, 21*(4), 387–404.
Ramirez, M. (1998). *Multicultural/multiracial psychology*. Maryland: Jason Aronson.
Rasmussen, K. M., Negy, C., Carlson, R., & Burns, J. M. (1997). Suicide ideation and acculturation among low socioeconomic status Mexican American adolescents. *Journal of Early Adolescence, 17*(4), 390–407.
Redfield, R., Linton, R., & Herskovitz, M. J. (1936). Memorandum for the study of acculturation. *American Anthropologist, 38*(1), 149–152.
Resnick, M. D., Bearman, P. S., Blum, R. W., Bowman, K. E., Harris, K. M., & Jones, J., et al. (1997). Protecting adolescents from harm: Findings from the national longitudinal study on adolescent health. *Journal of the American Medical Association, 278*, 823–833. doi:10.1001/jama.1997.03550100049038.
Rew, L., Thomas, N., Horner, S. D., Resnick, M. D., & Beuhring, T. (2001). Correlates of recent suicide attempts in a triethnic group of adolescents. *Journal of Nursing Scholarship, 33*(4), 361–367.
Rew, L., Resnick, M. D., & Blum, R. W. (1997). An exploration of help-seeking behaviors in female Hispanic adolescents. *Family and Community Health, 20*(3), 1–15.
Rivera, L. M. (2008). Acculturation and multicultural assessment: Issues, trends, and practice. In L. A. Suzuki & J. G. Ponterotto (Eds.), *Handbook of multicultural assessment* (3rd ed., pp. 73–91). San Francisco: Jossey-Bass.
Roberts, T. A., Auinger, P., & Klein, J. D. (2005). Intimate partner abuse and the reproductive health of sexually active female adolescents. *Journal of Adolescent Health, 36*, 380–385.
Roberts, R. E., & Chen, Y.-W. (1995). Depressive symptoms and suicidal ideation among Mexican-origin and Anglo adolescents. *Journal of the American Academy of Child and Adolescent Psychiatry, 34*(1), 81–90.
Roberts, R. E., Roberts, C. R., & Chen, Y.-W. (1997). Ethnocultural differences in prevalence of adolescent depression. *American Journal of Community Psychology, 25*(1), 95–110.
Rodriguez, C. E. (2000). *Changing race: Latinos, the census, and the history of ethnicity in the United States*. NY: New York University Press.
Rodriguez, N., Mira, C. B., Paez, N. D., & Myers, H. F. (2007). Exploring the complexities of familism and acculturation: Central constructs for people of Mexican origin. *American Journal of Community Psychology, 39*(1–2), 61–77.
Rodriguez, T., & Montané, D. (2007). *The daughters of Juarez: A true story of serial murder south of the border*. Los Angeles: Atria.

Rogler, L. H., Cortes, D. E., & Malgady, R. G. (1991). Acculturation and mental health status among hispanics: Convergence and new directions for research. *American Psychologist, 46*, 585–597.

Romero, A. J., Bracamonte Wiggs, C., Valencia, C., & Bauman, S. (2013). Latina teen suicide and bullying. *Hispanic Journal of Behavioral Sciences, 35*(2), 159–173.

Romero, A. J., Carvajal, S. C., Valle, F., & Oduña, M. (2007). Adolescent bicultural stress and its impact on mental well-being among Latinos, Asian Americans, and European Americans. *Journal of Community Psychology, 35*, 519–534.

Romero, A. J., Cuéllar, I., & Roberts, R. E. (2000). Ethnocultural variables and attitudes toward cultural socialization of children. *Journal of Community Psychology, 28*, 79–89.

Romero, A. J., Martinez, D., & Carvajal, S. C. (2007). Bicultural stress and adolescent risk behaviors in a community sample of Latinos and non-Latino European Americans. *Ethnicity and Health, 12*, 443–463.

Romero, A. J., & Roberts, R. E. (2003a). Stress within a bicultural context for adolescents of Mexican descent. *Cultural Diversity and Ethnic Minority Psychology, 9*(2), 171–184.

Romero, A. J., & Roberts, R. E. (2003b). The impact of multiple dimensions of ethnic identity on discrimination and adolescent's self-esteem. *Journal of Applied Social Psychology, 33*(11), 2288–2305.

Romero, A. J., & Ruíz, M. G. (2007). Does familism lead to increased parental monitoring?: Protective factors for coping with risky behaviors. *Journal of Child and Family Studies, 16*, 143–154.

Romero, A. J., Wiggs, C. B., Valencia, C., & Baumann, S. (2013). Latina teen suicide and bullying. *Hispanic Journal of Behavioral Sciences. 35*, 159–173

Rosaldo, R. (1993). Chapter 9: Border crossings. *Culture and truth: The remaking of social analysis* (pp. 196–217). Boston: Beacon Press.

Rosario, M., Schrimshaw, E. W., & Hunter, J. (2004). Ethnic/racial differences in the coming-out process of lesbian, gay, and bisexual youths: A comparison of sexual identity development over time. *Cultural Diversity and Ethnic Minority Psychology, 10*(3), 215–228.

Rosselló, J., & Bernal, G. (1999). The efficacy of cognitive-behavioral and interpersonal treatments for depression in Puerto Rican adolescents. *Journal of Consulting and Clinical Psychology, 67*, 734–745.

Rosselló, J., & Bernal, G. (2005). New developments in cognitive-behavioral and interpersonal treatments for depressed Puerto Rican adolescents. In E. D. Hibbs, P. S. Jensen (Eds.), *Psychosocial treatments for child and adolescent disorders: Empirically based strategies for clinical practice* (2nd ed., pp. 187–217). Washington, D.C.: American Psychological Association (xv, p. 839).

Rotheram-Borus, M. J., Piancentini, J., & Van Rossem, R., et al. (1999). Treatment adherence among Latina female adolescent suicide attempters. *Suicide and Life-Threatening Behavior, 29*(4), 319–331.

Rutter, M. (1979). Protective factors in children's responses to stress and disadvantage. *Annals of the Academy of Medicine, Singapore, 8*(3), 324–338.

Rutter, M. (1987). Psychosocial resilience and protective mechanisms. *American Journal of Orthopsychiatry, 57*, 316–331.

Sabogal, F., Marin, G., Otero-Sabogal, R., & Marin, B., et al. (1987). Hispanic familism and acculturation: What changes and what doesn't. *Hispanic Journal of Behavioral Sciences, 9*(4), 397–412.

Sameroff, A. J., Seifer, R., & Bartko, W. T. (1997). Environmental perspectives on adaptation during childhood and adolescence.

Santa Ana, O., & de Bustamante, C. G. (Eds.). (2012). *Arizona firestorm: Global immigration realities, national media, and provincial politics*. USA: Rowman and Littlefield Publishers.

Santisteban, D., Szapocznik, J., Perez-Vidal, A., Kurtines, W., Murray, E. J., & La Perriere, A. (1996). Efficacy of intervention for engaging youth and families into treatment and some variables that may contribute to differential effectiveness. *Journal of Family Psychology, 10*(1), 35–44.

Schmader, T., & Johns, M. (2003). Converging evidence that stereotype threat reduces working memory capacity. *Journal of Personality and Social Psychology, 85*(3), 440–452.

Schmader, T., Major, B., & Gramzow, R. H. (2001). Coping with ethnic stereotypes in the academic domain: Perceived injustice and psychological disengagement. *Journal of Social Issues, 57*(1), 93–111.

Schneider, S. K., O'Donnell, L., Stueve, A., & Coulter, R. W. (2012). Cyberbullying, school bullying, and psychological distress: A regional census of high school students. *Journal Information, 102*,1.

Schwartz, S. J., Unger, J. B., Zamboanga, B. L., & Szapocznik, J. (2010). Rethinking the concept of acculturation: Implications for theory and research. *American Psychologist, 65*(4), 237–251.

Seligman, M. E. P., & Csikszentmihalyi, M. (2000). Positive psychology: An introduction. *American Psychologist, 55*, 5–14.

Sellers, R. M., & Shelton, J. N. (2003). The role of racial identity in perceived racial discrimination. *Journal of Personality and Social Psychology, 84*(5), 1079–1092.

Sherrod, L., Flanagan, C. A., Kassimir, R., & Syversten, A. K. (2006). *Youth activism: an international encyclopedia*. Westport: Greenwood Press/Greenwood Publishing Group.

Shields, M. K., & Berhman, R. E. (2009). Children of immigrant families: Analysis and recommendations. *Children of Immigrant Families, 14*(2), 4–15.

Shweder, R. A. (1993). The cultural psychology of the emotions. In M. Lewis & J. M. Haviland (Eds.), *Handbook of emotions*. NY: The Guilford Press.

Silverman, J. G., Raj, A., & Clements, K. (2004). Dating violence and associated sexual risk and pregnancy among adolescent females in the United States. *Pediatrics, 114*, e220–e225.

Silverman, J., Raj, A., Mucci, L., & Hathaway, J. (2001). Dating violence against adolescent girls and associated substance use, unhealthy weight control, sexual risk behavior, pregnancy, and suicidality. *Journal of the American Medical Association, 286*, 572–579.

Smith, T., Domenech Rodriguez, M., & Bernal, G. (2011). Culture. *Journal of Clinical Psychology, 67*, 166–175.

Spears - Brown, C., & Bigler, R. S. (2005). Children's perceptions of discrimination: A developmental model. *Child Development, 76*(3), 533–553.

Spicer, R. S., & Miller, T. R. (2000). Suicide acts in 8 states: Incidence and case fatality rates by demographics and method. *American Journal of Public Health, 90*(12), 1885–1891.

Steele, M., & Doey, T. (2007). Suicidal behavior in children and adolescents. Part 1: Etiology and risk factors. *Canadian Journal of Psychiatry, 52*(6), 21.

Steidel, A. G. L., & Contreras, J. M. (2003). A new familism scale for use with Latino populations. *Hispanic Journal of Behavioral Sciences, 25*(3), 312–330.

Stephens, D. P., & Few, A. L. (2007). The effects of images of African American women in hip hop on early adolescents' attitudes toward physical attractiveness and interpersonal relationships. *Sex Roles, 56*(3–4), 251–264.

Suicide Prevention Resource Center. (2011). *Suicide among racial/ethnic populations in the U.S.: Hispanics*. Newton, MA: Education Development Center, Inc.

Szapocznick, J., Scopetta, M. A., Kurtines, W., & Aranaldi, D. (1978). Theory and measurement of acculturation. *Revista Interamericana de Psicologia, 12*, 113–130.

Szapocznik, J., & Coatsworth, J. D. (1999). An ecodevelopmental framework for organizing the influences on drug abuse: A developmental model of risk and protection. In M. Glantz & C. R. Hartel (Eds.), *Drug abuse: Origins and interventions* (pp. 331–366). Washington, D.C.: American Psychological Association.

Teten, A. L., Ball, B., Valle, L. A., Noonan, R., & Rosenbluth, B. (2009). Considerations for the definition, measurement, consequences, and prevention of dating violence victimization among adolescent girls. *Journal of Women's Health, 18*, 923–927.

Thompson, A. (2008). *A child alone and without papers: A report on the return and repatriation of unaccompanied undocumented children by the United States*. Austin, Texas: Center for Public Policy Priorities.

Trautman, E. C. (1961). The suicidal fit: A psychobiologic study on Puerto Rican immigrants. *Archives of General Psychiatry, 5*, 98–105.

Trimble, J. E. (1990–1991). Ethnic specification, validation prospects, and the future of drug use research. *International Journal of the Addictions, 25,* 149–170.
Turner, S. G., Kaplan, C. P., Zayas, L., & Ross, R. E. (2002). Suicide attempts by adolescent Latinas: An exploratory study of individual and family correlates. *Child and Adolescent Social Work Journal, 19*(5), 357–374.
U.S. Census Bureau. (2009). *Latinos in the United States*. Retrieved from http://www.census.gov/population/www/socdemo/Latino/Latino_pop_presentation.ht.
U.S. Census Bureau. (2010). *Hispanics in the United States.* Retrieved from http://www.census.gov/population/www/socdemo/hispanic/hispanic_pop_presentation.ht.
U.S. Census Bureau. (2011). *The Hispanic Population 2010*. Retrieved from http://www.census.gov/prod/cen2010/briefs/c2010br-04.pdf.
U.S. Census Bureau. (2012). *Facts for features: Hispanic heritage month 2012 Sept. 15–Oct. 15.* Retrieved from: http://www.census.gov/newsroom/releases/archives/facts_for_features_special_editions/cb12-ff19.html.
U.S. Department of Health and Human Services. (2001). *Mental health: Culture, race, and ethnicity—A supplement to mental health: A report of the surgeon general.* Rockville: U.S. Department of Health and Human Services, Substance Abuse and Mental Health Services Administration, Center for Mental Health Services.
Ulloa, E. C., Jaycox, L. H., Marshall, G. N., & Collins, R. L. (2004). Acculturation, gender stereotypes, and attitudes about dating violence. *Violence and Victims, 19,* 273–287.
Ulloa, E., Jaycox, L. H., Skinner, S., & Orsburn, M. (2011). Attitudes about violence and dating among Latino/a boys and girls. *Journal of Ethnic and Cultural Diversity in Social Work, 17*(2), 157–176. doi:10.1080/15313200801941721.
Umaña-Taylor, A. J., Alfaro, E. C., Bámaca, M. Y., & Guimond, A. B. (2009). The central role of familial ethnic socialization in Latino adolescents' cultural orientation. *Journal of Marriage and Family, 71*(1), 46–60.
Umaña-Taylor, A. J., Diversi, M., & Fine, M. A. (2002). Ethnic identity and self-esteem of Latino adolescents: Distinctions among the Latino populations. *Journal of Adolescent Research, 17*(3), 303–327.
Umaña-Taylor, A. J., & Fine, M. A. (2001). Methodological implications of grouping Latino adolescents into one collective ethnic group. *Hispanic Journal of Behavioral Sciences, 23*(4), 347–362.
Umaña-Taylor, A. J., & Updegraff, K. A. (2007). Latino adolescents' mental health: Exploring the interrelations among discrimination, ethnic identity, cultural orientation, self-esteem, and depressive symptoms. *Journal of Adolescence, 30,* 549–567.
Ungar, M., Brown, M., & Liebenberg, L., et al. (2007). Unique pathways to resilience across cultures. *Adolescence, 42*(166), 287–310.
Unger, J. B., Ritt-Olson, A., Teran, L., Huang, T., Hoffman, B., & Palmer, P. (2002). Cultural values and substance use in a multiethnic sample of California adolescents. *Addiction Research and Theory, 10*(3), 257–280.
Van der Wal, M. F., de Wit, C. A. M., & Hirasing, R. A. (2003). Psychosocial health among young victims and offenders of direct and indirect bullying. *Pediatrics, 111*(6), 1312–1317.
Van Orden, K. A., Witte, T. K., Cukrowicz, K. C., Braithwaite, S. R., Selby, E. A., & Joiner, J. E. (2010). The interpersonal theory of suicide. *Psychological Review, 117*(2), 575–600. doi:10.1037/a0018697.
Villarruel, F. A., Carlo, G., Grau, J., Azmitia, M., Cabrera, N., & Chahin, T. (2009). *Handbook of US Latino psychology.*
Wadsworth, T., & Kubrin, C. E. (2007). Hispanic suicide in US metropolitan areas: Examining the effects of immigration, assimilation, affluence, and disadvantage. *American Journal of Sociology, 112*(6), 1848–1885.
Walsh, M., Smith, R. A., Morales, A., & Sechrest, L. (2000). *Ecocultural research: A mental health researcher's guide to studying race, ethnicity and culture.* Cambridge: Human Research Services Center.
Walters, K., & Simoni, J. (2002). Reconceptualizing native women's health: An "Indigenist" stress-coping model. *American Journal of Public Health, 92,* 520–524.

Warren, C. S., Gleaves, D. H., Cepeda-Benito, A., Fernandez Mde, C., & Rodriguez-Ruiz, S. (2005). Ethnicity as a protective factor against internalization of a thin ideal and body dissatisfaction. *The International Journal of Eating Disorders, 37*(3), 241–249.
Wood, N. (1992). *Many winters: Prose and poetry of the pueblos.* NY: Doubleday for Young Readers.
Zayas, L. H. (1987). Toward an understanding of suicide risks in young Hispanic females. *Journal of Adolescent Research, 2*(1), 1–11.
Zayas, L. H. (2011). *Latinas attempting suicide.* NY: Oxford University Press.
Zayas, L. H., & Gulbas, L. E. (2012). Are suicide attempts by young Latinas a cultural idiom of distress? *Transcultural Psychiatry, 49*(5), 718–734.
Zayas, L. H., Gulbas, L. E., Fedoravicius, N., & Cabassa, L. J. (2010). Patterns of distress, precipitating events, and reflections on suicide attempts by young Latinas. *Social Science and Medicine, 70*(11), 1773–1779.
Zayas, L. H., Lester, R., Cabassa, L. J., & Fortuna, L. R. (2005). Why do so many Latina teens commit suicide? A conceptual model for research. *American Journal of Orthopsychiatry, 75*(2), 275–287.
Zayas, L. H., Kaplan, C., Romano, K., & Gonzalez-Ramos, G. (2000). Understanding suicide attempts by adolescent Hispanic Females. *Social Work, 45*(1), 53–63.
Zayas, L. H., & Pilat, A. M. (2008). Suicidal behavior in Latinas: Explanatory cultural factors and implications for intervention. *Suicide and Life Threatening Behavior, 38*(3), 334–342.

# Index




A. J. Romero et al., *Preventing Adolescent Depression and Suicide Among Latinas*,
SpringerBriefs in Psychology, DOI: 10.1007/978-3-319-01381-7,

MIX
Papier aus verantwortungsvollen Quellen
Paper from responsible sources
FSC® C105338

If you have any concerns about our products,
you can contact us on
**ProductSafety@springernature.com**

In case Publisher is established outside the EU,
the EU authorized representative is:
**Springer Nature Customer Service Center GmbH**
**Europaplatz 3, 69115 Heidelberg, Germany**

Printed by Libri Plureos GmbH
in Hamburg, Germany

# SpringerBriefs in Information Systems

**Series editor**

Jörg Becker

More information about this series at http://www.springer.com/series/10189

Jan vom Brocke · Armin Stein
Sara Hofmann · Sanja Tumbas

# Grand Societal Challenges in Information Systems Research and Education

## Ideas from the ERCIS Virtual Seminar Series

Springer

Jan vom Brocke
Department of Information Systems
University of Liechtenstein
Vaduz
Liechtenstein

Sara Hofmann
ERCIS
University of Münster
Münster
Germany

Armin Stein
ERCIS
University of Münster
Münster
Germany

Sanja Tumbas
Department of Information Systems
University of Liechtenstein
Vaduz
Liechtenstein

ISSN 2192-4929 ISSN 2192-4937 (electronic)
SpringerBriefs in Information Systems
ISBN 978-3-319-15026-0 ISBN 978-3-319-15027-7 (eBook)
DOI 10.1007/978-3-319-15027-7

Library of Congress Control Number: 2015932446

Springer Cham Heidelberg New York Dordrecht London


Printed on acid-free paper

Springer International Publishing AG Switzerland is part of Springer Science+Business Media (www.springer.com)

# Foreword

In the 1980s, the United States tackled the rise of Japanese computer research, particularly their "5th Generation Computer," by introducing "Grand Challenges." Engineers and scientists were called to action to address particular challenges to keep the Unites States' competitive advantage. These Grand Challenges were revised by the National Science Foundation in 2009 but have remained largely the same over time.[1]

Grand Challenges are defined for disciplines (like information and communication technology[2] and health[3]) and for regions (like Canada[4] and South Africa[5]). The White House provides the several attributes that Grand Challenges feature[6]:

- They help create the industries and jobs of the future.
- They expand the frontiers of human knowledge about ourselves and the world around us.
- They help to address important problems related to energy, health, education, the environment, national security, and global development.
- They serve as a "North Star" for collaboration between the public and private sectors.

Grand Challenges for the information systems (IS) discipline are not new: Key issues in management information systems were published in 1984 (Dickson et al.), 1987 (Brancheau and Wetherbe), 1990 (Niederman et al.), and 1996 (Brancheau et al.) in *Management Information Systems Quarterly* (*MISQ*). In 2006 the topic of Grand Challenges in IS was reprised at a panel discussion at the European Conference on Information Systems (ECIS) in Göteborg, Sweden, was

[1] http://www.nsf.gov/cise/aci/taskforces/TaskForceReport_GrandChallenges.pdf.

[2] http://www.futurict.eu/.

[3] http://www.grandchallenges.org.

[4] http://www.grandchallenges.ca.

[5] http://www.gov.za/documents/download.php?f=104227.

[6] http://www.whitehouse.gov/administration/eop/ostp/grand-challenges.

repeated in 2011 at the International Conference on Information Systems (ICIS) in Shanghai, and is now gaining new momentum.

The Millennium Project is an independent and ongoing effort (http://www.millennium-project.org) that addresses 15 global challenges society faces today that cannot be solved by individuals in the short term. Prominent examples of the 15 challenges include how everyone on the planet can have sufficient clean water and a decent education. The dimensions of such challenges indicate that the IS discipline alone cannot solve the challenges, just as other disciplines cannot solve them alone. Solutions call for transdisciplinary and interdisciplinary efforts to which diverse disciplines must contribute.

It is our firm belief that IS can play a major role in contributing to solutions to grand societal challenges. Focusing on environmental challenges, researchers like Richard T. Watson (Watson et al. 2010), Nigel Melville (2010), and Steve Elliot (2011) advocate that the IS discipline has both the responsibility and the historic chance to demonstrate societal value by contributing to solutions that lessen the negative environmental effects of people's behavior. For example, the field of energy informatics has been established (Watson and Boudreau 2011) to investigate the design and use of information systems to improve energy efficiency through sensor networks in both businesses and private lives. Seidel et al. (2013) show that information systems can play a major role in facilitating sustainability transformations in organizations of various types. A recent call from vom Brocke et al. (2013) presents directives for IS research to leverage this position in contributing to grand societal challenges.

In keeping with the spirit of IS as a science of solutions to grand societal challenges, the European Research Center for Information Systems' (ERCIS) headquarters in Münster, Germany, and the ERCIS partner at the University of Liechtenstein have jointly organized a seminar series in which students are invited to provide their thoughts on how IS can contribute to the 15 societal challenges identified by the Millennium Project. The course has been offered over 3 years and has involved students from more than 30 nations. The course creates virtual multinational teams, and each of these teams is given the task of envisioning the role IS could play in contributing to a selected challenge. The Association for Information Systems (AIS) gave the course the AIS Award for Innovation in Teaching at the International Conference on Information Systems 2013 in Milan (ICIS 2013).

The students from the winter term 2011/2012 decided to continue working on their papers after the course was over, and each paper underwent a substantive review by colleagues. This book compiles the best papers from this course to stimulate further contributions of this kind. Dominik Heddier and Agata Materek collaborate on how IS can contribute to environmental sustainability, considering eco-efficiency, eco-equity, and eco-effectiveness. Dominic Steffen and Rajesh Srinivasan deal with the exponential growth of the human population and ask how IS can be used to align the global resources with the growing demand. Adrian Dolensky and Stefan Laube discuss e-participation as a way to make policymaking more sensitive to long-term global perspectives. How the benefits of ICT, which is ubiquitous in some parts of the world but all but absent in others, can be made available to everyone is the topic of Tomal K Ganguly and Klaus Fleerkötter's paper.

How decision support systems, key elements of IS research, can be used to improve the capacity to decide as the nature of work and institutions change is discussed by Marina Maschler and Asin Tavakoli. Manuela Weiss and Ekaterina Tarchinskaya discuss how IT can support changing the status of women around the globe, especially in the STEM field (Science, Technology, Engineering, and Mathematics), where the underrepresentation of women in the workforce can have negative effects on the efficiency and quality of the work. Nadine Székely explains how IS can support law enforcement officials in fighting "the dark side" as industrialization and the appearance of the Internet provide companies and individuals, including those in organized crime, with means to communicate and synchronize in real-time. Stefan Debortoli addresses the growing demand for energy as we move away from fossil fuels and seek to reduce power consumption, while governments invest increasing amounts of money in alternative green energy. Sandro Weber sketches out how IS can accelerate the diffusion of these ideas and innovations into practice. Finally, Nebojsa Milic and Krzysztof Oleszkiewicz take a look at the ethical difficulties global decisions bring to a world with more than 7.2 billion people and how IS can contribute to resolving them.

Although we reviewed and supervised the work, we kept the papers as original student papers and did not edit them ourselves. We want to acknowledge the effort our students have put into their tasks and also hope to illustrate the contribution we all can make as we start to think about how IS can contribute to solving grand societal challenges through IS research and practice. We hope you will feel as inspired as we do!

Our thanks go to all contributors for their effort and enthusiasm in working on this exciting and important topic. In particular, we recognize all of the supervisors of our seminar: Katrin Bergener, Elena Gorbacheva, Marcel Heddier, Andrea Herbst, Marco de Marco, Oliver Mueller, Stefan Seidel, Alexander Simons, Paolo Spagnoletti, and Theresa Schmiedel.

If you would like to get involved in a future seminar, we encourage you to visit our website at http://virtual-seminar.ercis.org.

Vaduz, Liechtenstein, July 2014 Jan vom Brocke
Münster, Germany Armin Stein
Münster, Germany Sara Hofmann
Vaduz, Liechtenstein Sanja Tumbas

## References

Watson, R. T., Boudreau, M.-C., & Chen, A. J. (2010). Information systems and environmentally sustainable development: Energy informatics and new directions for the IS community. *MIS Quarterly*, 34(1), 23–38.

Melville, N. (2010). Information systems innovation for environmental sustainability. *Management Information Systems Quarterly*, 34(1), 1–21.

Elliot, S. (2011). Transdisciplinary perspectives on environmental sustainability: A resource base and framework for IT-enabled business transformation. *MIS Quarterly*, 35(1), 1–13.

Watson, R. T., Boudreau, M.-C., Chen, A. J., & Sepúlveda, H. H. (2011). Green projects: An information drives analysis of four cases. *The Journal of Strategic Information Systems*, 20(1), 55–62.

Seidel, S., Recker, J., & vom Brocke, J. (2013). Sensemaking and Sustainable Practicing: Functional Affordances of Information Systems in Green Transformations. *MIS Quarterly*, 37(4), 1275–1299.

vom Brocke, J., Watson, R., Dwyer, C., Elliot, S., & Melville, N. (2013). Green Information Systems: Directives for the IS Discipline. *Communications of the Association for Information Systems (CAIS)*, 33(30), 509–520.

# Preface

The primary concern of Information Systems (IS) research and education has been effective development and implementation of business IS. Without creating adequate benefits to business, investment in IS cannot be justified. However, as the Internet and smart phone technologies have proliferated and ERP has become standardized, the primary platform of IS is no longer internal business systems. It has become inevitable to research the societal impact of the global IS platform, and the term "IS" becomes to imply an "Information Society."

The benefits of IS have increased as the speed and capacity of information technology (IT) have increased, but the growth of IT's benefits has become marginal while the impact of malicious side effects have become more serious, potentially catastrophic and global. Therefore, it has become essential to change gears in IS research and education to pay special attention to establishing a fundamental foundation for sustainable progress.

First of all, IS should solve the problems it has caused, such as transnational cyber-crimes and cyber-terror like that Sony Pictures recently experienced. To identify the seriousness of IS's dark side and to develop preventive solutions, the Council of Association of Information Systems (AIS) has adopted the ICT-enabled Bright Society Initiative as a Grand Vision Project (in brief, the Bright ICT Initiative).

The key issues of the Bright ICT Initiative are related to the key global challenges, so the fifteen global challenges identified by the Millennium Project serve as a foundation for the Bright ICT investigation. Energy shortage and climate change are among these global challenges, and we must determine how IS can help to mitigate and adapt to climate changes. Next, the effect of IS on education and healthcare, particularly in underdeveloped regions, can be profound, and IT availability can be the platform of survival of people. We must also understand the impact of IS on employment and how it changes the nature of work in both developed and developing countries.

This book provides excellent discussions on these important issues.

2015–2016

Jae Kyu Lee
Chair Professor at Korea Advanced
Institute of Science and Technology
President of the Association
of Information Systems

# Contents

# Chapter 1
# How Can Sustainable Development Be Achieved for All While Addressing Global Climate Change?

**Dominik Heddier and Agatha Materek**

## 1.1 Introduction

According to the United Nations Millennium Project, unsustainable growth may be the greatest threat to the future of humanity since the danger of nuclear annihilation during the Cold War (Glenn 2001). The Millennium Project revealed that in May 2011 the atmospheric $CO_2$ was the highest it had been in at least two million years. Since environmental sustainability is one of today's most important public issues (Dick and Kuo 2009), There has been increasing discussion about the role of information systems (IS), which has transformed the nature of businesses and improved the environmental performance of the organizations that are now under increasing pressure to implement sustainable practices (Dick and Kuo 2009). For instance, software is a critical enabling technology in nearly all aspects of climate change, from the computational models used to increase awareness of the impacts of society's activities on climate to the IS required to establish an effective carbon-neutral society (Easterbrook 2010). However, IS and IT have also contributed to organizations' environmental degradation. In order to counteract this development, the area of Green IT, which uses IS and IT to improve environmental performance, has emerged (Dick and Kuo 2009). In the IT industry, major hardware companies emphasize reusability and elimination of toxic chemicals through environmentally friendly designs, while IT service companies offer Green IT consultation to clients. So far, less attention has been paid to the potential of IT to enable environmental gains in the broad economy (Dedrick 2010).

Our research question is "How can IS contribute to environmental sustainability?" This paper is based on a comprehensive literature review and structured along a conceptual model that consists of three milestones of environmental sustainability: eco-efficiency, eco-equity, and eco-effectiveness (Chen et al. 2008).



J. vom Brocke et al., *Grand Societal Challenges in Information Systems Research and Education*, SpringerBriefs in Information Systems,
DOI 10.1007/978-3-319-15027-7_1

## 1.2 Literature Review

### *1.2.1 Research Methodology*

We conducted a thorough literature review based on Vom Brocke et al. (2009) literature analysis framework. Our focus was on articles that deal with methods and concepts for which IS is either the source of or the solution to environmental sustainability problems. We scanned the 125 papers that matched our search criteria to verify that they matched the defined scope and conducted a forward and backward search, resulting in 34 relevant papers that we then synthesized and structured in the concept-driven framework (Webster and Watson 2002).

### *1.2.2 Literature Analysis and Synthesis*

We use Dyllick and Hockerts' (2002) concept model to analyse the relevant articles. They identify three goals of sustainability: *eco-efficiency*, *eco-equity*, and *eco-effectiveness*. The three dimensions serve as a common framework by which to identify and organize the IS concepts that contribute to environmental sustainability. (Compare, e.g., Chen et al. 2008; Schmidt et al. 2009; Jenkin et al. 2011).

*Eco-efficiency* is achieved "by the delivery of competitively priced goods and services that satisfy human needs and bring quality of life, while progressively reducing ecological impacts and resource intensity throughout the life-cycle to a level at least in line with the earth's carrying capacity" (Dyllick and Hockerts 2002). Literally, then, efficiency means "doing more with less" (Chen et al. 2008).

*Eco-equity* describes the relationship between business and society (Dyllick and Hockerts 2002) that needs to be incorporated into the organizational value system in order to increase the number of organizations that follow eco-friendly strategies and processes, even in the absence of cost-efficiency (Chen et al. 2008). It refers to the "equity between peoples and generations and, in particular, the equal rights of all peoples to environmental resources" (Gray and Bebbington 2000).

*Eco-effectiveness* "aims to stop contamination and depletion, instead of only slowing down their speed, by directing individual and organizational attention to the underlying and fundamental factors of environmental problems and to make possible long-term prosperity through a fundamental redesign of the system" (Chen et al. 2008). Eco-effectiveness refers to increasing output with fewer natural resources consumed, improving Earth's health, and reducing greenhouse gases (Melville 2010).

### *1.2.3 IS Concepts Related to Eco-efficiency*

*Optimization Technologies.* The huge information-processing capabilities of IS can be used to find optimal solutions to a wide range of problems. Energy savings—and, thus, reductions in carbon emissions—can be achieved by optimizing the design and execution of operational processes. Production and maintenance optimization systems are popular examples in this area (International Telecommunication Union 2007; Hasan et al. 2009). Sbihi and Eglese (2009) elaborate on how optimization techniques from the operations research field, supported by real time IS, can help to achieve green logistic processes.

*Intelligent Transport Systems (also related to eco-effectiveness)*. According to the World Resources Institute (2011), the transportation sector is responsible for 13.5 % of global greenhouse gas emissions. Most IS solutions in these fields are referred to as "Intelligent Transport Systems". An example is the smart parking applications that direct drivers to empty parking spaces, preventing long searches that use up unnecessary fuel and time. Dashboard displays can also be used to inform drivers about their fuel efficiency, prompting them to improve it (Deakin et al. 2009).

*Emissions Trading Systems*. Many countries ratified the directives of the Kyoto Protocol with market-based mechanisms for active trade of carbon emission certificates. A fixed amount of emission allowances forces $CO_2$ producers to reduce their emissions or buy costly allowances from the market (Department of Energy and Climate Change 2011; Melville 2010). An efficient and effective carbon market is a key element of a successful global climate regime (Bell and Drexhage 2005). IS can help by connecting local markets, leading to more efficient implementation of carbon emission limits and a reduction in carbon production.

*Early-warning and Mitigation Systems (also related to eco-equity)*. Predicting likely losses after natural disasters like tsunamis, earthquakes, and volcanic eruptions requires immense calculating power, which is possible only with appropriate IS. Simulation, which uses historical data and correlations of variables, is a possible solution to this complex problem. Using a combination of historical data, such as probabilities and simulated events, and a policy strategy, it is possible to evaluate the damage and estimate the economic consequences of a particular hazard strategy (International Telecommunication Union 2007).

*Decision Support Tools*. Decision-makers around the world need to balance climate-related targets with other social, economic, and environmental goals (Burger et al. 2009; Lindseth 2004; Schreurs 2008). Decision-support tools like carbon accounting calculate carbon emissions and are already used for regulatory and corporate decision-making (Easterbrook 2010). Most concepts improve the decision-making process by using, for example, simulations, optimization methods, climate prediction and modelling systems, and intelligent transport systems.

*Virtual Conferencing and Collaborating (also related to eco-effectiveness)*. Travelling (e.g., to meetings) has a significant carbon footprint. Information and communication technology (ICT) supports a relatively pollution-free virtual space for an extensive range of activities (Hasan et al. 2009). Potential IT solutions include teleconferencing, telecommuting, the virtual office, and group decision-support systems. Every per cent decrease in business travel that results from video- and audio-conferencing saves around a million tons of $CO_2$ emissions annually (International Telecommunication Union 2007).

*Going Paperless (also related to eco-effectiveness)*. ICT has greatly decreased the use of paper in many areas. Paperless processing methods like e-business, e-books, and online surveys (Hasan et al. 2009) help businesses to lower their carbon footprints by reducing the amount of paper used in daily operations and decreasing the need for physical storage space. Paperless processing enhances corporate sustainability practices and helps organizations to reach their corporate and environmental goals (Myers 2010).

### *1.2.4 IS Concepts Related to Eco-equity*

*Climate Modelling and Prediction.* Climate modelling systems enhance the understanding of the climate system and aid in predicting future climates. A large array of modelling systems has been developed that often use highly sophisticated and complex models, challenging even the fastest speeds of the most powerful supercomputers (McGuffie and Henderson-Sellers 2001). The most common IS for climate modelling are global circulation models, which simulate the atmosphere, cryosphere, biosphere, and oceans to study the processes of climate change on a global scale and make projections (Easterbrook 2010).

*Climate Data Collection and Remote Monitoring.* Polar ice caps, glaciers, volcanoes, the ocean, and the atmosphere are the typical locations for climate research, but they are also inhospitable to human life, so they require remote monitoring and data collection using, for instance, ICT-equipped sensors (telemetry). In the last two decades wildlife telemetry research using the technologies of, for example, satellite tracking and archival logging technologies has increased (Hart and Hyrenbach 2009). Other technologies, such as aerial photography, satellite imagery, grid technology, and global positioning by satellite (GPS), have been widely used to track the slow and long-term movement of glaciers. Without IS contribution climate data collection, analysis, and remote monitoring would not be possible.

*Educational Software.* Many types of educational software serve as educational material about the Earth, its climate, and climate change for schools, universities, and Internet users. In most cases, the software is free and downloadable from the Internet websites. Two examples of climate change software are SimClimat (Gama and Risi 2008) and Educational Global Climate Modelling (Educational Global Climate Monitoring 2012).

*Collective Intelligence Tools.* Organizations have adopted a new generation of web tools, such as social networking; sharing of photos, videos, tags, and bookmarks; and wiki-based editing, which are often referred to as collective intelligence tools (Convertino et al. 2010). They use crowd-sourcing techniques (e.g., Mechanical Turk, Wikipedia, and Yahoo Answers) to improve the quality of evidence and analysis in many areas, including environmental sustainability, development, and climate change (Easterbrook 2010). Collective intelligence tools support learning and development in the field of sustainability, but they also promote climate-related events and initiatives, such as flash mobs to call attention to climate change.

### 1.2.5 IS Concepts Related to Eco-effectiveness

*Sustainable Design and Management of IS.* While IS can support environmentally friendly design in other areas, the design and management of the IS themselves can have a considerable impact on the carbon footprint. Zhang et al. (2011) propose an environmentally driven strategic analysis approach in which the environmental aspects of any project are covered as a set of requirements in addition to the classic business objectives and concrete application scenarios. An approach by Watson et al. (2011) suggests replacing customer orientation in IS design with innovation orientation based on a framework of four innovation drives: ubiquity, uniqueness, unison, and universality. Following this "4U framework" in the design phase should increase the resulting IS's eco-friendly and sustainable properties.

*Measuring Environmental Sustainability.* One of the most important challenges for environmental sustainability is its correct measurement, which requires expertise in several fields. The principles of sustainability measurement, which elucidate accepted methods for managing measurement in practice, create the needed links between disciplines while help organizations protect the global environment (Corbett et al. 2011). IS provides a common knowledge base for the organizational evaluation of environmental performance. For instance, an enterprise resource planning system is a basis for sustainability assessment and reporting processes because it assists organizations with the collection, processing, and sharing of extensive environmental impact and performance data (Chen et al. 2010).

## 1.3 Discussion

The literature review highlights several opportunities for IS to have a positive impact on global climate change. In this chapter we discuss how these improvements stand in relation to other efforts and whether society is willing to put in the necessary effort to achieve any mentionable impact at all. We begin with the possible downsides of some of the approaches mentioned in the literature review.

The initiative to create a "paperless" office using IS for digital documents might reduce overall paper demand and prevent deforestation, but this approach can be hindered by national regulations concerning document storage and archiving. For example, German legislation requires companies to store documents like contracts and end results for up to 10 years in paper form (HGB § 257, 4).

The intelligent transport systems currently in use can help reduce driving time and, thus, the emission of $CO_2$ gas, but the implementation of these systems is costly, and the money it requires might be spent more effectively elsewhere. One example of a measure with a higher investment-impact ratio is the extinguishing of underground coal fires around the world. Stracher and Taylor (2004) point out that there are underground coal fires in northern China of huge dimensions that contribute 2–3 % to the annual global emission of atmospheric $CO_2$, and there are many more of those fires burning around the world.

The climate certificate trade has currently become effectively useless because governments have flooded the markets with certificates. Currently the emission allowance for 1 ton of $CO_2$ costs only 6.23 € (EEX 2012), while environmentalists estimate a price level of more than 30 € would be necessary for companies to want to invest in more eco-friendly methods (Spiegel 2011).

Besides the actual usefulness or economy of these approaches, there is also the problem of willingness to change current behaviour. Dedrick (2010) identifies several barriers to implementing Green IT initiatives: For most companies Green IT solutions are too expensive and have no visible business value. In addition, the lack of governmental incentives and business leadership on Green IT prevents many from committing resources to sustainable IT initiatives. The absence of the will to change current unsustainable behaviour is reflected in the decision of the UN Climate Conference in Durban to adopt a universal legal agreement on climate change no later than 2015 (UNFCCC 2012).

## 1.4 Research Agenda

Having in mind all the factors that limit positive impacts of IS on climate, there is still a broad research area to explore. First, in order to understand the effects of Green IS and IT, more accurate and comprehensive measures of environmental impacts must be developed (Jenkin et al. 2011), including new research methodologies and metrics (Melville 2010). IS researchers can indirectly contribute to the implementation of Green IS and IT by taking a wider view of the whole system and considering how organizations can optimize a set of decisions—such as those related to hardware selection, system architecture, data centre design and location, and insourcing versus outsourcing—to "optimize a combination of performance, cost, and GHG emissions" (Dedrick 2010).

Future research also needs to examine the dynamics of environmental orientation in organizations, which influences attitudes and the acceptance of carbon-reducing technologies. It should concentrate on absorptive capacity, organizational learning, and social marketing theories to support and align employees' environment-friendly attitudes and behaviours in order to initiate environmentally related changes (Jenkin et al. 2011). This area requires further research in order to clarify the relationship between social issues and technology (Hasan and Alony 2011).

Research is still needed on how to increase the utility, usability, and user acceptance of e-tools to reduce the use of paper (Hasan et al. 2009) and on processes for providing decision support, including the operation of networks and intermediaries between the manufacturers and users of information for decision support. This research should include consideration of the most effective channels and organizational structures that should be applied for delivering information for decision support, how such information can be prepared to fit into individual, organizational, and institutional decision routines, and the factors that determine whether potentially useful information is actually used. One of the most challenging areas for further research concerns how to overcome barriers to the use of decision-relevant information (National Research Council 2010).

Another future research field is the impact of government policy on IT investment and the consequences for carbon productivity (Dedrick 2010).

## 1.5 Conclusion

Green IS encompasses a large area of research and practice, with many methods, perspectives and measures. This paper addressed how IS can prevent or mitigate climate change that is due to global warming and several application areas in which IS are used to cope with climate change. Table 1.1 presents an overview of the concepts discussed in the paper and their roles as contributors to dealing

**Table 1.1** Overview of IS concepts contributing to sustainable development and mitigating climate change (Millennium Project Challenge 1)

| Sub concept: eco-dimension | IT artifact | Role of IT |
|---|---|---|
| Eco-efficiency | Intelligent transport systems | Reducing ecological impacts and resources intensity through energy-efficient and decision-supporting technologies |
| | Optimization technologies | |
| | Emission trading systems | |
| | Decision support tools | |
| | Early warning and mitigation systems | |
| | Virtual conferencing and collaborating | |
| | Going paperless | |
| Eco-effectiveness | Sustainable design and management of information systems | Increasing output with fewer resources consumed |
| | Intelligent transport systems | Supporting better environmental performance |
| | Virtual conferencing and collaborating | |
| | Going paperless | |
| | Environmental sustainability measures | |
| Eco-equity | Climate prediction and modeling | Providing comprehensive climate information chain |
| | Climate data collection and remote monitoring | Reducing risks of wealth losses and assuring sustainable development |
| | Early-warning and mitigation systems | Increasing awareness about climate change and environmental sustainability |
| | Educational software | |
| | Collective intelligence tools | |

with the challenge of how sustainable development can be achieved for all while addressing global climate change using IS.

The concepts listed in Table 1.1 are only a fraction of the ways IS can contribute to the overlying challenge of climate change. The biggest challenge now is to convince humanity to make use of this huge potential.

## References

Bell, W., & Drexhage, J. (2005). *Climate change and the international carbon market*. Winnipeg, Canada: International Institute for Sustainable Development.

Burger, N., Ecola, L., Light, T., & Toman, M. (2009). *Evaluating options for U.S. greenhouse-gas mitigation using multiple criteria*. Santa Monica, CA: RAND Corporation.

Chen, A. J. W., Boudreau, M.-C., & Watson, R. T. (2008). Information systems and ecological sustainability. *Journal of Systems and Information Technology, 10*(3), 186–201.

Chen, A. J., Watson, R. T., Boudreau, M. C., & Karahanna, E. (2010). An institutional perspective on the adoption of green IS. *Australasian Journal of Information Systems, 17*(1), 23–45.

Convertino, G., Grasso, A., Dimicco, J., Michelis, G. D., & Chi, E. H. (2010). Collective intelligence in organizations: Toward a research agenda. *Proceedings of ACM Conference on Computer Supported Cooperative Work* (pp. 613–614).

Corbett, J., Webster, J., Boudreau, M. C., & Watson, R. (2011). *Defining the role for information systems in environmental sustainability measurement*. Working Papers on Information Systems.

Deakin, E., Frick, T. K., & Skabardonis, A. (2009). Intelligent transport systems. *Access, 34*(1), 29–34.

Dedrick, J. (2010). Green IS: Concepts and issues for information systems research. *Communications of the Association for Information Systems, 27*(1), 143–185.

Department of Energy and Climate Change. (2011). *EU emissions trading system*. Retrieved April 01, 2012, from http://www.decc.gov.uk/en/content/cms/emissions/eu_ets/eu_ets.aspx.

Dick, G. N., & Kuo, B. N. (2009). The greening of organisational IT: What makes a difference? *Australasian Journal of Information Systems, 16*(2), 81–92.

Dyllick, T., & Hockerts, K. (2002). Beyond the business case for corporate sustainability. *Business Strategy and the Environment, 11*(2), 130–141.

Easterbrook, S. M. (2010). Climate change: A grand software challenge. *Proceedings of the FSE/SDP Workshop on Future of software Engineering Research* (pp. 99–104).

Educational Global Climate Monitoring. (2012). *Global climate model (GCM)*. Retrieved January 01, 2012, from http://edgcm.columbia.edu/.

EEX. (2012). *EU emission allowances, preise und handelsvolumen*. Retrieved April 01, 2012, from http://www.eex.com/de/Marktdaten/Handelsdaten/Emissionsrechte/EU%20Emission%20Allowances%20%7C%20Spotmarkt.

Gama, N., & Risi, C. (2008). *SimClimat: An educational software to simulate the climate*. Retrieved January, 2012, from http://www.lmd.jussieu.fr/~crlmd/simclimat/index_english.html.

Glenn, J. (2001). The millennium project: Challenges we face at the millennium. *Technological Forecasting and Social Change, 66*(2–3), 129–312.

Gray, R. H. & Bebbington, J. (2000): Environmental accounting, managerialism and sustainability: Is the planet safe in the hands of business and accounting? In: Advances in Environmental Accounting & Management. 1, p. 1–44.

Hart, K., & Hyrenbach, K. (2009). Satellite telemetry of marine megavertebrates: The coming of age of an experimental science. *Endangered Species Research, 10*, 9–20.

Hasan, H. M. & Alony, I. (2011): Suggesting a practical agenda for Green IS - Recent solutions within aframework of efficiency, information, and effectiveness. In N. Geri & Y. M. Kalman (Eds.), Proceedings of the 5th Israel Association for Information Systems (ILAIS) Conference (pp. 25–30).

Hasan, H., Ghose, A., & Spedding, T. (2009). IS solution for the global environmental challenge: An Australian initiative. *Proceedings of the Americas Conference on Information Systems (AMCIS)*, San Francisco.

International Telecommunication Union. (2007). ICT s and Climate Change. Tech. Report. Jenkin, T. A., McShane, L., & Webster, J. (2011). Green information technologies and systems: Employees' perceptions of organizational practices. *Business and Society, 50*(2), 266–314.

Jenkin, T. A., Webster, J., & McShane, L. (2011). An agenda for "Green" information technology and systems research. *Information and Organization, 21*(1), 17–40. Elsevier Ltd.

Lindseth, G. (2004). The cities for climate protection campaign (CCPC) and the framing of local climate policy. *Local Environment, 9*(4), 325–336.

McGuffie, K., & Henderson-Sellers, A. (2001). Forty years of numerical climate modeling. *International Journal of Climatology, 21*(9), 1067–1109.

Melville, N. (2010). Information systems innovation for environmental sustainability. *Management Information Systems Quarterly, 34*(1), 1–21.

Myers, R. (2010). *Going Paperless*. Retrieved May 02, 2012, from http://gettheinsideedge.com/cashmanagement/articles/going-paperless-randy-myers.aspx.

National Research Council. (2010). *America's climate choices. Panel on advancing the science on climate change.*. Washington DC: National Academy of Sciences.

Sbihi, A., & Eglese, R. W. (2009). Combinatorial optimization and green logistics. *Annals of Operations Research, 175*(1), 159–175.

Schmidt, N., Erek, K., Kolbe, L., & Zarnekow, R. (2009). Towards a procedural model for sustainable information systems management. *HICSS'09 42nd Hawaii International Conference on System Sciences* (pp. 1–10).

Schreurs, M. A. (2008). From the bottom up: Local and subnational climate change politics. *Journal of Environment & Development, 17*(4), 343–355.

Spiegel. (2011). *Schlapper Emissionshandel, Klimasünde zum Preis einer Pizza*. Retrieved April 1, 2012, from http://www.spiegel.de/wirtschaft/soziales/0,1518,802544,00.html.

Stracher, G., & Taylor, T.P. (2004). Coal fires burning out of control around the world: thermodynamic recipe for environmental catastrophe. *International Journal of Coal Geology, 59*(1–2), 7–17.

UNFCCC. (2012). *Durban Climate Change Conference—November/December 2011*. Retrieved April 1, 2012, from http://unfccc.int/meetings/durban_nov_2011/meeting.

Vom Brocke, J., Simons, A., Niehaves, B., Riemer, K., Plattfaut R., Cleven, A. et al. (2009). Reconstructing the giant: On the importance of rigour in documenting the literature search process. *Proceedings 17th European Conference on Information Systems 2009*, Verona.

Watson, R. T., Boudreau, M.-C., Chen, A. J., & Sepúlveda, H. H. (2011). Green projects: An information drives analysis of four cases. *The Journal of Strategic Information Systems, 20*(1), 55–62.

Webster, J., & Watson, R. T. (2002). Analyzing the past to prepare for the future: Writing a literature review. *MIS Quarterly 26*(2), xiii–xxiii.

World Resources Institute (WRI) in collaboration with United Nations Development Programme, United Nations Environment Programme, and World Bank. (2011). *World Resources 2010–2011: Decision Making in a Changing Climate - Adaptation Challenges and Choices*. Washington, DC: WRI.

Zhang, H., Liu, L., & Li, T. (2011). Designing IT systems according to environmental settings: A strategic analysis framework. *The Journal of Strategic Information Systems, 20*(1), 80–95.

# Chapter 2
# ICT-Applications to Align Global Resources with a Growing Population

**Dominic Steffen and Rajesh Srinivasan**

## 2.1 Introduction

More than 7 billion people populate the planet, and the UN (2012) projects that this number will rise to a staggering 9.4 billion or higher by 2050. Nearly a billion people are considered undernourished (FAO 2010), and sudden price spikes for food commodities in 2007 and 2008 alarmed the world (Piesse and Thirtle 2009; OECD/FAO 2011). Can we produce enough food to feed such a large number of people? It is imperative that we address this question.

In this context, information and communication technology (ICT) has transformed our way of life through innovations like personal computers, mobile telephony, the Internet, cloud computing, and social networks. Therefore, we look at the question from the perspective of Information Systems (IS) science and ask, "How can the application of ICT be used to align global resources with a growing population?" We performed a literature review to investigate how far and from which perspectives has this question been approached already.

This paper is organized as follows: The next section describes our literature review procedure. Then we analyse the underlying problem from the perspectives of demand and supply, introducing ICT applications that mitigate issues that arise from a growing population and investigating ICT applications that can help increase food production. As a result, we find that several types of ICT play a crucial role in: increasing the production efficiency and sustainability of agriculture, decision support, providing timely information about the market, geographical position and knowledge distribution.



J. vom Brocke et al., *Grand Societal Challenges in Information Systems Research and Education*, SpringerBriefs in Information Systems,
DOI 10.1007/978-3-319-15027-7_2

## 2.2 Literature Review

The best practice for conducting literature reviews is to document each step of the research process in order to demonstrate the methodology used so other researchers can recreate and replicate the process (vom Brocke et al. 2009; Webster and Watson 2002). With this process as a guiding principle, we discuss the research process for investigating our primary research question.

The paper addresses the question, "How can the application of ICT be used to align global resources with a growing population?" We began by investigating the underlying problem of food supply for a growing population, dividing the topic into supply and demand. For demand we concentrated on population-centric functions for growth and development, and for supply we concentrated on farmer-centric functions for resource growth. In the second stage of our research, we investigated how ICT could be applied to solving the underlying problem.

We used the academic search engines Business Source Premier, JSTOR, SpringerLink, and Google Scholar to find matches for the search query in titles, abstracts, keywords, and full texts. We performed searches using the search terms "population control", "population growth", "population growth food", "population ICT", "ICT agriculture", and "ICT food security". We assessed the first one hundred documents for relevance by analysing their titles and abstracts for a reference to indexing methodologies for population control and ICT (exact and approximate structure and substructure search). When this assessment was inconclusive, we analysed the full text. Articles that made no contribution to the problem were discarded.

## 2.3 The Challenge: Demand and Supply Side Perspective

***Demand-side perspective***. According to the most recent UN medium projections (UN 2012), the world's population will continue to grow at least until 2050, when the total number of people on the planet is expected to reach 9.4 billion, an increase of 2.4 billion over the 2011 population of 7 billion. Nearly all of this future growth will occur in the developing world—Africa, Asia (excluding Japan, Australia, and New Zealand), and Latin America—where the population is projected to increase from 4.5 to 8.2 billion between 1995 and 2050 (UN 2012).

In order to keep a balance between population and resource growth, each country must define population policies, defined as "deliberately constructed or modified institutional arrangements and/or specific programs through which governments influence, directly or indirectly, demographic change" (Demeney 2003). Malthus (1798) suggested solution was "to proportion the population to food, since the food could not be proportioned to population," but critics of this suggestion have argued that Malthus failed to foresee the potential technological improvements that would increase food production (Sen 1994). What's more, the nature of consumer demand is changing as consumers spend more, consume more, and have more choices than ever before. About 85 % of the increase in the global demand for cereals and meat between 1995 and 2020 will occur in developing countries (Pinstrup-Andersen et al. 1999).

***Supply-side perspective***. About 925 million people in the world are undernourished (FAO 2010). In the short to medium term, high volatility in agricultural commodity markets are of major concern, as they have major implications for food security (OECD/FAO 2011). Price spikes in commodity markets occur as markets fail to match demand in the short term as a result of supply shocks like crop production shortfalls or unexpected surges in demand (OECD/FAO 2011). In the long term, a declining trend in global food production growth is expected (Trostle 2008; OECD/FAO 2011). While the global potential for agricultural production has been estimated to be sufficient to produce an affluent diet for 16–24 billion people (Koning 2008) or even more (Chalkley 1997), such a production level would be unsustainable. Pimentel et al. (1994) put the population limit under sustainable agricultural production at less than 2.

Land, water, nutrient resources (fertilizer), and energy are important input factors for agricultural production (OECD/FAO 2011; Koning 2008; Evans 2011). Drivers that have a direct impact on production efficiency include access to agricultural knowledge and advanced farming technologies (Koning 2008; Balaji and Meera 2007; OECD/FAO 2011), investment (OECD/FAO 2011; Koning 2008), access to markets and market information (Rao 2007; Singh 2006), and improved biotechnology (Chrispeels 2000; Serageldin 1999; Bruce 2011). Investment decisions and farmers' production targets are the result of profitability considerations, as higher prices increase farmers' desires to increase production, making investment in agriculture more profitable and, therefore, more available (Koning 2008).

## 2.4 Research Results

### *2.4.1 Demand Side*

This section addresses population-centric functions and steps that can be taken to decrease population growth and inform the public about sustainable solutions and the need to make better decisions.

(1) ***ICT as a tool***. The main role of ICT is to enable communications and integrate technologies, enabling the user to create, access, store, and transmit information in the fields of economic and international development. The main function of ICT as a strategic tool for development is to facilitate access to and sharing of relevant information and knowledge. ICT strengthens the voices of poor, excluded, and disadvantaged groups and affects decision-making.

(2) ***Socially responsible functions***. As the population increases, a balance between the economy and the ecosystem is difficult to maintain. Socially responsibility benefits society at large through social economic development and poverty reduction. According to the demographic transition theory, rapid population growth happens in all societies at one time or another when improvements

in living conditions and health care lead to reduced death rates (Raleigh 1999). Economic growth also facilitates the reduction of poverty. The use of ICT in this function helps to create awareness and increase growth and development through improved decision-support systems.

(3) ***Universal healthcare functions***. Healthcare systems work on the principle of universal coverage for all members of society. When the composition of the population changes as a result of rapid growth, the needs of an increasing number of people of reproductive age must be met in order to enhance human potential and accelerate human achievement and development (Sachs and McArthur 2005). Meeting this need results in distribution of aid to future generations (e.g., maternal education and health in order to influence the well-being of the next generation and environmental sustainability) and the poor and disadvantaged segments of the population (e.g., addressing health inequalities and policies to improve the status of women) (Anand and Sen 2000).

(4) ***Environmental care functions***. With today's increasing human population, environmental changes and their costs have a pivotal role in growth and development. Changes to the ecosystem as a whole damage the environment in terms of climate, water shortages, deforestation, soil erosion, and decline in the level of biodiversity, so a slowing economy becomes inevitable. In addition, the rising danger of global warming from the ongoing greenhouse gas emissions is ever-increasing; according to most estimates, a rise in global temperature of four degrees Celsius is likely to occur during the twenty-first century (Knight 2010). What's more, deforestation not only adds to the $CO_2$ in the atmosphere but also destroys the plants and animals that occupy the ecosystem either permanently or temporarily.

## 2.4.2 Supply Side

Supply-side applications of ICT target production efficiency. We grouped applications that can improve production efficiency into two categories: Farmer-centric applications (points 1–3) and strategic applications (points 4–5).

(1) ***Knowledge distribution***. Sources of relevant knowledge for farmers are social and professional networks (Warren 2004; Rosskopf 2006), agricultural journals (Rosskopf 2006), private consultancy (Warren 2004; Rosskopf 2006), and public sector agricultural extension services (Richardson 2006; Munyua et al. 2008; Warren 2004; Rosskopf 2006). ICTs are a potentially convenient and efficient method of delivery of the knowledge contained in these sources (Warren 2004; Rosskopf 2006; Munyua et al. 2008). ICT-enabled knowledge distribution can take a number of forms. Time-tested methods of delivery are based on radio or television broadcasts of topics of relevance for the agricultural sector (Singh 2006), but knowledge can also be compiled into repositories and be distributed on static media like CDs and DVDs or made accessible through a website (Rao 2007;

Singh 2006; Munyua et al. 2008; Colle and Yonggong 2002; Flor 2002; Rosskopf 2006; Richardson 2006). Moreover, ICT-based systems can enable more interactive, less standardized queries in which farmers can send a query to an extension worker or agricultural expert (Singh 2006; Meera and Jhamtani 2004; Richardson 2006) using mobile text-message,[1] email,[2] telephone or VoIP,[3] website forms,[4] or in a video-conference[5] (Singh 2006; Balaji and Meera 2007; Rao 2007; Kumar 2005).

(2) ***Market information and access***. ICT-supported systems can connect farmers to markets efficiently (Rao 2007; Singh 2006) by providing accurate and timely market information on agricultural commodities and services that allow farmers to buy inputs and sell the outputs they produce (e-commerce) (Rao 2007). ICT can cut the transaction costs associated with information search (De Silva and Ratnadiwakara 2008), and the availability of market information empowers farmers against intermediaries through increased transparency (Singh 2006). E-commerce systems decrease costs for buyers and sellers by eliminating intermediaries and the costs attributed to them, increasing the profitability of the trades and the farmers' productivity as inputs become cheaper and outputs more profitable (Rao 2007). According to Rao (2007) and Singh (2006), another way e-commerce can benefit farmers is through demand aggregation, especially for rural, small-holder farmers who are not associated. Provision of market access to individual farmers may not be interesting to sellers of agricultural input because of the high cost and risk involved, but demand aggregation reduces risk and cost (). Therefore, e-commerce systems that provide demand aggregation connect farmers to input markets they could not access before and can increase their access to quality inputs at lower prices.

(3) ***Decision-support systems***. Decision-support systems enable farmers to make better, more informed decisions can lead to production-efficient use of input factors and increase production output (Newman and Lynch 2000). Fountas and Pederson (2005) described precision agriculture (PA) as a relatively recent farming technique that facilitates field management based on site-specific data, indicating targeted agricultural actions, such as fertilizer application or irrigation, on the sub-field level. The main activities within the PA framework are data collection, data processing, and determination of quantities of input factors, so the benefits associated with PA are improved economic returns and a reduced environmental impact of farming activities. As the efficiency of inputs improves, a global benefit is realized as reduced use of inputs increases sustainability.

ICTs and other modern technologies, such as GPS, sensor systems (e.g., yield monitors, soil sensors) and automated application technologies, are important components of PA (Fountas and Pedersen 2005; Gebbers and Adamchuk 2010).

---

[1] e.g. RDA's AIS (Singh 2006).

[2] e.g. RDA's AIS (Singh 2006).

[3] e.g. Kisan Call Centers (Kumar 2005), Tamil Market (Plauché and Prabaker 2006).

[4] e.g. eChoupal (Singh 2006).

[5] e.g. n-logue (Rao 2007).

The system works through ongoing collection of data on soil composition (Fountas and Pedersen 2005; Gebbers and Adamchuk 2010). Soil data can be collected by analysing soil samples in the lab, through locally deployed sensor systems, and remote sensing (Fountas and Pedersen 2005; Gebbers and Adamchuk 2010). Another advantage of sensor systems, which can be deployed as networks, is the ability to produce real-time information (Panchard et al. 2007; Gebbers and Adamchuk 2010).

Another important sensor class, yield monitors installed in harvesting equipment (Fountas and Pedersen 2005; Gebbers and Adamchuk 2010), analyse the quality of the output as it is harvested and relay the quality of the harvest with geographical information from GPS to facilitate input-output analyses (Fountas and Pedersen 2005; Gebbers and Adamchuk 2010). PA software typically generates maps from the data (e.g., yield-maps, soil-composition maps); performs data filtering, statistical analysis, and record-keeping; and computes variable application rates (Fountas and Pedersen 2005; Gebbers and Adamchuk 2010). Farmers can retain the data and use the software locally, but some companies and government extension services have introduced Internet-based PA services that allow farmers to send data to a service provider who analyses the data and produces maps, application rates, and other agronomic recommendations (Fountas and Pedersen 2005).

**(4)** ***Agricultural research information systems***. Agricultural research information systems (ARIS) integrate and coordinate the flow of information and access to knowledge resources on scientific, technical, and research matters among national agricultural research institutions (Maru 2002). The benefits of a common information system in agricultural research, which are based on realised synergies, include improvements in research, management and coordination of research activities, access to information in national research institutions, and improved protection of intellectual property rights (Maru 2002). According to Singh (2006), a successful example is South Korea's Agricultural Information Service (AIS), developed and maintained by the country's Rural Development Administration (RDA), which connected and integrated South Korea's national institutes into a high-speed computer network. AIS supports researchers by providing a knowledge portal and facilitating interaction between researchers and agricultural experts, which results in an improved quality of research and facilitates collaborative projects. Farmers can access a database on agricultural technology information through a website or take Internet-based training courses. The system also connects farmers to researchers and experts through a customer-relationship-management system (CRMS), where farmers can consult experts using e-mail, mobile text-messages, or in virtual meeting rooms. AIS is a good example of an ARIS that supports and improves agricultural research and efficiently and quickly disseminates generated knowledge to users.

**(5)** ***Geographic information systems***. GIS are automated systems that capture, store, retrieve, analyse, and display geospatial data (Clarke et al. 1996; Flor 2002). GIS can be used to monitor the environmental sustainability of farming techniques, evaluate the efficiency of agricultural techniques, assess the state of food security and vulnerabilities, and improve regional planning and crisis preparedness (Rao 2007; Stephenson 1997; Flor 2001). Data-gathering methods can range from crowd-sourced, distributed data-entry using mobile devices (Munyua et al. 2008; Arnquist 2009) to remote sensing.

The most important data sources are probably the data sets gathered in various institutions and organizations (Rao 2007), so access to the data and standardized formats to facilitate exchange are important issues (Rao 2007). GIS-based systems can be used to compile data on the use of input factors, farm management techniques, and output flows on the farm level, while aggregation of the data on a regional level allows the state of production and the efficiency of the farming techniques employed in a region to be assessed (Rao 2007). Another important area of use for GIS-based systems is in the assessment of food security vulnerability and crisis detection (Rao 2007; Stephenson 1997). Systems that help to identify potential risks and provide early warning allow policymakers to reduce vulnerabilities and plan a response to crises (Evans 2011).

On an international scale, the Food and Agriculture Organization of the UN and other international agencies operate and maintain a number GIS-based monitoring and early-warning systems (Rao 2007), the most important of which is probably the Global Information and Early Warning System on Food and Agriculture (GIEWS). Conceived and operated since 1975, it is the UN's major provider of information on food supply and demand (Food and Agriculture Organization n.d. 2011). The system continuously monitors food supply and demand in all countries, collecting and analysing information on global production of agricultural goods, stocks, food aid, and trade in agricultural commodities (Food and Agriculture Organization n.d.). The system uses this and other data to predict global food supply and demand and to calculate food import requirements.

## 2.5 Conclusion

Major changes in the global structure will follow as a consequence of population growth in the developing world, while the population of the developed countries will remain fairly stable. The essential question is whether the resources our planet provides can accommodate the growing number of people. When we last faced this problem, we had reached a production ceiling, but humankind was able to push the barrier further (Koning 2008). Since there is reason to believe that such a crisis may occur again, it is important to investigate how the application of ICT can mitigate the problems arising from a growing population by aligning resource consumption with rising demand.

We suggest that ICTs can support production growth in agriculture by providing farmers with access to knowledge on efficient production techniques and technologies and connecting them to agricultural experts. ICTs can also help farmers to use input factors efficiently and to optimize their production through information-driven decision-making.

Since access to the necessary technologies is expensive, and in the short-term bio-engineered crops and liberal uses of input factors may be better suited to increasing production quickly (Fountas and Pedersen 2005). However, with increasing factor costs and reduced hardware costs, information-driven decision-making will play an essential role in increasing the production efficiency and sustainability of agriculture in the future. ICT can be of strategic importance in improving agricultural research and streamlining the dissemination of generated knowledge to users. GIS-based solutions are already being used to monitor food security, evaluate production efficiency on a regional level, and provide early warning in order to detect and avert food security crises caused by unexpected shocks and improve reactions to crises.

All in all, ICTs, as catalysts of innovation, can play an important role in mitigating the problems that arise from an increasing demand for food and a production system that may have reached its limits of efficiency.

## References

Anand, S., & Sen, A. (2000). Human development and economic sustainability. *World Development, 28*, 2029–2049.

Arnquist, S. (2009). *In rural Africa, a fertile market for mobile phones*. New York Times.

Balaji, V., & Meera, S. (2007). ICT-enabled knowledge sharing in support of extension: Addressing the agrarian challenges of the developing world threatened by climate change, with a case. *E-Journal of SAT Agricultural, 4*, 1–18.

Bruce, T. J. A. (2011). GM as a route for delivery of sustainable crop protection. *Journal of Experimental Botany, 63*(2), 537–541.

Chalkley, K. (1997). Population growth and consumption. *Population Today, 25*(4), 4–5.

Chrispeels, M. J. (2000). Biotechnology and the poor. *Plant Physiology, 124*(1), 3–6.

Clarke, K. C., McLafferty, S. L., & Tempalski, B. J. (1996). On epidemiology and geographic information systems: a review and discussion of future directions. *Emerging Infectious Diseases, 2*(2), 85–92.

Colle, R. D., & Yonggong, L. (2002). ICT capacity-building for development and poverty alleviation, enhancing the role of agricultural universities in China. In *Third Asian Conference for Information Technology in Agriculture*, Oct 26 (pp. 67–74).

De Silva, H., & Ratnadiwakara, D. (2008). *Using ICT to reduce transaction costs in agriculture through better communication: A case study from Sri Lanka. LIRNEasia.*

Demeney, P. (2003). *Population policy, and concise summary*, no. 173, p. 62. Population council. Policy Research Division.

Evans, A. (2011). *Resource scarcity, climate change and the risk of violent conflict*. Background Paper, World Development Report (pp. 1–23).

FAO. (2010). *The state of food insecurity in the world 2010*. Rome, Italy: Addressing Food Insecurity in Protracted Crises.

Flor, A. G. (2001). ICT and poverty: The indisputable link. *Second Asian development forum* (pp. 11–14). Bangkok, Thailand: World Bank Publications.

Flor, A. G. (2002). *Information and communication opportunities for technology transfer and linkages*. Expert Consultation on Agricultural Extension, Research-Extension-Farmer-Market Linkages. FAORAP, Bangkok, 16.

The Global Information and Early Warning System on Food and Agriculture (GIEWS publications), 2011, Food and Agriculture Organization of the United Nations. Available at: http://www.fao.org/giews/english/giews_en.pdf [Accessed November 10, 2014].

Fountas, S., & Pedersen, S. (2005). *ICT in precision agriculture—diffusion of technology. ICT in agriculture* (pp. 1–15).
Gebbers, R., & Adamchuk, V. I. (2010). Precision agriculture and food security. *Science, 327*(5967), 828–831.
Knight, M. (2010) *Current emissions risk 'devastating' temperature rise, scientists warn—CNN.com. CNN.com—breaking news, U.S., world, weather, entertainment &video news.* http://www.cnn.com/2010/WORLD/europe/11/29/climate.four.degrees.warning/index.html.
Koning, N. (2008). Long-term global availability of food: Continued abundance or new scarcity? *NJAS—Wageningen Journal of Life Sciences, 55*(3), 229–292.
Kumar, D. (2005). *Information and communication technology (ICT) in Indian agriculture. General economics and teaching* (pp. 44–46).
Malthus, T. R. (1798). *An essay on the principle of population, as it affects the future improvement of society* (1st ed.). London: J. Johnson.
Maru, A. (2002). A normative model for agricultural research information systems. In *Proceedings of the Third Asian Conference for Information Technology in Agriculture. Asian Federation for Information Technology in Agriculture (AFITA)-Chinese Academy of Agricultural Sciences (CAAS)* (pp. 26–28).
Meera, S., & Jhamtani, A. (2004). *Information and communication technology in agricultural development: A comparative analysis of three projects from India.* AgREN Network Paper, no. 135.
Munyua, H., Adera, E., & Jensen, M. (2008). Emerging ICTs and their potential in revitalizing small scale agriculture in Africa. In *World conference on Agricultural Information and IT, IAALD AFITA WCCA 2008* (pp. 707–717).
Newman, S., & Lynch, T. (2000). Success and failure of decision support systems: Learning as we go. In *Proceedings of the American Society of Animal Science* (pp. 1–12). Indianapolis, Indiana, USA: American Society of Animal Science.
OECD/FAO (2011). OECD-FAO Agricultural Outlook 2011–2020.
Panchard, J., et al. (2007). Commonsense net: A wireless sensor network for resource-poor agriculture in the semiarid areas of developing countries. *Information Technologies & International Development, 4*(1), 51–67.
Plauché, M., & Prabaker, M. (2006). Tamil market: a spoken dialog system for rural India. In G. Olson & R. Jeffries (Eds.), *CHI EA '06 CHI '06 extended abstracts on Human factors in computing systems* (pp. 1619–1624). New York, NY, USA: ACM.
Piesse, J., & Thirtle, C. (2009). Three bubbles and a panic: An explanatory review of recent food commodity price events. *Food Policy, 34*(2), 119–129.
Pimentel, D., Harman, R., & Pacenza, M. (1994). Natural resources and an optimum human population. *Population and Environment, 15*(5), 347–369.
Pinstrup-Andersen, Per, RajulPandya-Lorch, & Rosegrant, M. W. (1999). *World Food Prospects: Critical Issues for the Early Twenty-First Century, Food Policy Report.* International Food Policy Research Institute, Washington, D.C.
Raleigh, V. S., (1999). World population and health in transition. *British Medical Journal, 319,* 981–984.
Ramamritham, K., Bahuman, A., Duttagupta, S., Bahuman, C., & Balasundaram, S. (2006). Innovative ICT Tools for information provision in agricultural extension. In *Information and Communication Technologies and Development,* 2006. ICTD'06. International Conference on. IEEE, pp. 34–38.
Rao, N. (2007). A framework for implementing information and communication technologies in agricultural development in India. *Technological Forecasting and Social Change, 74*(4), 491–518.
Richardson, D. (2006). ICTs-transforming agricultural extension? In *Proceedings of CTA's Observatory on ICTs.*
Rosskopf, K. (2006). Vom Daten-zum Wissensmanagement: Wofür verwenden Landwirte einen Computer. In K.-O. Wenkel et al. (Eds.) *Land- und Ernährungswirtschaft im Wandel: Aufgaben und Herausforderungen für die Agrar- und Umweltinformatik, Referate der 26. GIL Jahrestagung.* Potsdam, Germany, pp. 225–228.

Sachs, J. D, McArthur, J. W. (2005). The millennium project: A plan for meeting the millennium development goals. *Lancet, 365*, 347–353.

Sen, A. (1994). Population: Delusion and reality. *NY Review of Books, 41*, 62–71.

Serageldin, I. (1999). Biotechnology and food security in the 21st century. *Science, 285*(5426), 387–389.

Singh, S. (2006). Selected success stories on agricultural information systems. *Asia-Pacific Association of Agricultural Research (APAARI)*, Bangkok, Thailand.

Stephenson, R. (1997). Disasters and the information technology revolution. *Disasters, 21*(4), 305–334.

Thysen, I. (2000). Agriculture in the information society. *Journal of Agricultural Engineering Research, 76*(3), 297–303.

Trostle, R. (2008). *Global agricultural supply and demand: Factors contributing to the recent increase in food commodity prices*. Washington, DC, USA: DIANE Publishing.

vom Brocke, J., Simons, A., & Niehaves, B. (2009). Reconstructing the giant: On the importance of rigour in documenting the literature search process. In *17th European Conference on Information Systems*, pp. 1–13.

Warren, M. (2004). Farmers online: drivers and impediments in adoption of Internet in UK agricultural businesses. *Journal of Small Business and Enterprise Development, 11*(3), 371–381.

Webster, J., & Watson, R. T. (2002). Analyzing the past to prepare for the future: Writing a literature review. *Management Information Systems Quarterly*, *26*(2), 3.

# Chapter 3
# How Can Information Systems Help to Make Policymaking Be More Sensitive to Global Long-Term Perspectives?

**Adrian Dolensky, Stefan Laube and Elena Gorbacheva**

## 3.1 Introduction

In the last decades several positive world trends have indicated that, in general, the world is getting healthier, better educated, richer, more peaceful, and better connected. However, many major global challenges remain, including climate change, organized crime, terrorism, corruption, the widening gap between rich and poor, and increasing economic insecurity. According to the Millennium Project, one way to tackle these problems is to improve decision-making on the individual, group, national, and institutional levels (Glenn et al. 2011). Governments and politicians play a major role in achieving long-term global goals by setting and implementing the right policies on a global scale, so the Millennium Project asks, "How can policymaking be made more sensitive to long-term global perspectives?"

Technological developments like the ongoing evolution of information and communication technology (ICT) and expansion of the Internet have accelerate the complexity of these issues, along with the number and extent of changes and globalization. Therefore, it is important to understand how technology, especially information systems (IS), can be used to inform policymaking and improve its long-term perspective.

This chapter focuses on the research question, "How can IS help to make policymaking more sensitive to long-term global perspectives?" and addresses, among other questions, how *e-participation* can be used to make policymaking more sensitive to long-term global perspectives, how IS can support *global cooperation in policymaking*, and the chance that such ongoing research topics as *resilience thinking* and *collective intelligence* will affect policymaking. In addition to summarizing the topic-related concepts identified in the literature, this study proposes a research agenda.



J. vom Brocke et al., *Grand Societal Challenges in Information Systems Research and Education*, SpringerBriefs in Information Systems,
DOI 10.1007/978-3-319-15027-7_3

## 3.2 Research Method

The research question (RQ) for our literature review was derived from challenge number five of the Millennium Project (Glenn et al. 2011) and adapted to the context of IS. The process by which papers were selected consisted of four steps, in accordance with the methodologies proposed by Randolph (2009) and Vom Brocke et al. (2009).

In the first step, we defined the search terms and created a mind map on the RQ, where we summarized our ideas. As a result, a set of thirteen search strings related to IS and policymaking resulted.

In the second step, we defined the scope of our search. We used such academic journal databases as Google Scholar, EBSCOhost, and Springer Link to perform the search and chose not to limit our search to top-ranked peer-reviewed IS journals because of the interdisciplinary nature of our RQ.

Next, we defined the population of results in order to make the overall number of articles for reviewing manageable. We evaluated the first twenty pages of Google Scholar and Springer Link and the first ten pages of EBSCO Host to limit the hits for each keyword to 700.

**Table 3.1** Selected papers focusing on the IS-supported concepts contributing to making policymaking more sensitive to global long-term perspectives

| Paper | Paper Nr. |
|---|---|
| Braess et al. (2005) | [P1] |
| Cockerill et al. (2009) | [P2] |
| Dolowitz and Marsh (2000) | [P3] |
| Edwards (1996) | [P4] |
| Gordon (2003) | [P5] |
| Gordon (2009) | [P6] |
| Guiraudon (2000) | [P7] |
| Keen (1981) | [P8] |
| Kreileman et al. (1996) | [P9] |
| Lampathaki et al. (2010) | [P10] |
| Larrain et al. (2000) | [P11] |
| Leach (2008) | [P12] |
| Lempert et al. (2004) | [P13] |
| Lempert et al. (2009) | [P14] |
| Macintosh (2004) | [P15] |
| McCartney (1988) | [P16] |
| Misuraca et al. (2011) | [P17] |
| Parr et al. (2003) | [P18] |
| Phang and Kankanhalli (2008) | [P19] |
| Rotmans et al. (2001) | [P20] |
| Tumer and Wolpert (2000) | [P21] |

Finally, we evaluated the literature, identifying the potentially relevant papers as those that dealt with the RQ. We assumed that the term "long-term perspective" referred to "a time frame of one, two or three generations" (Rotmans et al. 2001), as even 30 years might not be long enough for our purposes. In the first round of the literature evaluation process, we read the papers' titles and, when a title was promising, we read the abstract. If both the title and the abstract suggested a paper's inclusion was justified, we read its conclusion as well. We excluded many papers that addressed policymaking from a long-term global perspective because they did not consider IS's contribution to the challenge. The papers we selected in the end either stated explicitly how IS contribute to the challenge or derived IS's contribution indirectly. We also extended the literature through a backward search of the articles chosen. The resulting list of twenty-one papers is presented in Table 3.1.

Analysis of the selected articles included identification of the papers' central issues and examination of the authors' lines of argumentation. As a result, we found seven IS-supported concepts that contribute to making policymaking more sensitive to long-term global perspectives. These concepts and related papers are presented in Fig. 3.1.

The next section discusses each concept in detail.

| Paper | E-participation in e-democracy | Counter-counter-implementation | Policy transfer | Cooperation in policymaking | Resilience thinking | Collective intelligence | Predictive methods | | | |
|---|---|---|---|---|---|---|---|---|---|---|
| | | | | | | | Policy context analysis | Policy modelling | Policy simulation | Policy evaluation |
| [P1] | | | | | | X | | | | |
| [P2] | | | | | | | | X | X | |
| [P3] | | | X | X | | | | | | |
| [P4] | | | | | | | | X | X | |
| [P5] | | | | | | | X | | | |
| [P6] | | | | | | | X | | | |
| [P7] | | | X | X | | | | | | |
| [P8] | | X | | | | | | | | |
| [P9] | | | | | | | | X | X | X |
| [P10] | | | | | | | X | X | X | X |
| [P11] | | X | | | | | | | | |
| [P12] | | | | | X | | | | | |
| [P13] | | | | | | | X | X | X | X |
| [P14] | | | | | | | X | X | X | X |
| [P15] | X | | X | | | | | | | |
| [P16] | | X | | | | | | | | |
| [P17] | X | | | | | X | X | X | X | |
| [P18] | | | | | X | | | | | |
| [P19] | X | | | | | | | | | |
| [P20] | | | X | X | | X | | | | |
| [P21] | | | | | | X | | | | |

**Fig. 3.1** IS-supported concepts contributing to making policymaking be more sensitive to global long-term perspectives and the related papers

## 3.3 Results and Discussion

The first IS-supported concept that contributes to making policymaking more sensitive to long-term global perspectives refers to *e-participation*, the inclusion of the general public into the policymaking process with the help of IS. A growing number of government organizations use technology for this purpose, providing access to policy information and requesting comments from the public about it (Macintosh 2004). In order to ensure that a policy fulfils the requirements of as many people who are affected by it as possible, policymakers must reach the targeted group. IS offer a broad variety of solutions to meet this need (e.g., e-participation in online communities) by removing time and space constraints (Phang and Kankanhalli 2008). Such solutions should take into account how deeply citizens should be involved in the policymaking processes (be informed about a policy or be actively involved in its creation or implementation) and what information policymakers need from citizens and within what time frame. It is also important to evaluate the outcome of each e-participation project (See the passage on *Policy evaluation*, below).

The second IS-supported concept that contributes to making policymaking more sensitive to long-term global perspectives deals with policy *counter-implementation* (CI) and *counter–counter-implementation* (CCI). CI occurs when a policy implementation is sabotaged. As policymaking is deeply linked to political decisions (in both the public and the private sector), there will always be some kind of resistance. In relation to IS, counter-implementation might happen if, for example, people involved in policy implementation feel threatened by the change (Keen 1981). CCI is the tactic used to combat CI by analysing the scenarios of potential CIs and how they can be avoided before a project breaks down. Information is the key element of CCI, and IS can be assist considerably by reducing communication costs and enhancing information flow (Larrain et al. 2000). Setting up knowledge bases, which include the best practises on measures that have been applied successfully, can also help.

The third concept is related to *policy transfer*. Before introducing a policy, decision-makers usually determine whether a comparable policy in a similar environment (e.g., another country with a similar political and/or economic background) already exists, in which case all or part of it can be transferred and used. Technological developments make such a policy transfer process easier by offering new ways of communication and storing and accessing information, which has already resulted in increased occurrences of policy transfer (Dolowitz and Marsh 2000). Since such policy transfer can also fail, as there are many internal and external factors that influence its potential for success, the important questions concern which factors influence a policy's outcome and who should be involved in implanting a policy. Information can reduce the risk of policy transfer failures significantly, so the contribution of modern communication technologies supported by IS to the exchange of information on a global scale should not be underestimated.

The fourth IS-supported concept that contributes to making policymaking more sensitive to long-term global perspectives is that of *cooperation in policy-making*. One example of a Europe-wide long-term-oriented policy is the Schengen agreement. This agreement not only facilitates the creation of a single European market and travelling without border controls but also introduces within the agreement the Schengen Information System, which contains information about people who have violated the law (Guiraudon 2000). The availability of the system in all member states significantly eases the process of catching criminals. Other technological cooperation delineated in the agreement includes the finger-printing of asylum-seekers, which the IS-supported means of world-wide communication makes easier. The Schengen Information System, which does not require large investments of effort and money investments, helps to make cooperation in policymaking easier and cheaper by, for example, providing opportunities to conduct virtual meetings, which allows geographically separated teams to work together.

The fifth concept is *resilience thinking*, which refers to a society's resilience or response to shocks and ability to face change and reduce such uncertainties as risks, ambiguity, and ignorance (Leach 2008). Resilience thinking includes policies that affect how a society deals with natural disasters, catastrophes, and crises from a long-term global perspective. One important aspect of this concept is the need to raise society's awareness about the environmental change caused by humans and the need for more sustainable production and consumption. Here, IS can reveal the consequences of society's behaviour by means of rapid analyses of huge data sets and communication of the research results to policymakers and the wider public (Parr et al. 2003). Advanced IS tracking tools can also help to prevent or manage disasters.

The sixth concept considers *collective intelligence* (COIN), which refers to a large system that includes many connected actors with no centralized communication and control but a utility function that rates the histories of the full system (Tumer and Wolpert 2000). The goal of the COIN approach is to determine how the people involved should update their behaviour in order to get the best results possible (e.g., to avoid traffic jams). By using mass collaboration platforms and real-time visualisation, policymakers can create, learn, engage, and share group knowledge and adapt policies to make them more efficient. A good deal of information is already available, but it remains to be mapped to usable applications, and this is where IS can be of help in supporting global sustainable policy development.

The last concept reviews *prediction-based methods* that are relevant to (1) *policy context analysis,* (2) *policy modelling,* (3) *policy simulation, and* (4) *policy evaluation* (Lampathaki et al. 2010). These four domains were proposed within the CROSSROAD project, a support action project funded by the European Commission that is aimed at determining the key research challenges in the field of ICT for policy modelling and governance. The concepts of resilience thinking and COIN rely on real-time information that, when combined with information about the past, can be used to predict the future.

(1) *Policy context* analysis focuses on analytical and descriptive analysis of policy or analysis for the formulation of policies and proposals. IS facilitates such statistical methods as foresight, forecasting, and backcasting, which can analyse the past in order to develop short-term forecasts. One useful approach for policy analysis on issues within a 10-year horizon is the State of the Future Index (SOFI), which forecasts the results of today's actions (Gordon 2003) with the goal of improving the sensitivity of decision makers at the local, national, and global levels to long-term global perspectives.

The Real-Time Delphi Method is another forecasting approach for analysing policy contexts. In contrast to traditionally expensive and time-consuming Delphi studies, the Real-Time Delphi Method uses software to provide each participant with the responses of other participants and the reasoning for these responses in real time. This method supports mass collaboration and can be used to evaluate policy options from remote locations without unnecessary expenditure of time. Such IS tools play an important role in future policymaking, as IS facilitates policy intelligence (comparable with business intelligence), IS-driven decision analysis, gaming-based simulations in the context of policymaking (Misuraca et al. 2011), and process optimisation and control. Lempert et al. (2004) pointed out that long-term policy analysis using IS tools should help policy-makers to make systematic, grounded decisions.

(2) *Policy modelling* is another method that help to increase policy-makers' sensitivity to long-term global perspectives. Policy-makers can use models (made by researchers working with policy-makers) as simplified, abstract views of reality. Several studies have mentioned the importance of such collaboration for the success of policy-modelling (Cockerill et al. 2009; Kreileman et al. 1996; Lampathaki et al. 2010). Edwards (1996) suggested that such comprehensive modelling informs policy-makers about the structure and the extent of global problems, thereby supporting the decision-making process.

(3) The aim of IS-supported policy-making in the context of *policy simulation* is to gain feedback from artificial, yet realistic tests in order to choose the best policy option. For example, computer-based simulation models enable policy-makers to play 'what if' games, where they can evaluate various scenarios related to policy implementation (Cockerill et al. 2009). Simulation of social and political systems can be done via agent-based modelling, where agents are (for example) governments (Lempert et al. 2009).

(4) Several IS tools that support qualitative and quantitative methodologies for *policy evaluation* have already been developed. Policy evaluation uses scenario techniques to support the policy-making process at a strategic level (Lampathaki et al. 2010). Robust decision-making (RDM) refers to the practice of policy simulation and policy evaluation, where computers are used to create plausible long-term paths (scenarios), after which robust short-term policy options are identified using statistical computer algorithms. As a result of RDM, policy-makers are able to gain a systematic understanding of various scenarios so they can consider the differing consequences of feasible paths before they decide on a policy (Lempert et al. 2009).

## 3.4 Research Agenda

The literature review process made clear that the proposed concepts cannot address our main RQ fully (although they still provide valuable ideas on how IS can contribute to making policy-making more sensitive to long-term global perspectives). As a result, we developed additional RQs from the literature.

*To what extent is the integration of population into political decisions beneficial for the policy-making process? E-participation* in *e-democracy* is important for integrating the population into political decisions, but with the help of IS, such integration can become so easy that too many participants in policy-making would lead to endless debates. Another challenge arises with the need to ensure the security of IS that supports online voting during governmental elections.

*Can policy transfer be simulated using (for example) modelling techniques?* Further reviews on the best practices supported by IS are needed for such areas as *counter–counter-implementation* and *policy transfer*. Creating a policy-transfer database might be a reasonable idea, but we found no such idea during the literature review. However, several other IS for political cooperation have already been implemented, among which is the Schengen Information System. Since policy transfer can be misleading, prior *simulation* might be useful.

*Can a global IS that supports politics and is implemented in multiple countries become a basis for making decisions in the context of global resilience?* Our main finding from the literature is that cooperation between governments with the help of IS yields significant possibilities. IS can be of great value in the communication and coordination of policy-making between countries. As a next step, we suggest implementation of a global IS to support policies that are useful for groups of countries. However, further research is required on how to collect and integrate data from multiple countries into a single IS. A global IS could be an integrated collective intelligence system that supports cooperation in politics and policy-making. However, data security and the influence of national cultures remain problems and require further investigation.

We found four sub-areas of research on using IS to make policy-making more sensitive to long-term global perspectives in the literature: *policy context analysis*, *policy modelling*, *policy simulation*, and *policy evaluation* (Lampathaki et al. 2010).

*Policy context analysis* deals with predicting the future and evaluating its impacts and implications using such methods as foresight, forecasting, and backcasting. *Policy modelling* incorporates findings from various scientific domains and research areas related to the creation of computer models. It usually refers to specific fields of application (e.g., climate change) and uses such approaches as system dynamics, game theory, and mathematics. *Policy simulation* is a powerful method for making short- and long-term predictions. It involves the creation of computer models that stand in for social systems in order to evaluate social interdependencies. Such computer-assisted social simulations are influenced by physics, computer science, and mathematics. Finally, *policy evaluation* involves the assessment of policy impacts in order to explain why policy instruments perform differently than they were supposed to perform. Policy evaluation can be used to derive best practices or feedback for policy-making. There are several frameworks on how policy evaluation should be implemented (e.g., Mark et al. 2000), but research on policy evaluation is still in progress (Lampathaki et al. 2010).

We believe that predictive methods supported by IS can be a powerful driver in making policy-making more sensitive to long-term global perspectives. The ongoing research in such fields as mathematics and computer science regularly delivers new prediction-based methods. Now it is important to identify which of them predict the future in the context of policy-making in the most accurate way.

## 3.5 Conclusion

A long-term perspective and a global context are the two most vital aspects of modern policy-making. There are numerous challenges associated with these requirements, and it is important to know whether IS can support the concepts that are intended to address these problems. This study analysed several areas of global and long-term policy-making and the role of IS in their support in order to derive three primary conclusions.

First, IS can support global and long-term policy-making that deals with particularly complicated issues, such as disasters and climate change. *Resilience*, one of the keywords proposed in this paper, shows how policies influence how society deals with unexpected situations. Resilience requires deep insights about the future, and IS can enhance prediction-based methods for *policy context analysis, policy modelling, policy simulation,* and *policy evaluation.*

Second, until now policy decision-makers have been a small group of people. We propose that approaches like *e-participation* could help decision-makers to get feedback from all relevant target groups on various policy activities, which would ensure that a policy meets the requirements of those who are most affected by it.

Third, there are few areas in which no policy has been established in at least one country. It is logical to reuse or adapt these policies in a process called *policy transfer*. One example of how IS support the use of a policy in many countries is the Schengen Information System developed within the Schengen agreement. The core idea of the Schengen Information System is to help all of the signatory countries to know what the others are doing in certain areas (e.g., what criminals they are looking for), which makes the decision-making process more efficient. Such *collective intelligence* helps security organizations but can be also applied in areas like transport organisation and optimisation.

We find the creation of one integrated IS for several countries to be the most exciting topic for future research. We believe that integrated policies are needed in order to face long-term global problems. These integrated policies can be created only if there is an integrated view on the world, which can be facilitated by IS.

By using IS for cooperation and coordination, the time and space usually required for policy-making can all but disappear. Thus, IS can be of great value in making policy-making more sensitive to long-term global perspectives.

## References


Braess, D., Nagurney, A., & Wakolbinger, T. (2005). On a paradox of traffic planning. *Transportation Science, 39*(4), 446–450. doi:10.1287/trsc.1050.0127.

Cockerill, K., Daniel, L., Malczynski, L., & Tidwell, V. (2009). A fresh look at a policy sciences methodology: Collaborative modeling for more effective policy. *Policy Sciences, 42*(3), 211–225. doi:10.1007/s11077-009-9080-8.

Dolowitz, D. P., & Marsh, D. (2000). Learning from abroad: The role of policy transfer in contemporary policy-making. *Governance, 13*(1), 5–23. doi:10.1111/0952-1895.00121.

Edwards, P. N. (1996). Global comprehensive models in politics and policymaking. *Climatic Change, 32*(2), 149–161. Retrieved from http://www.springerlink.com/index/M950J6027250521Q.pdf.

Glenn, J., Gordon, T., & Florescu, E. (2011). *2011 State of the future* (1st ed.). The Millennium Project.

Gordon, T. J. (2003). State of the Future Index (SOFI): A method for improving decision making that affects the future. *21st century opportunities and challenges: an age of destruction or an age of transformation* (p. 37).

Gordon, T. J. (2009). The Real-Time Delphi Method. *Futures Research Methodology Version, 3.*

Guiraudon, V. (2000). European Integration and migration policy: vertical policy-making as venue shopping. *JCMS: Journal of Common Market Studies, 38*(2), 251–271. doi:10.1111/1468-5965.00219.

Keen, P. G. W. (1981). Information systems and organizational change. *Communications of the ACM, 24*(1), 24–33.

Kreileman, E., Leemans, R., & Alcamo, J. (1996). Global models meet global policy: How can global and regional modellers connect with environmental policy makers? What has hindered them? What has helped? *Science, 6*, 255–259.

Lampathaki, F., Koussouris, S., Charalabidis, Y., Asjounis, D., Mouzakitis, S., Passas, S., et al. (2010). State of the art. In F. Lampathaki, S. Koussouris, Y. Charalabidis, & D. Asjounis (Eds.), *ICT for governance and policy modelling*. CROSSROAD White Paper.

Larrain, F. B., Sachs, J. D., & Warner, A. (2000). A structural analysis of Chile's long-term growth: History, prospects and policy implications*. *Growth (Lakeland)*, (January).

Leach, M. (2008). Re-framing resilience: a symposium report. In *Symposium a quarterly journal In modern foreign literatures.*

Lempert, R., Popper, S., & Bankes, S. (2004). *Shaping the next one hundred years: new methods for quantitative, long-term policy analysis. Technological forecasting and social change* (Vol. 71). Santa Monica: RAND. doi:10.1016/j.techfore.2003.09.006.

Lempert, R., Scheffran, J., & Sprinz, D. F. (2009). Methods for long-term environmental policy challenges. *Global Environmental Politics, 9*(3), 106–133. Retrieved from http://www.mitpressjournals.org/doi/abs/10.1162/glep.2009.9.3.106.

Macintosh, A. (2004). Characterizing e-participation in policy-making. In *Proceedings of the 37th Annual Hawaii International Conference on System Sciences, 2004. 00*(C), p. 10. doi:10.1109/HICSS.2004.1265300.

Mark, M. M., Henry, G., & Julnes, G. (2000). *Evaluation: an integrated framework for understanding, guiding, and improving policies and programs* (p. 400). San Francisco: Jossey-Bass.

McCartney, G. (1988). Implementing computer-based systems. In *Proceedings of the 16th Annual ACM SIGUCCS Conference on User Services—SIGUCCS '88* (pp. 117–122). doi:10.1145/62548.62581.

Misuraca, G., Broster, D., & Centeno, C. (2011). Digital Europe 2030: Designing scenarios for ICT in future governance and policy making. *Government Information Quarterly*. doi:10.1016/j.giq.2011.08.006.

Parr, T. W., Sier, A. R. J., Battarbee, R. W., Mackay, A., & Burgess, J. (2003). Detecting environmental change: science and society-perspectives on long-term research and monitoring in the 21st century. *The Science of the Total Environment*, *310*(1–3), 1–8. doi:10.1016/S0048-9697(03)00257-2.

Phang, C. W., & Kankanhalli, A. (2008). A framework of ICT exploitation for e-participation initiatives. *Communications of the ACM*, *51*(12), 128. doi:10.1145/1409360.1409385.

Randolph, J. (2009). A guide to writing the dissertation literature review. *Practical Assessment Research Evaluation*, *14*(13). Retrieved from http://www.mendeley.com/research/guide-writing-dissertation-literature-review-1/.

Rotmans, J., Kemp, R., & Van Asselt, M. (2001). More evolution than revolution. *World*, *03*(01), 17.

Tumer, K., & Wolpert, D. H. (2000). Collective Intelligence and Braess' Paradox. In *Proceedings of AAAI 2000*. Morgan Kauffman.

Vom Brocke, J., Simons, A., Niehaves, B., Riemer, K., Plattfaut, R., & Cleven, A. (2009). Reconstructing the giant: On the importance of rigour in documenting the literature search proccess. In *ECIS 2009* (pp. 1–13). Verona.

# Chapter 4
# Making the Global Convergence of ICT Available for Everyone

**Tomal K Ganguly and Klaus Fleerkötter**

## 4.1 Introduction

In 2011, the pace in the development of information and communication technology (ICT) seemed to peak, but the convergence phenomenon can still be described as a first-world trend. While the majority of citizens of North America, Australia, and Europe have Internet access, the world average is only every third person. In Africa, Internet penetration plunges as low as 11 %.[1]

While access to the Internet may not be as directly valuable as goals like eliminating hunger or curing disease, the converging applications in information and communication can accelerate the improvement of the human condition. Across the globe, agile protestors are overthrowing non-democratic governments by organizing via social media that uses ICT channels, and national legislations are facing empowered citizens while struggling with regulating the Internet. Inhabitants of the world grow together by means of digital communication and information and establish the first genuine digital democracies. Unrestricted information access supports and enables democratization, learning environments, and free speech, but this global convergence renders national regulation almost powerless. While, on one hand, this development routinely defies governmental censorship, it also imposes new kinds of threats that defy current law enforcement structures. Governments of leading countries in technological development criticize regimes that try to inhibit their citizens' communication connections and information access while also trying to establish similar censorship infrastructures under the cover of preserving intellectual property rights.

This outline demonstrates how the global convergence of information and communication affects a wide range of technological and social topics. This article

[1] http://www.internetworldstats.com/stats.htm.



J. vom Brocke et al., *Grand Societal Challenges in Information Systems Research and Education*, SpringerBriefs in Information Systems,
DOI 10.1007/978-3-319-15027-7_4

assesses the emerging themes in this area and the opportunities, threats, and challenges that these developments create.

The remainder of the article is structured as follows: first we describe our literature review approach. Then we discuss the emergent themes regarding global convergence. Finally, we propose a research agenda that could guide further research.

## 4.2 Literature Review

The Millennium Project is a report that identifies trends and issues that lead to global challenges. The present paper analyses how ICT can contribute to the challenge of making ICT available to everyone.

The framework (Fig. 4.1) explains the procedure that our research followed and shows the bridge between information and communication regarding the opportunities and threats of internet access and its availability to citizens. Moreover, it highlights which technological developments might help to close the gap and build an expedient bridge (Fig. 4.1).

A literature review seeks to uncover the sources relevant to a topic under study, so it makes a vital contribution to the relevance and rigor of research (vom Brocke et al. 2009). We searched among several databases: ACM Digital Library, Business source premier, EBSCOhost, Elsevier, JSTOR, and IEEE. The review can be characterized as focusing on research application, integrating various research areas, and having a neutral perspective and a target audience of general scholars (Cooper 1988). The literature review followed a procedure in which search results matching the respective topic ("hits") were saved to a project database; the hits were then analysed based on evaluation of their titles, abstracts, and summaries and, if

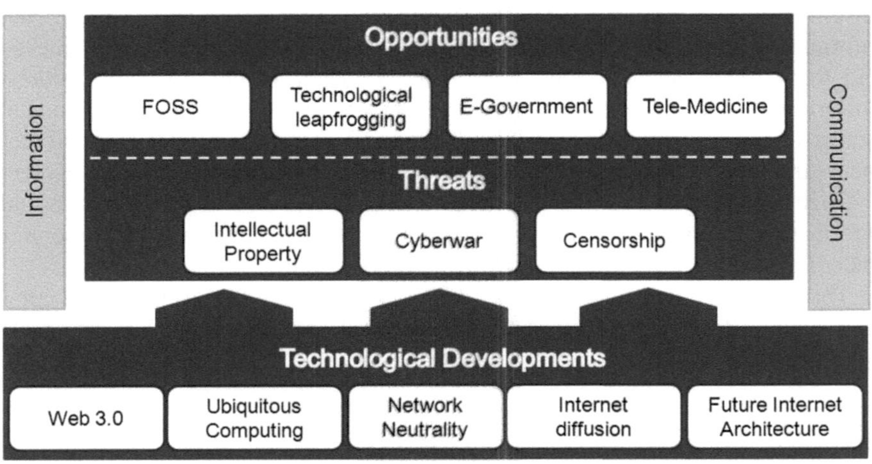


**Fig. 4.1** Conceptual framework

they were valuable for the purposes of this review, they were catalogued based on a tag-based taxonomy ("evaluated"); and articles that addressed the core research interest were synthesized and marked as "reviewed".

## 4.3 Results and Discussion

(1) ***Opportunities***. Developing countries set trends in applications, revenue models, and cost-saving approaches, especially for mobile networks (Khalil and Kenny 2008). ICT has immense potential to facilitate this development through, for example, *technological leapfrogging*, which bypasses some of the processes of accumulation of human capabilities and fixed investment in order to narrow the gaps in productivity and output that separate industrialized and developing nations (Steinmueller 2001). The developing world is well-equipped to use modern technologies or even to become the lone adopter of such products (Chen 1999) because of the unavailability of an established technology and the ambitions and aspirations of the developing world (Prakash 2005; Malairaja 2003). However, opinions vary concerning whether technological leapfrogging is possible beyond the mobile phone phenomenon and whether it is key to the successful development of countries that sometimes lack even basic sanitation facilities (Steinmueller 2001).

Digital government or *electronic government* (e-Government) has emerged as a new form of interaction between the state and the citizens with the purpose of improving government performance and processes (Andersen et al. 2010). The key principles of e-Government initiatives include transparency, participation, and collaboration. A certain level of trust in the electronic interaction with the government must underpin these key principles and requirements. The intention to use e-Government depends on access and skills, since skills are an important element in terms of how the population uses the Internet—whether for general use, online purchases, or online information searches.

The healthcare industry claims to be one of the first service industries to introduce the use of ICT in electronic health services (e-Health), which began with telephone-based medical consultations and grew from there (Houghton 2002). e-Health is also the focus of the European health policy, which has led to a significant increase in health insurance cards and e-services like reimbursement and the electronic transmission of prescriptions. Telemedicine, an important part of the European healthcare agenda, has attracted considerable attention from the participating countries (Denz 2008).

(2) ***Threats and Requirements***. The regulation of cyberspace via *Internet censorship* is an important topic that has been linked to the topics of democracy, free speech, and privacy. Internet censorship is primarily defined as state-controlled Internet filtering (e.g., South Korea's blocking of pro-North Korean sites, India's blocking of sites of groups that foment domestic conflict, blocking search results from the Chinese Google website or Wikileaks). Actually, access to the unrestricted Internet cannot be found in any country in the world (Tariq 2006). The

states find new and more sophisticated techniques to restrict some of the Internet activities, and the motives, scope, and effectiveness of Internet censorship vary widely from country to country (Boyle 1997).

The convergence of media and digital content into a realm of common communication tends to exacerbate the *intellectual property* problem because of issues like the impermanence, multiplication, and heterogeneity of communication sources (ICT 2005). The property of ideas and their expression is hard to maintain in cyberspace in the forms in which they are usually protected (e.g., patents, trademarks, copyrights). Dynamic content can be modified in minutes, so it is difficult to preserve its integrity in order to claim property rights to it. The same applies to the varying formats in which digital content may be available. Legal issues in this area are complicated by the fact that some media formats, such as sound files, are covered by rules specific to the media (Okediji 2004). Another example of the complex nature of such issues is the legal claims of websites in respect to the display of their content (e.g., images) through search engines. The heterogeneous nature of digital collections often leads to confusion in terms of copyright infringements, such as the when one considers whether a print publication is allowed to add the work of freelance writers to an electronic database.

Because of its relative novelty, the vulnerability of the Internet has yet to undergo a complete assessment in terms of evaluating the threat of *cyberwar*. Each target infrastructure requires a much more detailed analysis of normal rates of failure and response, the degree to which critical functions are accessible from public networks, and the level of human control, monitoring and intervention in critical operations (Sommer 2001). The amount of damage caused by the launch of cyber-attacks imposes high economic costs in relation to the comparatively minor cost of launching such an attack (Ramsaroop 2003). Opportunity costs (e.g., lost sales or lower productivity) make up a large proportion of the cost of cyber-attacks and viruses (Lewis 2002).

**(3)** ***Technological Developments***

*Ubiquitous Computing.* Ubiquitous computing is based on Weiser's (1991) vision of the human world merging with virtual content, and computers vanishing into the background. Making this vision a reality involve a multitude of purpose-specific devices, such as communication micro-chips, pads, wall screens, and smartphones, that are connected in a ubiquitous network and provide content in ways suited for their specific purposes. Therefore, devices must be aware of their and their user's environment and be physically and cognitively available (Banavar and Bernstein 2002; Waller and Johnston 2009). This development is enabled by inexpensive hardware, small-form factors, and advanced human interfaces.

For global convergence, ubiquitous computing serves the goal of availability and provides access to digital information content and communication channels that are embedded in the contexts of prospective users, lowering entry barriers and learning curves. Its real-world application would be driven by falling prices, improved hardware performance, and the ability to connect to services through a variety of networks (Islam and Fayad 2003; Lyytinen and Yoo 2002). However, it requires a large-scale, diffuse service infrastructure that can integrate a multitude of requirements and a large-scale understanding of how to adapt and understand technology (Lyytinen et al. 2004).

*Web 3.0.* Although researchers have not yet agreed upon a precise scientific definition, "Web 3.0 represents an evolutionary shift in how people interact with the web, and vice versa" (Green 2011). Certain themes are emerging to form the next-generation web, which provides information tailored to the individual user's needs, location, and identity. This web is likely to be a semantic Web that is not only readable by humans but also understandable and traversable for machines (Green 2011). Manual definitions of the relationships between types of content make complex information searches and aggregations more feasible than do text mining and AI algorithms, which are comparably slow and long-running (Berners-Lee and Hendler 2001).

While Web 2.0 is user-centric, Web 3.0 will be personal, as the correspondence between the retrieved information and the user's need is improved by leveraging the social graph—that is, the user's contacts. Web 3.0 also takes into account the properties of ubiquitous computing (embedded virtuality, mobility of computing; Green 2011) and may become a "web of things", where not only users and services but also a multitude of autonomous devices communicate transparently, connecting appliances that share and consume data. Rich Internet applications form the heart of the dynamic, data-centric Web 2.0, but the mobile web emerges in Web 3.0 with interconnected applications, mobile websites, and a swarm of new types of devices that act as user interfaces. With the properties of collective intelligence, Web 3.0 may also have a positive influence on learning and organizational development (Green 2011).

*Information Infrastructure.* Network neutrality refers to the principle of an information network that treats all data packets equally (Schuett 2010) and does not favour one application over others (Wu 2003). Network neutrality is important if long-term public interest in having an innovative communication infrastructure, where services can build in natural competition, is to be achieved (Wu 2003). Standing against this effort are the short-term interests of ISPs, which (because of their control over the "last mile" to the customer) act as selective gatekeepers.

A recent example is the mobile Internet market: Because wireless network providers block voice-over-IP packets by employing deep packet inspection, they use their control power over the end user's technology to block competitors (e.g., Skype). Ensuring that the tools that enable information and communication are available to as many humans as possible poses several requirements on a feasible Internet architecture. We derived requirements from the emerging themes introduced in the literature and aligned them to six architecture goals, classified according to Paul et al. (2011).

In the next-generation Internet, security will be a part of the architecture rather than being overlaid on top of the original architecture (Paul et al. 2011). Confidential message exchange between parties (e.g., humans, devices, and services) that can trust each other is a necessary prerequisite for global communication, which requires user-centric privacy and identity management. Research suggests the use of secure cryptographic identities (e.g., in resource-oriented networking) that enables identity-based access control on the level of packets (per-packet attribution) to be enforced throughout a network (e.g., Security Architecture for Networked Enterprises (SANE)).

Another architecture requirement is reliable *content delivery mechanisms* that ensure that information content is persistent, available, authenticated, and provided with preserved integrity for all users. The scarce bandwidth available at the transnational backbone must also be used effectively—that is, with the highest throughput and lowest latency. On the state-of-the-art Internet, content distribution networks (CDN) serve static (i.e., cacheable) content to fasten load times (Paul et al. 2011). Future Internet Assembly (FIAs) will make use of highly distributed content nodes (e.g., next-generation CDNs) and peer-to-peer delivery (e.g., next-generation peer-to-peer, swarming architecture), both of which will be aware of their underlying topology and the location of peers (e.g., swarming architectures, provider-aided distance information systems), resulting in fewer costly ISP transits and shortened content routes (Paul et al. 2011; Poese et al. 2012; Bell and Walker 2011).

Other requirements include architectures like delay-tolerant networks (DTN) and mobile ad hoc networks (Kumar 2009) for areas where Internet connections and power supply are flaky and routes between mobile devices are temporarily unavailable, fault-prone, or subject to long delays. These *challenged network environments* are defined as "heterogeneous network environments where continuous end-to-end connectivity cannot be assumed" (Paul et al. 2011). In addition, *management control frameworks* that prohibit censorship and ensure network neutrality must be found. Finally, in order to achieve network neutrality, manageable content policies and network path optimization, with regard to overall network efficiency, the *internetworking layer* must be redesigned to improve scalability, mediation between user-defined content policies, and ISP-based performance optimizations.

Basic ICT markets must be established in order to facilitate competition and enable foreign competitors to invest in local infrastructure. To ensure that ICT is used to its full potential, regulatory institutions are needed for broadcast infrastructure and content—either combined or split (Khalil and Kenny 2008)—that examine and leverage the impact of ICT and drive change and privatization (Jussawalla 1999). Such regulations may include the liberalization of state-owned monopolies or policies of open economies and export-oriented investment in technology, providing the basis for long-term economic growth, as has been observable in Asian countries (Jussawalla 1999). Regulation also includes the reformation of mechanisms of spectrum allocation for wireless communication, which is best handled through property rights (Khalil and Kenny 2008).

The objective of this paper is to answer the question concerning how the global convergence of ICT can work for everyone. While seeking to close the gap between information and communication (Fig. 4.1), one must find the opportunities, threats, and requirements and define which solutions fit best. While one cannot say that there is only one right answer, focusing and narrowing down the opportunities and threats reveals some of the most plausible topics.

In defining the opportunities, threats, and requirements, we find that the IT area is important but should not be the sole focus. We try to link the thoughts of the Millennium project with other fields, such as the political view, the judicial view, and the socio-cultural aspects of the issues. Combining those four areas, we identified the topics of Free and open-source-software (FOSS), technological leapfrogging, e-Government, and Tele-Medicine. While looking for best-practice methods among these areas, we must also consider the threats that are related to the four areas and that are also linked to other problems throughout the project work. Intellectual property, cyberwar, and censorship were identified as threats, while intellectual property can also be seen as an opportunity.

We narrowed the topics to five that may give an answer to the question posed for the literature review. The main point was to define areas that are important to our ability to connect the world. We listed critical issues within these points, such as issues within the topics of internet diffusion and network neutrality, as well as technological concepts that can address them. Although we are setting a good pace in achieving global connectivity, future technologies—such as a new internet architecture that optimizes bandwidth usage but also installs centralized censorship possibilities—may also slow our ability to overcome these critical issues. We highlighted the cross-disciplinarily of these issues by bringing together the views and aspects of IT, politics, law, and socio-culture.

We developed an overall method and framework to add the technological developments of Web 3.0, ubiquitous computing, network neutrality, internet diffusion, and future internet architecture. We combine these developments with the opportunities of FOSS, technological leapfrogging, e-Government and tele-medicine while considering at the same time the threats and requirements of intellectual property, cyberwar, and censorship.

## 4.4 Research Agenda

Performing a literature review on the topic presented here uncovered a gap in the research on the global convergence of information and communication: Although the areas presented here are already adequately covered, research on the core nature of global convergence is missing. Therefore, our conceptualization can serve as a basis from which to outline future research's attempts to strengthen the links between the socio-cultural factors and information systems to improve the future global state. Interdisciplinary collaboration from diverse disciplines, such as politics, law, computer science, and socio-culture, will be necessary in this effort.

Many questions are yet to be answered, including how to find a working balance between preventing censorship and governmental control of access to content against the need to preserve value for cultural minorities (which is likely to require a filtering infrastructure). The question of internet diffusion and the role of governments in this process will become even more important as Internet penetration increases in the near future.

A possible model for a diffusion process that is cost-effective and that preserves the public interest in an innovative infrastructure could be *technology transfer through governmental incentives for open source*. Openness of interfaces and platforms is an essential prerequisite for establishing a global infrastructure like the internet. Therefore, developed countries should set incentives for the domestic technology markets to pursue open source. The more open source components are developed (and, consequently, the more open interfaces are used), the more developing countries can benefit from this technology, save costs, and leapfrog their development by extending on the state of the art. In addition, open source teams can help to achieve the transfer of knowledge into developing regions, support the externalization of first-world knowledge, and enable a global virtual and physical exchange of professional talent. Finally, private competition in developing countries' markets is enhanced because suppliers can build on available commons like the FOSS infrastructure components.

## 4.5 Conclusions

According to Khalil and Kenny (2008), and as confirmed by the findings of our literature review, developing countries will set the trend in applications, revenue models, and cost-saving approaches, especially for mobile networks. We created a framework that can serve as a guideline to answering the question concerning how the global convergence of information and communication technologies can work for everyone. We focused not only on IT but also on aspects of politics, law, and socio-cultural issued, combining them with current affairs. In seeking to close the gap between information and communication (Fig. 4.1), we had to consider other important areas, such as threats and requirements, to give a shape to the framework.

Future research should improve the framework and detail it to the point at which it can be evaluated and include other findings to round it out. The developed framework should be seen as a basis from which to provide an answer to the questions concerning to connect the people all around the world using tools, considering the opportunities and threats that are present, and taking into account the challenges and complexity of the task.

## References

Andersen, K. N., Henriksen, H. Z., Medaglia, R., Danziger, J. N., Sannarnes, M. K., & Enemærke, M. (2010). Fads and facts of e-government: A review of impacts of e-government (2003–2009). *International Journal of Public Administration, 33*(11), 564–579.

Banavar, G., & Bernstein, A. (2002). Software infrastructure and design challenges for ubiquitous computing applications. *Communications of the ACM, 45*(12), 92–96.

Bell, S., & Walker, S. (2011). Futurescaping infinite bandwidth, zero latency. *Futures, 43*(5), 525–539. Elsevier Ltd.

Berners-Lee, T., & Hendler, J. (2001). Publishing on the semantic web. *Nature, 410*(6832), 1023–1024.

Boyle, J. (1997). Foucault in cyberspace: surveillance, sovereignty, and hard-wired censors. *University of Cincinnati Law Review 66*, 177.

Chen, Z. (1999). Adoption of new technology by a lagging country: Leapfrogging or no leapfrogging? *Pacific Economic Review, 4*(1), 43–57.
Cooper, H. M. (1988). Organizing knowledge syntheses: A taxonomy of literature reviews. *Knowledge in Society, 1*(1), 104–126. Springer: Netherlands.
Denz, M. D. (2008). Sustainable telemedicine: Paradigms for future-proof healthcare. *European Health Telematics Association, 1.*
Green, M. (2011). Better, smarter, faster: web 3.0 and the future of learning. *Training + Development, 65*(4), 70–72.
Houghton, J. (2002). *Information technology and the revolution in healthcare.* Victoria University of Technology, Centre for Strategic Economic Studies, Melbourne. www.cfses.com.
Islam, N., & Fayad, M. (2003). Toward ubiquitous acceptance of ubiquitous computing. *Communications of the ACM, 46*(2), 89–92. Association for Computing Machinery, Inc, One Astor Plaza, 1515 Broadway, New York, NY, 10036–5701, USA.
Jussawalla, M. (1999). The impact of ICT convergence on development in the Asian region. *Telecommunications Policy, 23*(3–4), 217–234.
Khalil, M., & Kenny, C. (2008). The next decade of ICT development: Access, applications, and the forces of convergence. *Information Technologies & International Development, 4*(3), 1–6.
Kumar, M. (2009). Distributed computing in opportunistic environments. *Ubiquitous Intelligence and Computing*, 1–1. Springer.
Lewis, J. A. (2002). Assessing the risks of cyber terrorism, cyber war and other cyber threats. *Center for Strategic & International Studies.*
Lyytinen, K. J., & Yoo, Y. (2002). Issues and challenges in ubiquitous computing. *Communications of the ACM, 45*(12), 62–65.
Lyytinen, K. J., Yoo, Y., Varshney, U., Ackerman, M., Davis, G., Avital, M., et al. (2004). Surfing the next wave: design and implementation challenges of ubiquitous computing. *Communications of the Association for Information Systems, 13*(1), 40.
Malairaja, C. (2003). Learning from the Silicon Valley and implications for technological leapfrogging the experience of Malaysia. *International Journal of Technology Management and Sustainable Development, 2*(2), 73–95.
Okediji, R. L. (2004). Development in the information age: issues in the regulation of intellectual property rights, computer software and electronic commerce. *UNCTAD-ICTSD, 9.*
Paul, S., Pan, J., & Jain, R. (2011). Architectures for the future networks and the next generation internet: A survey. *Computer Communications, 34*(1), 2–42. Elsevier B.V.
Poese, I., Frank, B., Ager, B., Smaragdakis, G., Uhlig, S., & Feldmann, A. (2012). Improving content delivery with PaDIS. *IEEE Internet Computing, 16*(3), 46–52.
Prakash, G. (2005). Leapfrogging into the knowledge era: Use of ICT for development. *IMR Conference* (pp. 47–56).
Ramsaroop, P. (2003). Cybercrime, cyberterrorism and cyberwarfare. Pan American Health Organization.
Schuett, F. (2010). Network neutrality: A survey of the economic literature. *Review of Network Economics, 9*(2). doi:10.2202/1446-9022.1224.
Sommer, J. S. (2001). Against cyberlaw. *Berkeley Technology Law Journal, 15*(3), 1145–1232.
Steinmueller, E. (2001). ICTs and the possibilities for leapfrogging by developing countries. *International Labour Review, 140*(2), 193–210.
Tariq, O. (2006). Internet censorship: the end of digital libertarianism? *London School of Economics.*
vom Brocke, J., Simons, A., Niehaves, B., Riemer, K., Plattfaut, R., & Cleven, A. (2009). Reconstructing the giant: On the importance of rigour in documenting the literature search process. *17th European Conference on Information Systems* (pp. 1–13).
Waller, V., & Johnston, R. B. (2009). Making ubiquitous computing available. *Communications of the ACM, 52*(10), 127.
Weiser, M. (1991). The computer for the 21st century. *Scientific American, 265*(3), 94–104. New York.
Wu, T. (2003). Network neutrality, broadband discrimination. *Journal on Telecommunications & High Technology Law, 2*(1), 141–175.

# Chapter 5
# How Can the Capacity to Decide Be Improved as the Nature of Work and Institutions Change?

**Marina Maschler and Asin Tavakoli**

## 5.1 Introduction

Trans-institutional decisions are often slow, inefficient, and ill-informed. The Gartner Group (Gartner 2011) identified the "top 10 strategic technologies for 2011", which included social analytics, context-aware computing, and ubiquitous computing. Human decision-making is already dealing with these IT trends. Especially during the last 3 years of the 2008 financial crisis, decision-making by computer systems gained a foothold in financial markets. These automated decisions and trades are driven by the interplay among legislative measures, increased competition between execution venues, and improvements in information technology (Avellaneda and Stoikov 2008). High-frequency trading, one of the key developments stemming from this technological advance, has become a significant feature of modern financial markets (OICV-IOSCO 2011). However, the automated decisions and actions of such systems can lead to unforeseen events, like the 'flash-crash' that took place on 6 May 2010 (Fischer 2011). The Millennium Project's challenge number nine deals with this conflict between the human decision-making process and the use of information technology to make decisions.

Such grand challenges have not yet been addressed in the field of information systems (IS) research. The IS community is more focused on individual researcher's agendas, deprioritizing such large joint research endeavours as global challenges (Winter and Butler 2011). This article identifies the contributions to the ninth global challenge that have already been made and, based on these contributions, discussions two research questions:

1. What are the major topics in IS that can contribute to the ninth global challenge?
2. How should IS researchers address the problems described in the ninth global challenge?



J. vom Brocke et al., *Grand Societal Challenges in Information Systems Research and Education*, SpringerBriefs in Information Systems,
DOI 10.1007/978-3-319-15027-7_5

## 5.2 Literature Review

### *5.2.1 Pre-identification of Relevant Topics*

Technology offers a huge range of opportunities for connecting computers and humans. Decision support systems (DSS), as a special kind of IS, organize data, information, and models to help improve human decisions. DSS support the decision-maker with the objective facts that may be needed. For example, in large organizations or nations like the UN, decision-making can be improved by showing how visions, missions, goals, strategies, and objectives are linked. An initial summary of the ninth global challenge shows that many areas must be addressed to solve it. Table 5.1 provides an overview of the sub-problems contained in the ninth challenge.

### *5.2.2 Research Approach*

A literature review has been conducted based on vom Brocke et al. (2009) to analyse the available contributions of the IS discipline to these sub-problems.

**Table 5.1** Sub-problems of the ninth challenge

| Global challenge sub-problems | Sub-problem description | Source |
|---|---|---|
| Limited capacity to decide | The human ability to analyze and synthesize the available information in a timely fashion is limited. A result of this is the requirement to reduce decision-making time. Besides, the ability to make timely strategic decisions becomes necessary | The Millennium Project (2011) |
| Information overload | The increasing amount of available information leads to excessive choice proliferation. Besides, the number and complexity of choices increases. In addition, more and more information sources are made accessible | The Millennium Project (2011) |
| Increasing number of stakeholders | The number of stakeholders involved in a single decision is steadily increasing. Topics of interest are therefore group decision-making, trans-institutional decision-making, and from a political context also governmental decision-making | Glenn and Gordon (2001) |

The focus of the search words is on the decision-making realm and related areas. A concept-centric approach to the literature review is aligned with Webster and Watson (2002). The sub-sections are organized around the sub-problems, after which major concepts and artefacts (if possible) that are addressed by the IS realm are presented.

### 5.2.3 Literature Search Results

#### 5.2.3.1 Sub-problem: Limited Capacity to Decide

In general, discussing and analysing DSS was popular among IS researchers during the 1980s and 1990s. Sixteen of thirty-six relevant papers describe at least the architecture, the functions, the implementation or a framework for DSS (e.g., Adelman 1991; Alter 1977; Finlay and Forghani 1998; Kersten 1985; Wang and Shen 1989; Sauter and Schofer 1988). In addition, there are a few publications about the theory behind DSS (Arnott 2004; Kasper 1996).

Another group of papers deals with specific types of DSS (Mendonça 2003; Xu et al. 2008), and there are also studies about special types of DSS, such as knowledge-based DSS (Elam and Henderson 1983; Jacob and Pirkul 1992). This special type of IS takes specific knowledge to simulate human expert consulting and uses the concepts of knowledge engineering (Elam and Henderson 1983).

The sub-problem 'limited capacity to decide' refers to the challenge of accessing the most relevant information among irrelevant information and information pollution in order to take the right decision (Orman 1984). Decision-making, as well as its representation, is often an intricate process (Srinivas and Shekar 1997). As Yadav and Khazanchi (1992) mentioned, decision-makers are always involved with understanding, formulating and understanding problems, but they are constantly confronted with limited information capabilities (Yadav and Khazanchi 1992). As a solution, decision-makers could use external IS to access external data. To make the right decision, it is also important to get internal data (Walters et al. 2003).

The results of the studies that deal with the question concerning the degree to which a DSS influences and limits managers in their capacity to decide have shown that managers with high-quality DSS make better decisions with a generic tool like Microsoft Excel (Lilien et al. 2004; Riedel and Pitz 1986).

#### 5.2.3.2 Sub-problem: Increasing Number of Stakeholders

The literature identified as relevant to this topic can be divided into five groups. In the first group are publications that provide a general overview of the topic (Buchanan and O'Connell 2006; Fjermestad and Roxa 1998). The second group contains publications that deal with group characteristics like communication synchronicity (Benbunan-Fich et al. 2003), or that deal with the effects of group size and spatial arrangements (Cummings et al. 1974) or group composition (Daily et al. 1996) on group decision-making performance or decision-making quality.

Papers in the third group analyse techniques that could be used in the group decision-making process to enhance performance. Examples of such techniques are problem structuring and modelling (Barki and Huff 1985), decisional guidance (Limayem and DeSanctis 2000), and aspiration-led and quasi-satisficing paradigms for eliciting user's preference (Lewandowski 1989). Researchers in this group also explore techniques to overcome certain problems in decision-making groups. For example, researchers have looked at the personal preferences problem in group settings (Farris et al. 1975), decisions under threats (Gladstein and Reilly 1985), and situations that require fast and flexible decision-making (Mendonça 2003). From a leadership perspective, research has been conducted on group decision approaches (Miner 1979). Technology support is also a prominent research area in the area of group decision-making. Groups that use technology performed better than those that did not (Daily et al. 1996). Others had similar results using adaptive structuration theory building on the research findings of a twenty-year-long project (DeSanctis et al. 2008). Moreover, it has been empirically shown that technology-supported groups develop project agendas that are more closely related to the requirements of the team members (Dennis et al. 2003). The additional use of expert systems outperforms the use of systems in individual settings (Nah et al. 1999). Zigurs et al. (1988) took a closer look at group processes, identifying influence behaviour as a major variable, while El-Shinnawy and Vinze (1998) identified polarization and persuasive argumentation in group settings as important factors.

The fourth group of publications deals with GDSS. From a historical perspective some authors began with computer-supported decision-making in groups using a UNIX-based PDP11 to support the interactive process with technology (Steeb and Johnston 1981). Others showed that GDSS help groups that are communicating face-to-face perform better (Smith and Vanecek 1990). The concept of GDSS has also been aligned with other available systems, such as by combining them with knowledge management systems to improve results (Jacob and Pirkul 1992). The prevailing finding of the literature search deals with media choice in GDSS (Sarker et al. 2010).

The fifth set of publications focuses on practically relevant topics, such as specific professions and industries. (Choudhury et al. 2006; Finnegan and O'Mahony 1996).

#### 5.2.3.3 Sub-problem: Information Overload

The literature that is relevant to this topic can be divided into two groups. In the first group are publications that deal with information overload that affects individual people (Jones et al. 2004), how the complexity of messages affects user behaviour (O'Reilly III 1980), how decision-makers' perceived information overload is associated with higher satisfaction and lower performance (Turetken and Sharda 2001, 2004), how visualization of web search results leads to less information overload, how information overload affects decision quality

(Paul and Nazareth 2010), and the relationship among information processing, information complexity, and time pressure.

In the second group are publications that deal with groups of people who are collaborating. Grise and Gallupe (1999) designed an idea regulator to regulate the flow of information ideas with higher levels of complexity, while Paul and Nazareth (2010) concluded that there is a relationship among information processing, information complexity, and time pressure.

## 5.3 Discussion

Referring to the first research question, the IS discipline deals with the capability of IS to improve the effectiveness and efficiency of human organizations. The discipline seeks to obtain knowledge about the productive application of IT in the organization and to perform research on the development and use of IS for managerial purposes. Hevner et al. (2004) argued that both behavioural and design science research are necessary for this purpose. From the global challenges perspective, Sidorova et al. (2008) identified IS as a supportive academic discipline because DDS, among other research areas, has been extensively researched).

The sub-problem *limited capacity to decide* was a significant topic during the 1980s and 1990s. Publications about architectures and frameworks for developing DSS are the most frequent, and some researchers aligned DSS with the social and political situation. As for *information overload*, the second sub-problem, the literature search showed that the topic has been addressed from both an individual and a group perspective and that some researchers have focused on specific information-overload situations, such as those that occur when one performs a web search. The literature that is relevant to the sub-problem *increasing number of stakeholders* can be divided into six groups: topic overview, group characteristics, group decision-making techniques, group decision approaches, GDSS, and practice.

Clearly, the IS discipline addresses many topics in the decision-making realm. Many publications, as a detailed analysis of the sub-problems has shown, have focused on technical or methodological approaches and frameworks to develop such systems. Some researchers also included human beings as individuals or as groups in order to determine which of their characteristics influence the decision-making process or need to be adjusted to the DSS. From a methodological perspective, the majority of studies analysed were conducted by empirical research, while only a small were intended to contribute to theory-building.

## 5.4 Research Agenda

To address the second research question, a research agenda is presented to guide future IS research in supporting the resolution of the ninth global challenge. The aim of the research agenda is to derive sharp and insightful questions for research to consider that are aimed at solving the ninth challenge from an IS perspective (vom Brocke et al. 2009). As the IS discipline has addressed the first three sub-problems—limited capacity to decide, increasing number of stakeholders, and information overload—they are not be taken into consideration for a future research agenda.

Governmental decision-making is a general topic that IS research has not covered. One research question could be "How can governments collaborate and come to joint decisions quickly?" Referring to the example of the financial crisis, the ability of governments or larger groups to come to faster decisions could be an advantage in such circumstances.

Other examples of apparently never-ending discussions that lead to no result are the ongoing debate about the national debt of some states in the European Union and the issue of the education sector, where every European country has its own study programmes, regulations, and fees. Politicians are discussing the solutions or actions to take for these issues, but the decision-making process is slow, opaque, and inefficient. One problem in this decision process is that every state wants to have the best solution for its own situation, which usually implies disadvantages for the other participants in the discussion (Egeberg 1999). Part of this challenge could be addressed in research that seeks to answer questions like "How can groups like the European Union or NGOs collaborate in the decision-making process to avoid locally optimal decisions in favour of globally optimal decisions?" Thinking in a larger, trans-institutional context, the European Union could benefit from scale effects, to which IS research could contribute meaningfully in support of global joint decision-making.

A closer look at research about trans-institutional decision-making reveals a gap in the research about trending technologies like cloud computing. A possible research question is "To what degree can IT trends like cloud computing and real-time analysis support trans-institutional decision-making?"

Another topic is the cultural background of decision-making, which has to be considered when taking joint decisions. Whether a group decision is to be made by decision-makers from different countries with different backgrounds must be considered during the discussion phase. In this area, research questions like "What role does the cultural background of a decision-maker in a group decision process play?" or "How can the cultural and social aspects of decisions be implemented in DSS?" can provide input to the ninth global challenge.

In summary, IS can help master the problems of complexity and speed in the decision-making process, and IS research is on track to resolve the ninth global challenge. The discipline should use the plethora of empirical findings in the extant literature to create theories that will strengthen the core of this research area. While researchers began with a focus on the technical and informational support of these processes, they have also started to take the specific characteristics of individuals and groups into account. This is the right direction to take to cover challenge number nine as a whole.

## 5.5 Conclusion

This article took a close look at how the IS discipline can support the ninth global challenge, identified by the Millennium Group. The literature review provides an overview of the sub-problems of the challenge and describes the most relevant IS-related papers. However, only a limited set of journals and conferences was searched, and most of the resulting papers refer to core IS values. To take a broader view, a broader set of journals and conferences should be reviewed using a quantitative content analysis, rather than the qualitative content analysis employed for this paper, in order to identify additional gaps in the research (Indulska et al. 2011).

This paper proposed a research agenda based on the literature review and the subsequent discussion of the topics identified in the review. While the IS discipline has addressed the sub-problems analysed here, the research area of trans-institutional decision-making is of particular import as an area that could affect the response to such crises as the 2008 financial crisis by providing better information quality and transparency without being influenced by automatic decisions made by poorly programmed systems.

Clearly, the IS discipline addresses a plethora of topics related to the ninth global challenge, although some areas, such as trans-institutional decision-making and governmental decision-making, have not been addressed. The IS discipline can help to resolve the ninth global challenge more comprehensively by widening its scope to include these problems.

## References


Adelman, L. (1991). Experiments, quasi-experiments, and case studies: A review of empirical methods for evaluating decision support systems. *IEEE Transactions on Systems, Man, and Cybernetics, 21*(2) 293–301.

Alter, S. (1977). A taxonomy of decision support systems. *Sloan Management Review, 19*(1) 39–56.

Arnott, D. (2004). Decision support systems evolution: Framework, case study and research agenda. *European Journal of Information Systems, 13*(4), 247–259.

Avellaneda, M., & Stoikov, S. (2008). High-frequency trading in a limit order book. *Quantitative Finance, 8*(3), 217–224.

Barki, H., & Huff, S. L. (1985). Change, attitude to change, and decision support system success. *Information & Management, 9*(5), 261–268.

Benbunan-Fich, R., Hiltz, S. R., & Turoff, M. (2003). A comparative content analysis of face-to-face vs. asynchronous group decision making. *Decision Support Systems, 34*(4), 457–469.

Buchanan, L., & O'Connell, A. (2006). A brief history of decision making. *Harvard Business Review, 1*, 32–41.

Choudhury, A. K., Shankar, R., & Tiwari, M. K. (2006). Consensus-based intelligent group decision-making model for the selection of advanced technology. *Decision Support Systems, 42*(3), 1776–1799.

Cummings, L. L., Huber, G. P., & Arendt, E. (1974). Effects of size and spatial group on arrangements decision making. *The Academy of Management Journal, 17*(3), 460–475.

Daily, B., Whatley, A., Ash, S. R., & Steiner, R. L. (1996). The effects of a group decision support system on culturally diverse and culturally homogeneous group decision making. *Information & Management, 30*(6), 281–289.

DeSanctis, G., Poole, M. S., Zigurs, I., DeSharnais, G., D'Onofrio, M., & Gallupe, B., et al. (2008). The Minnesota GDSS research project: Group support systems, group processes, and outcomes. *Journal of the Association for Information Systems, 9*(10/11), 551–608.

Dennis, A. R., Garfield, M. J., & Adoption, G. (2003). The Adoption and use of GSS in project teams: Toward more participative processes and outcomes. *Management Information Systems Quarterly, 27*(2), 289–323.

Egeberg, M. (1999). Transcending intergovernmentalism? Identity and role perceptions of national officials in EU decision-making. *Journal of European Public Policy, 6*(3), 456–474

El-Shinnawy, M., & Vinze, A. S. (1998). Polarization and persuasive argumentation: A study of decision making in group settings. *Management Information Systems Quarterly, 22*(2), 165–198.

Elam, J. J., & Henderson, J. C. (1983). Knowledge engineering concepts for decision support system design and implementation. *Information & Management, 6*(2), 109–114.

Farris, D. R., Sage, A. P., & Member, S. (1975). Introduction and survey of group decision making with applications to worth assessment. *IEEE Transactions on Systems, Man, and Cybernetics, 1*(3), 346–358.

Finlay, P. N., & Forghani, M. (1998). A classification of success factors for decision support systems. *Journal of Strategic Information Systems, 7*, 53–70.

Finnegan, P., & O'Mahony, L. (1996). Group problem solving and decision making: An investigation of the process and the supporting technology. *Journal of Information Technology, 11*, 211–221.

Fischer, J. (2011). *Am Börsenparkett dominieren heute Software-Programme das Geschehen. Format*. Retrieved December 20, 2011, from http://www.format.at/articles/1133/528/304498/am-boersenparkett-software-programme-geschehen.

Fjermestad, J., & Roxa, S. (1998). An assessment of group support systems experimental research: Methodology and results. *Journal of Management Information Systems, 15*(3), 7–149.

Gartner. (2011). *Gartner identifies the top 10 strategic technologies for 2011*. Retrieved December 20, 2011, from http://www.gartner.com/it/page.jsp?id=1454221.

Gladstein, D. L., & Reilly, N. P. (1985). Group decision making under threat: The tycoon game. *Management, 28*(3), 613–627.

Glenn, J. C., & Gordon, T. J. (2001). The millennium project: Challenges we face at the millennium. In *Technological Forecasting and Social Change* (Vol. 66). New York: Elsevier.

Grise, M.-L., & Gallupe, R. B. (1999). Information overload: Addressing the productivity paradox in face-to-face electronic meetings. *Journal of Management Information Systems, 16*(3), 157–185.

Hevner, A. R., March, S. T., Park, J., & Ram, S. (2004). Design science in information systems. *MIS Quarterly, 28*(1), 75–105.

Indulska, M., Hovorka, D. S., & Recker, J. (2011). Quantitative approaches to content analysis: Identifying conceptual drift across publication outlets. *European Journal of Information Systems, 21*(1), 49–69. Nature Publishing Group.

Jacob, V. S., & Pirkul, H. (1992). A framework for supporting distributed group decision-making. *Decision Support Systems, 8*(1), 17–28.

Jones, Q., Ravid, G., & Rafaeli, S. (2004). Information overload and the message dynamics of online interaction spaces: A theoretical model and empirical exploration. *Information Systems Research, 15*(2), 194–210.

Kasper, G. M. (1996). A theory of decision support system design for user calibration. *Information Systems Research, 7*(2), 215–232.

Kersten, E. (1985). NEGO—group decision support system. *Information & Management, 8*, 237–246.

Lewandowski, A. (1989). SCDAS—Decision support system for group decision making: Decision theoretic framework. *Decision Support Systems, 5*, 403–423.

Lilien, G. L., Rangaswamy, A., Van Bruggen, G. H., & Starke, K. (2004). DSS effectiveness in marketing resource allocation decisions: Reality vs perception. *Information Systems Research, 15*(3), 216–235.

Limayem, M., & DeSanctis, G. (2000). Providing decisional guidance for multicriteria decision making in groups. *Information Systems Research, 11*(4), 386–401.

Mendonça, D. (2003). Decision support for improvisation: Prospects, challenges and opportunities. In *Proceedings of the International Conference on Information Systems* (pp. 1–15). Seattle.

Miner, F. C, Jr. (1979). A comparative analysis of three diverse decision making approaches. *The Academy of Management Journal, 22*(1), 81–93.

Nah, F. H. (Fiona), Mao, J., & Benbasat, I. (1999). The effectiveness of expert support technology for decision making: individuals versus small groups. *Journal of Information Technology, 14*(2), 137–147.

OICV-IOSCO. (2011). Regulatory issues raised by the impact of technological changes on market integrity and efficiency. In *Regulation* (pp. 1–60). Madrid.

Orman, L. (1984). Fighting information pollution with decision support systems. *Journal of Management Information Systems, I*(2), 64–71.

O'Reilly, C. A, I. I. I. (1980). Individuals and information overload in organizations: Is more necessarily better? *The Academy of Management Journal, 23*(4), 684–696.

Paul, S., & Nazareth, D. L. (2010). Input information complexity, perceived time pressure, and information processing in GSS-based work groups: An experimental investigation using a decision schema to alleviate information overload conditions. *Decision Support Systems, 49*(1), 31–40. Elsevier B.V.

Riedel, S. L., & Pitz, G. F. (1986). Utilization-oriented evaluation of decision support systems. *IEEE Transactions on Systems, Man, and Cybernetics, 16*(6), 980–996.

Sarker, S., Sarker, S., & Valacich, J. S. (2010). Media effects on group collaboration: An empirical examination in an ethical decision-making context. *Decision Sciences, 41*(4), 887–932.

Sauter, V. L., & Schofer, J. L. (1988). Evolutionary development of decision support systems: Important issues for early phases of design. *Journal of Management Information Systems, 4*(4), 77–92.

Sidorova, A., Evangelopoulos, N., Valacich, J. S., & Ramakrishnan, T. (2008). Uncovering the intellectual core of the information systems discipline. *Management Information Systems Quarterly, 32*(3), 467–482.

Smith, J. Y., & Vanecek, M. T. (1990). Dispersed group decision making using nonsimultaneous computer conferencing: A report of research. *Journal of Management Information Systems, 7*(2), 71–92.

Srinivas, V., & Shekar, B. (1997). Strategic decision-making processes: Network-based representation and stochastic simulation. *Decision Support Systems, 21*(2), 99–110.

Steeb, R., & Johnston, S. C. (1981). A computer-based interactive system for group decision making. *IEEE Transactions on Systems, Man, and Cybernetics, 11*(8), 544–552.

The Millennium Project. (2011). *Global challenges facing humanity*. Retrieved October 20, 2011, from http://www.millennium-project.org/millennium/Global_Challenges/chall-09.html.

Turetken, O., & Sharda, R. (2001). Visualization support for managing information overload in the web environment. *Proceedings of the International Conference on Information Systems* (pp. 1–12). New Orleans.

Turetken, O., & Sharda, R. (2004). Development of a fisheye-based information search processing aid (FISPA) for managing information overload in the web environment. *Decision Support Systems, 37*(3), 415–434.

vom Brocke, J., Alexander, M., Niehaves, B., Riemer, K., Plattfaut, R., & Cleven, A. (2009). Reconstructing the giant: On the importance of rigour in documenting the literature search proccess. In *ECIS 2009* (pp. 1–13). Verona.

Walters, B. A., Jiang, J. J., & Klein, G. (2003). Strategic information and strategic decision making: The EIS/CEO interface in smaller manufacturing companies. *Information & Management*, *40*, 487–495.

Wang, H. F., & Shen, S. Y. (1989). Group decision support with MOLP applications. *IEEE Transactions on Systems, Man, and Cybernetics, 19*(1), 121–126.

Webster, J., & Watson, R. T. (2002). Analyzing the past to prepare for the future: Writing a literature review. *MIS Quarterly*, *25*(2), xiii–xxiii.

Winter, S. J., & Butler, B. S. (2011). Creating bigger problems: Grand challenges as boundary objects and the legitimacy of the information systems field. *Journal of Information Technology, 26*(2), 99–108.

Xu, Z., Yuan, Y., & Ji, S. (2008). A decision analysis framework for emergency notification. In *Proceedings of the 41st Annual Hawaii International Conference on System Sciences (HICSS 2008)* (pp. 1–9). IEEE.

Yadav, S. B., & Khazanchi, D. (1992). Subjective understanding in strategic decision making. *Decision Support Systems, 8*, 55–71.

Zigurs, I., Poole, M. S., & DeSanctis, G. L. (1988). A study of influence in computer-mediated group decision making. *Management Information Systems Quarterly, 12*(4), 625–644.

# Chapter 6
# The Role of Information Technologies in Changing the Status of Women to Improve Human Conditions

**Manuela Weiss and Ekaterina Tarchinskaya**

## 6.1 Introduction

Experiences all over the world have shown that women drive development, and holding them back has an adverse impact on economic growth (Dlodlo 2009). The level of social and political development is also affected by the level of women's contribution to society (Best and Maier 2007). Therefore, those societies that discriminate by gender pay a high price in terms of their ability to grow and reduce poverty (Sharma 2003).

The meaning of quality differs in different parts of the world, especially between developed countries and developing countries (Glenn and Gordon 2001). In developing countries equality refers to reducing the number of women who are living in poverty and situations of violence, increasing women's literacy and education rates, providing access to resources and financial aid, and increasing women's participation in public and professional life (Glenn and Gordon 2001). Best and Maier (2007) argue that long-standing gender inequalities in developing countries lead to a gap that is difficult to close. As for developed countries, despite great progress, industrialized nations are also far from reaching gender equality, as wage gaps and limited access to executive power pose unique problems (Glenn and Gordon 2001). Even in developed countries, women represent far less than half of a nation's use of intellectual capital, skilled labor, and economic contribution (Leahy and Yermish 2002).

The international community sees information and communication technologies (ICT) as the most effective tool in the context of strengthening gender development and economic development almost simultaneously (Best and Maier 2007). Many results have demonstrated that it can be a vital tool in triggering a revolutionary change that is beneficial to both women and men (Dighe and Reddi 2006; Hafkin and Huyer 2006). Hence, the goal of this paper is to identify existing theories and the state-of-the art in this field.



J. vom Brocke et al., *Grand Societal Challenges in Information Systems Research and Education*, SpringerBriefs in Information Systems,
DOI 10.1007/978-3-319-15027-7_6

## 6.2 Literature Review

The objective of our literature review was derived from challenge number five of the Millennium Project (Glenn et al. 2001) and adapted to the context of IS. The review follows a method presented by Watson and Webster (2002), which focuses on journal and conference publications because of their traditionally high quality and timeliness. The review followed three steps:

(1) *Defining a research basis.* We chose Science Direct and EBSCO Business Source Premier as a starting point for the analysis. The search covers an unrestricted timeframe and sought the keywords "information technology" and "woman" in the papers' titles, abstracts, and keywords (Table 6.1).
(2) *Extracting and categorizing relevant research.* Our search yielded 297 articles, of which 245 articles were published in Science Direct and 52 were published in EBSCO Business Source Premier. Among these articles we identified 20 candidate articles that focus on the role of women in correlation with IT and that cover either the aspects of women's roles that are supported by IT or the aspects of women's roles in IT. Among these 20 articles, 12 were from Science Direct, and 8 were from EBSCO Business Source Premier.
(3) *Analyzing the research.* It was apparent that, among these 20 relevant studies, ICT was usually described as a diverse set of technological tools and resources (Nath 2001; Pattanaik 2005) and as a tool to get human voices heard (Shirazi 2011), challenge long-standing inequalities, open up new perspectives, and raise awareness about inequity (Mbarika et al. 2007; Nath 2001; Pattanaik 2005; Sandys 2005; Shirazi 2011).

Three main perspectives are prevalent in the IS literature: *gender and technology, gender in the IT workplace,* and *empowerment.* The gender and technology category includes papers that deal with problems of digital divide, female-friendly IT design, and women's perceptions of IT. The Gender in the IT workplace category covers a broad range of barriers that women experience at their workplaces as well as barriers that prevent them from entering the IT industry. The empowerment category groups the kinds of empowerment that come from IT: social-economic, political, economic, and psychological empowerment.

**Table 1** Research approach

| | Science direct | EBSCO business source premier |
|---|---|---|
| Source | All sources | All sources |
| Time frame | All years | All years |
| Search term 1 | Woman | Woman |
| Search field 1 | TITLE-ABSTR-KEY | TITLE |
| Search term 2 | [AND] "information technology" | [AND] "information technology" |
| Search field 2 | TITLE-ABSTR-KEY | TITLE |

## 6.3 Results and Discussion

Having identified papers that are relevant to the role of women and IT in conjunction with the three broad categories to which the papers belong, the next step is to describe the current state of our research topic in the literature. A qualitative approach to the categories of *gender and technology*, *gender in the IT* workplace, and *empowerment* of women identifies subtopics for each of these categories. Since the categorization of our results has a hierarchical structure, our descriptions begin with the high-level category (e.g., *gender and technology*) before delving more deeply into the category.

### *6.3.1 Gender and Technology*

*Women's perception of IT.* New technology is often portrayed as a male domain (Dlodlo 2009; Wasserman and Richmond-Abbott 2005). Researchers explain that those gender-specific differences have their origins in the fact that women have traditionally constructed a more distant relationship to technology than men have and have underestimated their usage skills (Kelan 2007; Rasmussen and Håpnes 1991; Wasserman and Richmond-Abbott 2005). Socio-cultural and religious influences and preconceptions about women's ability to understand and manipulate technology hinder women from actively engaging in the use of ICT (Bimber 2000; Dlodlo 2009; Wasserman and Richmond-Abbott 2005).

*Female-friendly IT design.* Researchers argue that technology may be "gendered by design" because men's values have been institutionalized in the technology through its creators, embedding a cultural association with masculine identity in the technology itself (Bimber 2000; Kaminski and Reilly 2004; Kelan 2007). Especially the Internet is biased toward men because of gender inequality in the professions that produce Internet technologies, the commercial success of male-oriented pornography, and so on (Bimber 2000). ICT and its content—designed by men in the English language and without considering women's interests, concerns, perspectives, and required information—create a significant obstacle for women, especially in the developing world (Dlodlo 2009; Parmentier and Huyer 2008; Sandys 2005).

Within this context are two main perspectives on female-friendly IT design: that which views women as users of IT design and that which is concerned with women as designers of IT. Fountain (2000) highlights that women as users have the power to shape IT design. However, the influence of users, although important and far-reaching, is limited; designers fashion technology more deeply, pervasively, and fundamentally. In this regard, women are poorly represented as IT designers and experts, so research should pay attention to the potential role women might play as designers in the information-based society (Fountain 2000; Wajcman and Lobb 2007).

*Gender digital divide.* Studies on the digital divide address who (e.g., individuals, organizations/communities, societies/countries/world regions) uses what attributes (e.g. income, education, geography, age, profitability, sector) to connect to what kind of technology (e.g., phone, Internet, computer) and how they do it (e.g., pain access, usage, real impact) (Hilbert 2010, 2011). Adoption of gender analysis regarding ICT developments sheds light not only on the digital divide between developed and developing nations but also on a gender divide between men and women in terms of access to, usage of, and benefits of ICTs (Best and Maier 2007; Brännström 2011; Dighe and Reddi 2006; Hilbert 2011; Sandys 2005; Wasserman and Richmond-Abbott 2005). As available statistical data show, in developing countries women are less likely than men are to use ICTs (Brännström 2011; Dighe and Reddi 2006; Huyer 2005; Sandys 2005), leading related research to conclude that a gender digital divide exists and poses a severe threat to women (Hafkin and Huyer 2006; Hilbert 2011). However, our findings also indicate that the digital divide is closing, and in some countries like Kenya the process is rapid (Brännström 2011). Other case studies and research papers show that ICTs hold the promise of empowering women (Hafkin and Huyer 2006; Hu 2003; Sandys 2005; Sharma 2003). In any case, the extant literature is inconclusive, arguing either that ICT is a threat to women or that it is a unique opportunity for empowerment.

### 6.3.2 Gender in the IT Workplace

*Barriers to entering the IT field.* The literature presents two dominant viewpoints about the barriers to women's entering the IT field. As Trauth (2002, p. 99) argues, one perspective "focuses on the presumption of inherent differences between women and men to explain the perception of IT as a male domain [...]." Social influences during childhood and adolescence, such as parents, media, race, ethnicity, and socio-economic status, influence girls' choices in many ways and constrain their education and career decisions later (Fountain 2000; Kaminski and Reilly 2004; Kelan 2007; Sanders 2005). Research has found that women perceive the IT workplace negatively, as a field for "geeks and nerds" (Gürer and Camp 2002; Kelan 2007; Rasmussen and Håpnes 1991; Wentling and Thomas 2009). Women also frame the work as difficult, isolated, unattractive, and lacking the necessary social interaction and work-family balance (Armstrong et al. 2007; Kelan 2007; Wentling and Thomas 2009). Other impediments are the general lack of role models and networking opportunities (Armstrong et al. 2007; Ballard et al. 2006; Fountain 2000; Medeiros 2005; Riemenschneider et al. 2006; Wentling and Thomas 2009).

*Barriers while working in the profession.* The fact that the number of women in IT field is declining has triggered discussions about how to retain those who are currently employed (Armstrong et al. 2007). The literature that explores workplace conditions has identified several barriers for women who work in the IT profession, including discrimination (Armstrong et al. 2007; Panteli et al. 2001;

Riemenschneider et al. 2006; Stevens 2009), lack of respect (Riemenschneider et al. 2006; Stevens 2009), lack of consistency (e.g., lack of consistency at the organizational level; lack of consistency in how supervisors treat employees) (Riemenschneider et al. 2006), promotion barriers (Armstrong et al. 2007; Riemenschneider et al. 2006), work-family conflict (Armstrong et al. 2007; Gürer and Camp 2002; Riemenschneider et al. 2006; Stevens 2009), work stress (Riemenschneider et al. 2006), difficulty into keeping up with the pace of the IT industry (Kelan 2007), and lack of confidence (Stevens 2009). Only a little research has been conducted about male IT employees, their perceptions, and opinions in regard to the shortage of women in their profession (Kelan 2007).

*Work conditions for women in IT.* There is a tendency for women to be at a disadvantage compared to men in terms of their terms and conditions of employment (Panteli et al. 2001). Researchers claim that women are consistently paid less than men are for similar work, even after years of education and experience (Armstrong et al. 2007; Baroudi and Igbaria 1995; Dubnoff and Kraft 1986; Klawe and Leveson 1995; Panteli et al. 2001; Roan and Whitehouse 2007; Wright and Jacobs 1994) and that it is more difficult for women than men to advance in their careers, especially to managerial jobs, because they are more likely than men are to interrupt their careers, to work part-time, and to work under short-term contracts. (Armstrong et al. 2007; Fountain 2000; Panteli et al. 2001). Exclusion from male-dominated social networks in the organization is a particular obstacle in terms of gaining skills, job leads, and mobility in the organization (Trauth et al. 2003), and it limits women's ability to build technical skills and knowledge (Trauth et al. 2003). Working long hours and the need for continuous learning in order to keep up with changes in the field and required skills make it nearly impossible to balance work and family obligations (Armstrong et al. 2007; Fountain 2000). We also included in this sub-topic research that identifies women's attitudes toward such work conditions and barriers. Trauth et. al. (2003) distinguish three types of women in this regard: those who tend not to focus on the fact that they are women, those who accept the uneven playing field, and those who have experienced the uneven playing field and are willing to speak up about it.

### *6.3.3 Empowerment of Women*

Addressing the gender dimensions of access, need, and use of ICT is central to women's empowerment in the area of ICT. As Nath (2001) explains, effective initiatives and projects have to take women's needs and concerns regarding ICTs into account in order to ensure that women and men benefit equally.

*Socio-economic empowerment.* Relevant research about socio-economic empowerment deals with a variety of topics in the context women's use of ICTs. Under the topic of socio-economic empowerment, we included articles that discuss women's empowerment in the realms of education, health care, and violence against women. Combinations of traditional and new ICTs are applied in education in order to support the education and training of women and girls in formal and informal learning, distance education, and establishing e-learning centers. Sandys (2005) argues that education initiatives that focus on women in poor communities or on computer literacy have already demonstrated the value of ICT for women.

There is vast potential for ICT to facilitate global, regional, and national health initiatives for women. Organizations like Satellife[1] and HealthNet[2] have successfully provided medical information and connected medical specialists to developing countries. ICT is an effective tool in disseminating information about women's health, including sexual and reproductive rights and health (Sandys 2005). ICT can also help women to fight against violence. For example, ICT can be used to foster awareness about the many forms that violence against women can take on the Internet and to develop a community that can respond to these issues. By using ICT, women's groups can participate in the development of policies, legislation, and other actions to combat the exploitation of women and girls (Sandys 2005).

*Political empowerment.* Women and their organizations have pioneered strategic and empowering uses of ICT to promote women's rights (Sandys 2005). For example, Shirazi observes that "blogging in Iran has helped repressed and marginalized groups reach out, including women's and human rights activists, ethnic and religious minorities and Iranian youth to get their voices heard and to challenge the long-standing univocal government and Islamic religious authorities [...]." Women in Zambia use mobile phones to keep informed about court hearings in order to mobilize women to meet at the courthouse to advocate for women's rights (Keifer-Boyd 2011). ICT in this context is used as a tool to promote women's rights, keep them informed, and help them to challenge authorities.

*Economic empowerment.* There is an evidence that internationally outsourced jobs like medical transcription work and software services have made a considerable difference in women's work opportunities in developing countries (Sandys 2005). For instance, women hold positions in most telephone operating companies in India (Valk and Srinivasan 2011). ICT also has the power to make time and distance less significant, so women can work from anywhere and at any time, even if they have young children, and become financially independent and empowered (Nath 2001). Jiyane and Mostert (2010) point out that ICT supports businesswomen in South Africa by connecting them with business partners. Another powerful application of ICT is as a part of knowledge networking—that is, electronic commerce that refers not only to "selling products and services on-line, but also to the promotion of a new class of ICT-savvy women entrepreneurs in both rural and urban areas" (Nath 2001).

*Psychological empowerment.* The psychological dimension of empowerment refers to women's sense of self-value and self-appreciation (Keifer-Boyd 2011). Hu (2003) defines psychological empowerment as "a motivational construct

[1] http://www.satellife.org.

[2] http://www.healthnet.org.

manifested in four cognitions: meaning, competence, self-determination, and impact [...]." This definition suggests that frequent use of IT at work raises women's confidence, self-assurance, and innovativeness and provides them with a significant amount of influence and control at their workplaces. ICT's psychological impact applies not only to the work environment but also to everyday life (Keifer-Boyd 2011).

## 6.4 Research Agenda

The results of the literature review suggest the need for future research on the impact of the changing status of women with regard to information systems on the improvement of the human condition. Generally speaking, there is little research on how the changing status of women in conjunction with IT can improve overall human conditions. Most of the papers in our review are concerned with the conditions women are currently under in the IT field, or they provide information about problems and challenges with regard to the role of women and IT. The impact of empowering women with IT about the human condition and about how entire human communities can benefit from the changing status of women by means of IT is only a side note in these articles. Hence, based on the current status of the literature, we propose three questions to further the research.

(1) *How can the changing status of women by means of ICT improve human conditions?* Most of the extant research investigates the current status of women in IT or how IT can change their current status. However, a shift should be made to question how women's changing status by means of ICT can improve human conditions.
(2) *How can the changing status of women by means of ICT improve human conditions in accordance with regional considerations?* The Millennium Project (2011) distinguishes the problems that women experience according to whether they live in Africa, Asia and Oceania, Europe, Latin America, or North America. Our literature review also showed that the role of women and IT issues can be differentiated based on whether they live in developed countries or developing countries. Further elaboration is needed in order to clarify how improvement of certain regional problems can affect overall human conditions.
(3) *What negative impact might the changing status of women by means of IT have on human conditions?* Every coin has two sides, and so far most papers have focused on the positive influence of IT. Additional research on the problems that might occur with regards to the role of women and IT topic would be useful because it has far reaching consequences on team work, virtualization of working practices and work-life balance.

## 6.5 Conclusion

This literature review provides both quantitative and qualitative insights into the current state of research on how IT contributes to the changing role of women. We believe that understanding the current body of knowledge is the first step to realizing how the impact of IT on the status of women might affect the human condition. Unfortunately, the extant literature illustrates the possible consequences for society but does not provide sufficient empirical evidence. This literature review provides evidence that research has not answered the main question of how the changing status of women can help improve the human condition and how information systems can influence this process. Thus, this study serves as a starting point for future research in this field.

This review has some limitations. First, the scope of our review is a limitation, as articles found in Science Direct and EBSCO Business Source Premier tend to have a technical focus. Therefore, initial research that considers a non-technical focus—that is, research in the social or psychological sciences—is necessary in order to validate and, if needed, to broaden the extracted discussions and trends. Second, other researchers may have come up with different classifications than ours, especially in relation to the category of empowerment. We do not claim that the present literature review is exhaustive but only that it presents an initial categorization of the trends and ongoing discussions.

We see the category of *gender and technology* as a comparatively new trend in the overall discussion of women and IT, whereas *gender in the IT workplace* and *empowerment* have been topics of discussion since the early 1990s, although the results of empowerment initiatives have not been discussed in depth. Further research in the areas of gender and technology and the results of empowerment initiatives is needed.

## References

Armstrong, D., Riemenschneider, C., Allen, M., & Reid, M. (2007). Advancement, voluntary turnover and women in IT: A cognitive study of work–family conflict. *Information & Management, 44*(2), 142–153.

Ballard, J., Scales, K., & Edwards, M. A. (2006). Perceptions of information technology careers among women in career development transition. *Information Technology, Learning, And Performance, 24*(2), 1–9. Springer.

Baroudi, J. J., & Igbaria, M. M. (1995). An examination of gender effects on career success of information systems employees. *Journal of Management Information Systems, 11*(3), 181–201.

Best, M. L., & Maier, S. G. (2007). Gender, culture and ICT use in rural south India. *Gender, Technology and Development, 11*(2), 137–155.

Bimber, B. (2000). Measuring the gender gap on the Internet. *Social Science Quarterly, 81*(3), 1–10.

Brännström, I. (2011). Gender and digital divide 2000–2008 in two low-income economies in Sub-Saharan Africa: Kenya and Somalia in official statistics. *Government Information Quarterly, 29*(1), 60–67. Elsevier Inc.

Dighe, A., & Reddi, U. V. (2006). *Women's literacy and information and communication technologies: Lessons that experience has taught us*. New Delhi: Commonwealth Educational Media Centre for Asi.

Dlodlo, N. (2009). Access to ICT education for girls and women in rural South Africa: A case study. *Technology in Society, 31*(2), 168–175. Elsevier Ltd.

Dubnoff, S., & Kraft, P. (1986). *Gender discrimination in the computer industry' (mimeo), Dept of Sociology*. Binghamton: SUNY.

Fountain, J. (2000). Constructing the information society: Women, information technology, and design. *Technology in Society, 22*(1), 45–62. doi:10.1016/S0160-791X(99)00036-6.

Glenn, J. C., & Gordon, T. J. (2001). The Millennium project: Challenges we face at the Millennium. *Technological Forecasting and Social Change, 66*(2–3), 129–312.

Gürer, D., & Camp, T. (2002). An ACM-W literature review on women in computing. *ACM SIGCSE Bulletin, 34*(2), 121–127. ACM.

Hafkin, N., & Huyer, S. (2006). *Cinderella or Cyberella? empowering women in the knowledge society*. Bloomfield: Kumarian Press.

Hilbert, M. (2010). The manifold definitions of the digital divide and their diverse implications for policy responsibility. In *The 38th Research Conference on Communication, Information and Internet Policy* (pp. 1–16).

Hilbert, M. (2011). Digital gender divide or technologically empowered women in developing countries? A typical case of lies, damned lies, and statistics. *Women's Studies International Forum, 34*(6), 479–489. Elsevier Ltd.

Hu, S. (2003). Effects of expectancy-value, attitudes, and use of the Internet on psychological empowerment experienced by Chinese women at the workplace. *Telematics and Informatics, 20*(4), 365–382.

Huyer, S. (2005). Women, ICT and the information society: Global perspectives and initiatives. In *CWIT 05 Proceedings of the International Symposium on Women and ICT Creating Global Transformation* (pp. 5–9).

Jiyane, V., & Mostert, J. (2010). Use of information and communication technologies by women hawkers and vendors in South Africa. *African Journal of Library, Archives and Information Science, 20*(1), 53–61.

Kaminski, J. A. M., & Reilly, A. H. (2004). Career development of women in information technology. *SAM Advanced Management Journal, 69*, 20–30. SAM society for advancement management.

Keifer-Boyd, K. (2011). African women and ICTs: Investigating technology, gender and empowerment. *Journal of International Women's Studies, 12*(1), 212–217.

Kelan, E. K. (2007). "I don't know why"—Accounting for the scarcity of women in ICT work. *Women's Studies International Forum, 30*(6), 499–511.

Klawe, M., & Leveson, N. (1995). Women in computing: Where are we now? *Communications of the ACM, 38*(1), 29–32.

Leahy, K. B., & Yermish, I. (2002). Information and communication technology: Gender issues in developing nations. In *Informing Science + IT Education Conference Proceedings*. Santa Rose: Informing Science Institute.

Mbarika, V. W., Payton, F. C., Kvasny, L., & Amadi, A. (2007). IT education and workforce participation: A new era for women in Kenya? *The Information Society, 23*(1), 1–18.

Medeiros, C. (2005). From subject of change to agent of change: women and IT in Brazil. In *CWIT 05 Proceedings of the International Symposium on Women and ICT Creating Global Transformation*.

Nath, V. (2001). Empowerment and governance through information and communication technologies: Women's perspective. *The International Information & Library Review, 33*(4), 317–339.

Panteli, N., Stack, J., & Ramsay, H. (2001). Gendered patterns in computing work in the late 1990s. *New technology, Work and Employment, 16*(1), 3–17.

Parmentier, M. J. C., & Huyer, S. (2008). Female empowerment and development in latin america: Use versus production of information and communications technology. *Information Technologies & International Development, 4*(3), 13–20.

Pattanaik, D. (2005). Engendering knowledge networks—empowering women through ICT. In *CWIT 05 Proceedings of the International Symposium on Women and ICT Creating Global Transformation* (pp. 2–6).
Rasmussen, B., & Håpnes, T. (1991). Excluding women from the technologies of the future?: A case study of the culture of computer science. *Futures, 23*(10), 1107–1119. Elsevier.
Riemenschneider, C.K., Armstrong, D. J., Allen, M. W., & Reid, M. F. (2006). Barriers facing women in the IT work force. *ACM SIGMIS Database, 37*(4), 58–78. ACM.
Roan, A., & Whitehouse, G. (2007). Women, information technology and "waves of optimism": Australian evidence on "mixed-skill" jobs. *New Technology, Work and Employment, 22*(1), 21–33. Wiley Online Library.
Sanders, J. (2005). Gender and technology in education: A research review.
Sandys, E. (2005). Gender equality and empowerment of women through ICT. *United Nations, Division for the Advancement of Women.* Retrieved from http://www.un.org/womenwatch/daw/public/w2000-09.05-ict-e.pdf.
Sharma, U. (2003). *Women empowerment through Information Technology.* New Delhi: Authors Press.
Shirazi, F. (2011). Information and communication technology and women empowerment in Iran. *Telematics and Informatics, 29*(1), 45–55. Elsevier Ltd.
Stevens, H. (2009). The professional fate of woman engineers in the computer sciences: Unexpected reversals. *Sociologie du Travail, 51*, e15–e33.
The Millennium Project. (2011). *The millennium project global futures studies & research.*
Trauth, E. M. (2002). Odd girl out: An individual differences perspective on women in the IT profession. *Information Technology & People, 15*(2), 98–118.
Trauth, E. M., Nielsen, S. H., & von Hellens, L. A. (2003). Explaining the IT gender gap: Australian stories for the new millennium. *Journal of Research and Practice in Information Technology, 35*(1), 7–20.
Valk, R., & Srinivasan, V. (2011). Work–family balance of Indian women software professionals: A qualitative study. *IIMB Management Review, 23*(1), 39–50. Elsevier Ltd.
Wajcman, J., & Lobb, L. A. P. L. (2007). The gender relations of software work in Vietnam. *Gender, Technology and Development, 11*(1), 1–26.
Wasserman, I. M., & Richmond-Abbott, M. (2005). Gender and the internet: Causes of variation in access, level, and scope of use. *Social Science Quarterly, 86*(1), 252–270.
Watson, R. T., & Webster, J. (2002). Analyzing the past to prepare for the future: Writing a literature review. *MIS Quarterly, 26*(2), xiii–xxiii.
Wentling, R. M., & Thomas, S. (2009). Workplace culture that hinders and assists the career development of women in information technology. *Information Technology, Learning, and Performance Journal, 25*(1), 25–42.
Wright, R., & Jacobs, J. (1994). *Gender inequality at work.* London: Sage.

# Chapter 7
# How Can Transnational Organized Crime Networks Be Stopped from Becoming More Powerful and Sophisticated Global Enterprises?

**Nadine Székely**

## 7.1 Introduction

In 1994 the Secretary-General of the United Nations, Boutros Boutros-Ghali, referred to transnational crime as having been "organized along relatively simple lines and involv[ing] a limited number of activities" (Boutros-Ghali 1994, p. 1). Today, however, their areas of activity have grown to include trade in nuclear technology and human organs (Boutros-Ghali 1994).

Shelly (1995) emphasized that transnational organized crime organisations undermine the laws of the nation-state through corruption and that trafficking in nuclear materials threatens the state. In addition, Shelly emphasized that the transnational organized crime organisations of the twenty-first century are no longer interested in stable states since they generate their profits from the chaos of wars and conflicts, so they cooperate with terrorist organisations (Shelley 2006). In fact, the support of terrorist groups can be seen as one of the main activities of transnational organized crime organisations (Guymon 2000).

This paper investigates how transnational organized crime networks can be defeated through the use of information systems (IS). The paper starts with a literature review to determine the current state of research on the topic. The first part of the literature review, which shows how transnational organized crime organisations can be characterised, is followed by an overview of the typical activity areas of transnational organized crime. The last part of the literature review deals with the question concerning known ways to defeat transnational organized crimes. Then the results of the literature review are discussed to show the current contribution of IS to the topic, and the last part of the paper presents suggestions for future research in this field.



J. vom Brocke et al., *Grand Societal Challenges in Information Systems Research and Education*, SpringerBriefs in Information Systems,
DOI 10.1007/978-3-319-15027-7_7

## 7.2 Literature Review

### 7.2.1 Research Method

This paper focuses on the question concerning what transnational organized organisations are doing and how they can be precluded from getting more power. I conducted a structured literature research with search words that stem from the domains of IS and organised crime, followed by a backward search. To get a better overview of the papers I found and to cluster them, I created a framework based on the papers' topics (cf. Table 7.1), which framework is reflected in the structure of this paper.

### 7.2.2 Characteristics of Transnational Organized Crime Organisations

Donald R. Cressey described organized crime in 1969. Based on his work, von Lampe (2003) developed a model (Fig. 7.1) that shows how organized crime stands in relation to the government, society, and (illegal) markets. Today's research sees networks as the ideal structure for transnational organized crime organisations (Williams and Godson 2002), since network structures can adapt to

**Table 7 1** Clustering of relevant papers

| Topic | | Relevant paper |
|---|---|---|
| Characteristics | | 5 |
| Activity areas | | 3 |
| Means against TOCs | Legal regulations | 1 |
| | Crime detection | 5 |
| | Preventive methods | 6 |
| | Cooperation | 1 |

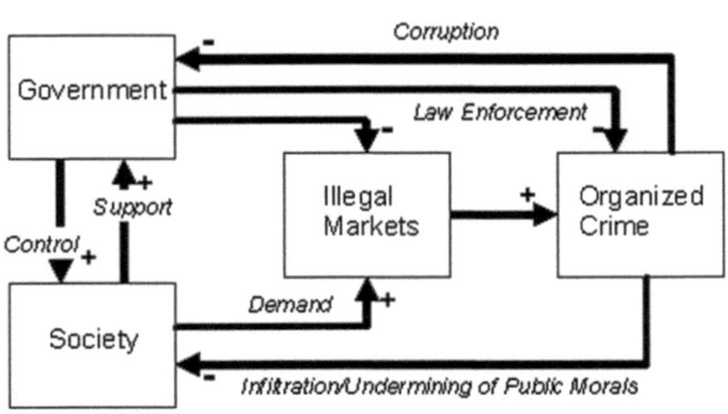


**Fig. 7.1** "Cressey Model" (von Lampe 2003)

a continuously changing world (Klerks 2001) and they minimize the risk for the organisations since they are difficult to defeat (Williams and Godson 2002).

Through globalisation, global opportunities and pressure in traditional areas of operation forced organized crime organisations to become transnational (Williams and Godson 2002). Criminal organisations take advantage of globalisation's opening of frontiers and national economies (Williams and Godson 2002; Boutros-Ghali 1994), so their success depends "increasingly on their capacity to act transnationally" (Williams and Godson 2002, p. 342).

Transnational organized crime organisation have high levels of flexibility and the ability to adapt methods and techniques to changes quickly (Williams and Godson 2002), which requires that they use the latest information and communication technologies (ICT) (Etges and Sutcliffe 2008) to, for example, identify leaks in their own contexts and to find potential police informants (Etges and Sutcliffe 2008).

### 7.2.3 Main Activity Fields

Transnational organized crime organisations are active in several fields. According to Europol (2006), *violence* against persons or property and *corruption* are often connected to other areas of crime Other areas of activity include the *facilitating crime*, which includes counterfeiting of documents in support of illegal immigration; *money laundering*; trafficking in *illegal markets*, the focus of which is goods that are not available legally, such as prohibited or regulated goods (e.g., drugs) and stolen goods; and *fraud*, which includes credit card fraud and non-payment of VAT (Europol 2006). Other areas in which organized crime organisations are active are *human trafficking*, especially the trafficking of women for commercial sex, and *intellectual property theft* and *counterfeiting* (especially Chinese criminal enterprises), while criminal Turkish groups focus on the trafficking of drugs, people, and nuclear material (Williams and Godson 2002). The Internet provides a contact platform for organized crime organisations to offer their services, including the exploitation of children and human trafficking (Etges and Sutcliffe 2008). The services are paid for with e-currencies, which provide anonymity to both buyer and seller and are also used by the crime organisations to launder money (Etges and Sutcliffe 2008).

### 7.2.4 Means Against Transnational Organized Crime

There are several ways to combat transnational organized crime, the first of which is clear laws and regulations that declare which activities are unlawful and what punishment transgressors can expect. Methods to investigate and prevent crimes must recognize that transnational crimes often affect several states, so cooperation between the states and their law enforcement agencies plays an important role.

The United Nations convention against transnational organized crime (2004) was signed in 2000. This convention asks the signatory states to ensure that domestic laws against transnational organized crime are established as a way to combat transnational organized crime (Guymon 2000).

Forensic methods play an important role in investigating crimes, especially when there are no eyewitnesses (Afanasyev et al. 2011). P. Kevin Smith, VP of North American sales at LTU Technologies defined digital forensics as "the investigation of digital data, be it on a hard drive or over a network or the web used for investigative purposes" (Miller 2007, p. 40). Although there are already tools in this area, new tools are needed since the current tools cannot cope with the enlarging targets (Golden and Vassil 2006).

In addition to the states, software companies can assist in combatting crime. The Business Software Alliance (BSA) fights against piracy (Nykodym and Taylor 2004) using ICT, and companies support the Child Exploitation Tracking System (CETS), which helps in the search of exploited children (Etges and Sutcliffe 2008).

Combatting crime means not only reacting to crime but also being proactive. Therefore, law enforcement agencies need early information about possible crimes in order to establish good strategies (Verfaillie and Beken 2008; Williams and Godson 2002) through broad data collection and analysis and the extensive use of intelligence sources (Williams and Godson 2002).

The characteristics of transnational organized crimes, such as networked crime, transnational crime, and complex crime, are similar to the nature of the new threats against states that appeared after the Cold War (Dunn Cavelty and Mauer 2009), so the methods used to forecast such threats can also be used in the fight against transnational organized crime. Methods of monitoring criminal activities include "geospatial predictive analysis, data-mining technologies, project management-based approaches and social network analysis" (Dunn Cavelty and Mauer 2009, p. 130). Through the use of data-mining technologies, links and patterns in data sets can be identified. For example, traffic and travel analyses can detect patterns and identify likely areas of new activity (Williams and Godson 2002). The success of social network analysis was evident in 2010, when a boss of an Italian mafia organization was traced down through his Facebook logons (cf., e.g., Wise 2010).

Transnational organized crime organisations use the Internet as a contact platform for their businesses (Etges and Sutcliffe 2008). The use of the Internet for criminal activities can be defeated through regulation. Franklin (2010) discussed the difficulties inherent in implementing such regulation and summarized several possible approaches, including the kind of state regulation that China imposes, where Internet use within China's boarders are monitored and censored by the government (Franklin 2010). The main question here concerns how to regulate the Internet without the state's being accused of being a 'big-brother' (Franklin 2010, p. 68). Franklin also suggested other multi-stakeholder participatory models, keeping in mind that the Internet is not created by an external institution but is built by its users (Franklin 2010). Franklin also questioned, however, whether it was even possible to regulate the Internet within any borders.

Afanasyev et al. (2011) presented an approach to combatting transnational organized crime using digital forensic methods. In the digital world, there is nothing like DNA or fingerprints in the physical world. They proposed to introduce a new network-layer capability that would allow authorized parties to identify the physical machine that sent a particular packet via the Internet so law-enforcement agencies could map traffic over the internet back to the physical source device.

## 7.3 Discussion on Results

The previous section summarized results of the literature review on the topic of transnational organized crime. The literature review revealed that IS play three roles in this area: as a communication tool for transnational organized crime organizations, as a new field for criminal activities and a support for existing activities, and as a support for law enforcement agencies that investigate and prevent crimes.

Etges and Sutcliffe (2008) mentioned the use of the IT by transnational crime organisations. Combating these organizations' use of IT is difficult, but once participants are found, their communications can be monitored to prevent future crime. Etges and Sutcliffe (2008) also wrote about how the Internet makes it possible for criminals to offer their services transnationally and how they use the Internet to launder money through the use of e-currencies. Franklin (2010) discussed several ways to regulate the Internet but found no perfect solution and even asked whether it is possible.

Information technology, particularly calculation power, is also used by law enforcement. For example, geospatial predictive analysis is possible only when the required calculations and simulations are done by a computer.

Figure 7.2 summarizes the areas in which IS plays a role in connection with transnational organized crime. Next, these areas are analysed and ideas for further research are developed.

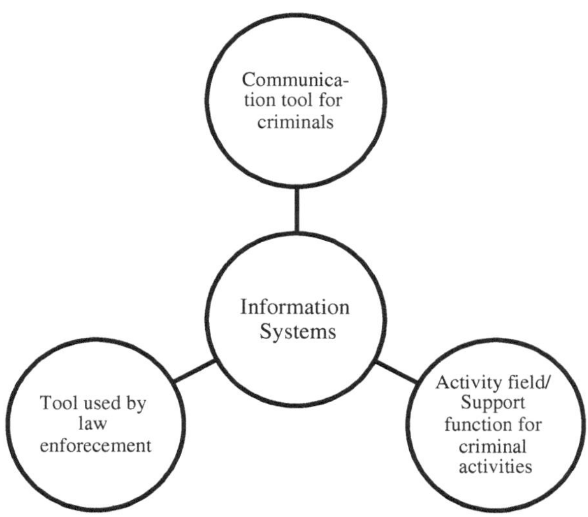


**Fig. 7.2** Information systems in connection with transnational organized crimes

## 7.4 Research Agenda

### 7.4.1 *Areas of Further Research*

IS influences transnational organized crime as a communication tool for criminal organisations, as a new field of activity and a support for existing fields of activity for criminal organisations, and as tool to combat criminal organisations. These areas suggest useful approaches for future research.

### 7.4.2 *Communication Tool for Criminals*

The communication tools transnational organized crime organizations use are the same tools that legal enterprises use. Therefore, the question concerning how to regulate things that are available to everybody so they cannot be used for criminal purposes is a particularly difficult one. Future research should investigate to what degree it is possible use ICT to identify criminal communication in order to disable it.

If regulation is not possible, monitoring becomes more important. Therefore, the approach Afansayev et al. (2011) identified should be examined in more detail to find methods to monitor criminals while securing the privacy rights of the general public. In connection with data privacy is the topic of data security, which must also be covered by the chosen tools.

### 7.4.3 *Support of Criminal Activities*

The Internet supports all kinds of activities, and criminal enterprises are no exception. If regulation of the Internet is not possible, ways of scanning the Internet and finding suspicious pages using search algorithms should be examined.

The broader field of information technology and its role as a support for criminal activities was not examined in detail as part of this paper, so further research should determine whether special means to combat one or more fields of criminal activity are already being used and how they can be enhanced to support the fight against transnational organized crime.

### 7.4.4 *Supporting Law Enforcement*

Section 7.4 discussed several methods for predicting criminal actions through pattern analysis and simulations, but further research is needed to focus on the algorithms that are necessary to do such analysis. Future research should also determine which new technologies, such as in-memory technology, can be used to enhance the success of such methods and what new technologies may include new possibilities to create predictions. In general, then, future research in this area should focus on how the existing analysis can be executed faster and how it can be enhanced through better technical solutions.

Besides supporting the prediction of crimes, IS are used to investigate crimes. In this area digital forensics and the tools that are used to examine digital evidence are important, but the tools that currently exist are not sufficient (Golden and Vassil 2006). Future research should look into the requirements of such new tools and how to build them.

## 7.5 Conclusion

The aim of this paper was to review the existing research results on the question concerning how IS can help to keep transnational organized crime networks from becoming more powerful. The paper examined how transnational organized crime organisations are characterized in the literature and summarized the main fields of activity in which these organizations engage. The last part of the literature review focused on papers that have addressed way to combat transnational organized crime and identified legal regulations, crime detection, preventive methods, and cooperation.

The results of the literature review were summarized to three main areas where IS play a role: as a communication tool for the criminals, as new field of activity for them, and as a support tool for the law enforcement. Based on these three areas, several areas for future research were developed.

## References

Afanasyev, M. et al. (2011). Privacy-preserving network forensics. *Communications of the ACM*, *54*(5), 78. Retrieved September 6, 2011 from http://portal.acm.org/citation.cfm?wdoid=1941487.1941508.

Boutros-Ghali, B. (1994) Transnational crime. *Vital Speeches of the Day*, *61*(5), 130.

Dunn Cavelty, M., & Mauer, V. (2009). Postmodern intelligence: Strategic warning in an age of reflexive intelligence. *Security Dialogue*, *40*(2), 123–144. Retrieved September 29, 2011 from http://sdi.sagepub.com/cgi/doi/10.1177/0967010609103071.

Etges, R., & Sutcliffe, E. (2008). An overview of transnational organized cyber crime. *Information Security Journal: A Global Perspective*, *17*(2), 87–94. Retrieved October 21, 2011 from http://www.tandfonline.com/doi/abs/10.1080/19393550802036631 .

Europol, (2006). *EU organised crime threat assessment 2006* (p.26). Retrieved January 3, 2012 from http://scholar.google.com/scholar?hl=en&btnG=Search&q=intitle:EU+Organised+Crime+Threat+Assessment+2006#0.

Franklin, M. I. (2010). Digital dilemmas: Transnational politics in the twenty-first century. *Security*, *xvi*(ii), 68.

Golden, G. R. I., & Vassil, R. (2006). Next-generation digital forensics. *Communications of the ACM*, *49*(2), 76–80. Retrieved from http://www.ncbi.nlm.nih.gov/pubmed/20715682.

Guymon, C. L. D. (2000). International legal mechanisms for combating transnational organized crime: The need for a multilateral convention. *Berkeley Journal of International law. 18*, 53. Retrieved October 21, 2011 from http://heinonlinebackup.com/hol-cgi-bin/get_pdf.cgi?handle=hein.journals/berkjintlw18&section=7.

Klerks, P. (2001). The network paradigm applied to criminal organisations : Theoretical nitpicking or a relevant doctrine for investigators? Recent developments in the Netherlands. *Defense, 24*(3), 53–65.

von Lampe, K. (2003). *The Use of Models in the Study of Organised Crime* (pp. 1–14). Retrieved from www.organized-crime.de.

Miller, R. (2007). The truth is in there. *EContent, 30*, 30–43.

Nykodym, N., & Taylor, R. (2004). The world's current legislative efforts against cyber crime. *Computer Law & Security Report, 20*(5), 390–395. Retrieved October 21, 2011 from http://www.sciencedirect.com/science/article/pii/S0267364904000706.

Shelley, L. (2007). The unholy trinity: Transnational crime, corruption and terrorism. In U. Gori and I. Paparela. *Invisible Threats: Financial and Information Technology Crimes and National Security*. Amsterdam, Berlin, Oxford, Tokyo, Washington, D.C. (NATO series): IOS Press, 2007, pp. 100–107.

Shelley, L. (1995). Transnational organized crime: An imminent threat to the nation-state. *Journal of International Affairs, 48*. Retrieved October 21, 2011 from http://www.questia.com/googleScholar.qst?docId=5000275997.

Verfaillie, K., & Beken, T. V. (2008). Proactive policing and the assessment of organised crime. *Policing: An International Journal of Police Strategies & Management, 31*(4), 534–552.

Williams, P., & Godson, R. O. Y. (2002). Anticipating organized and transnational crime. *Crime, Law and Social Change, 37*, 311–355.

Wise, A. (2010). Mafia Boss betrayed by Facebook. In *ABC News*.

# Chapter 8
# How Can Growing Energy Demands Be Met Safely and Efficiently?

**Stefan Debortoli and Nadine Székely**

## 8.1 Introduction

A United Nations survey identified environmental sustainability as the most important issue of the future (Watson et al. 2010). In particular, meeting the world's growing energy demands safely and efficiently is one of fifteen global challenges identified by The Millennium Project (2011).

While technology provides the electricity we need to live an affluent lifestyle, technology creates carbon emissions that affect the environment (Watson et al. 2008). The information technology (IT) sector contributes to the growing demand for energy especially with the ongoing production of new mobile devices, but information systems (IS) has also been one of the major enablers of productivity growth in many countries and, as such, has helped save resources (Watson et al. 2008). Since the root cause of global warming is fossil fuel consumption (Hoffert et al. 2002), Watson et al. (2010) establish a new subfield of IS, energy informatics, that they define as being "concerned with analyzing, designing, and implementing systems to increase efficiency of energy demands and supply systems" (p. 24). They express the core idea with a simple formula:

$$Energy + Information < Energy$$

Watson et al. (2010) present an integrated framework that incorporates all elements of an energy supply and demand system with an information system at its heart, containing flow networks, sensor networks, and sensitized objects. They identify nine IS research questions, all of which deal with dimensions of the energy informatics framework. This paper elaborates on the IS issues that are related to the supplier side, as depicted in Watson's (2010) energy informatics framework, by focusing on a single research question:

*How can an information system integrate supply and demand data to increase energy efficiency?*



J. vom Brocke et al., *Grand Societal Challenges in Information Systems Research and Education*, SpringerBriefs in Information Systems,
DOI 10.1007/978-3-319-15027-7_8

To address this question, a literature review to identify existing concepts and the current state of the research in this field is conducted. Then the findings are discussed and gaps in the research are identified and evaluated. Next, a future research agenda is presented which the IS academic community should carry out in order to change the world for the better in terms of meeting its growing energy demands safely and efficiently.

## 8.2 Literature Review

### *8.2.1 Literature Search*

There are always two parties involved in an energy consumption transaction, a supplier and a consumer (Watson et al. 2010), both of which must be involved equally in order to manage energy efficiency effectively. This paper focuses on the supplier side by investigating the possible effects if the supplier uses various sources of data to understand the customer.

To address the body of knowledge on this topic, all articles published from 2000 to 2011 in the top six journals in the field of IS (*MIS Quarterly*, *European Journal of Information Systems*, *Information Systems Journal*, *Information System Research*, *Journal of AIS*, and *Journal of MIS*) listed by the Association for Information System (AIS) have been searched. Seventeen papers were published between 2000 and 2011, among which only one was considered relevant after applying a full text search on "energy efficiency," "energy demand data," "energy supply data," "demand side management," and "smart grid." Therefore, the search strategy was extended to papers outside the six journals. A second search wave with the same keywords led to a huge amount of results, so the search terms were changed to "energy efficiency AND green IS," "energy efficiency AND information systems," and "energy efficiency AND smart metering." These searches resulted in more than 800 papers, among which not even a dozen were considered relevant. However, two major concepts for leveraging IS to increase energy efficiency were extracted from these papers: demand side management and smart grid technologies. Another literature review on only these two topics refined the results and provided more detailed insights.

The following sections discuss the identified concepts in depth and suggest other concepts that may be researched in the future.

### *8.2.2 Demand Side Management*

During the height of the oil crisis of the 1970s, many electricity utilities introduced demand side management (DSM) with the intention of lowering the customers' energy consumption (Bloom 2010). In a broad sense, DSM refers to modifying customer energy use, while in a more narrow sense, DSM programs are designed to lower energy consumption (Masters 2004). Even though it is against the utilities' logical strategy of selling more energy, they have since promoted energy-efficient products and technologies intensively. The utilities did not change their strategy because of their benevolence, but lowering the overall energy consumption was an essential step for keeping the Kyoto protocol targets and was less expensive than building new power plants to keep up with the growing demand (Bloom 2010; Geller and Attali 2005). A study by the US Department of Energy revealed that DSM helped to reduce energy consumption in the 1990s by a third.

The first introduction of DSM was in the 1970s, but experts agree that we are currently again in time of aggressive DSM activity (Bloom 2010). There has never been such a high level of governmental funding for the purpose of energy independence, the use of clean green power, and the shift to low-power-consuming electronic devices.

Andrew (2010) states that peak energy periods are the major and most cost-intensive issue in energy production. The problem with electrical energy is that it cannot be stored effectively for times when higher capacities are required; peak times have to be supplied by the power plants directly, without the possibility of using energy produced previously. Andrew (2010) identifies three fundamental components for an effective DSM program (p. 30):

Education and communication

Smart grid technologies

Voluntary curtailment of use

Education and communication is essential if a DSM program is to work effectively, as the best energy-efficient technology does not save the same amount of energy as a well-founded education (Andrew 2010). Today's information systems play a major role in educating consumers by providing internet-based resources for learning about energy-saving practices through program websites, frequently asked questions (FAQs), e-mail newsletters, discussion forums, social media initiatives, and more. Besides providing how-to instructions, such information systems can motivate lower energy consumption by providing specific and challenging goals in combination with public commitment (Graml et al. 2011). Other elements of education and communication include support for repeated behavior through prompts and reminders and learning by doing through immediate feedback—a combination of descriptive and injunctive feedback (Graml et al. 2011). Through social networks, consumer can connect to other people who are also concerned about saving energy and learn about their consumption behavior (Zhang 2012).

Smart grid technologies are another component that can lead to a drastic reduction in energy consumption. Since smart grid technologies have already been identified as one of the major concepts in the literature review, Sect. 8.2.3 is fully dedicated to this topic.

While DSM has many positive aspects and effects (Auffhammer et al. 2007; Geller and Attali 2005). Geller and Attali (2005) identified nine major criticisms and collected responses to them. One of the most common criticisms was the claim that the rebound effect abolishes all energy savings—that is, that reduced energy costs that are due to low-power devices and subsidized prices increase the demand for other energy services, such as heating, refrigeration, and lighting, when consumers change their behavior because of the lower costs. Several studies have tried to accept or reject the hypothesis of the rebound effect (Auffhammer et al. 2007; Loughran and Kulick 2004), but researchers have not yet concluded whether the effect exists.

DSM is a means for increasing energy efficiency that has been around for more than 40 years. The core idea does not necessarily require information systems in order to function, but IS heavily supports the educational component of DSM.

### *8.2.3 Smart Grid Technologies*

"Smart grid" describes the use of information technologies in energy supply systems (Potter et al. 2009). Smart grid technologies are advanced sensors and communication technologies that improve the use of assets, improve reliability, and enable a wider range of services (Potter et al. 2009). Examples of smart grid technologies are smart meters, which are "advanced meters that identify consumption in more detail than conventional meters and communicate via a network back to the utility for monitoring and billing purposes" (Darby 2010; Farnworth and Castilla-Rubio 2008). One key feature of smart grid technologies their ability to leverage demand side data in order to make better operational supply decisions. This concept is highly related to and interconnected with DSM, but when DSM was introduced forty years ago, the required technology was not available to integrate demand and supply data in almost real-time.

Potter et al. (2009, p. 1) collects some defining features for smart grids:

*A smart grid provides an interface between consumer appliances and the traditional assets in a power system (generation, transmission, and distribution).*

This two-way communication allows individual customers to control their energy consumption and empowers the operator to control the households' loads, which enables a more agile system behavior.

*A smart grid is at least semi-autonomous.*

The use of intelligent systems allows the utilities to respond automatically to power-related changes observed by sensor networks, without any human interaction. This capability facilitates much faster reactions to interruptions and other emergency situations. Jansen and Green (2011) describe this behavior as an outage management system (OMS).

*A smart grid optimizes the assets of the power system.*

Smart grids help to ensure optimal power flows and improve reliability. By communicating peak demand periods and applying dynamic pricing (i.e., charging higher prices during peak hours) based on the system load, smart grids ensure that consumers reduce energy consumption during peak demand periods. This trend lowers overall costs for the utilities since the need for flexible and expensive power generation is reduced dramatically.

*A smart grid supports the integration of distributed energy generation into the conventional centralized power system.*

Improved communication from two-way metering allows a more intelligent incorporation of decentralized power production (i.e., by connecting residential, commercial, or industrial local energy-generation equipment) into the centralized power system. Thus, the customer could be a supplier as well as a consumer.

Smart grid technologies collect data about the reality in energy distribution systems, measuring the energy that has been consumed and capturing power quality and power outages (Johnson 2010). This capability implies the potential for deriving a huge volume of data from the energy network that must then be turned into meaningful information by means of a central hub for information exchange. Johnson (2010) envisions this hub as a kind of meter data manager, the core responsibilities of which are reading data from the network and feeding information to the network operations and business management by developing the context that surrounds the data. An example is the collection of data about power quality from the smart meter, including issues related to power quality, such as voltage sag problems or overstressed sections of the infrastructure, which would be vital data for the utility (Johnson 2010). The meter data manager also encourages customer interaction by providing personalized power-usage information and mechanisms for economical energy choices through web-based portals.

Johnson's (2010) vision has the meter data manager acting as a central, real-time data transfer and information exchange system that uses all available data sources to provide information to multiple operational and business applications, thereby providing a number of benefits (Johnson 2010, p. 48):

Proactive incident management

Proactive management of power quality

Improvement in the efficiency of asset use

Integration of green energy sources into any part of the network

Consumer engagement in the energy efficiency value chain

Many researchers have identified the use of smart grid technologies as a requirement for lowering the carbon footprint by integrating renewable energy sources into the supply stream (Gobble 2011). Realizing this vision is a challenge by itself, as many studies show. Several countries already have smart grid pilots installed and running since many of the required technologies—such as low-cost sensors to measure network performance and low-cost communications to route that information to consumers, utilities, and distributors—already exist (Gobble 2011).

Another challenge is the integration of smart grids into aging transmission networks while ensuring continued system reliability and security. Although new wind and solar power plants have been integrated into the existing infrastructure, the re-use of currently existing cables is desired, but Gobble (2011) cites an interview with Greenpeace Germany as saying that this is possible only with intelligent systems.

Even though most of the required technology and many pilots exist, more intensive research in the area of intelligent control systems is required, as is research on the security-related aspects of these systems (Gobble 2011). For instance, an attack by a Trojan horse called Stuxnet targeted communications and controls at energy infrastructures (Davies 2010), which some reports said targeted a nuclear power plant in Iran. Because of those kinds of occurrences, researchers seek to determine whether the benefits of smart grids are worth the associated risk (Davies 2010).

Smart grid technologies are essential for increasing energy efficiency since these technologies close the gap between demand data and supply data. Other effects are new business models like dynamic pricing, which are not currently possible to any detailed extent. However, unresolved issues remain regarding data security in the smart grid that must be the subject of future research initiatives.

## 8.3 Discussion and Research Agenda

Based on current knowledge about the two fundamental concepts for bringing demand and supply data together, IT can make two contributions: measuring data directly on the customer side and distributing it to the supplier to managing the offer as part of the overall idea of smart grid technologies, and generating education and awareness. DSM can leverage IT to meet these goals.

Smart grid technologies can consider only real-time data in applying new revenue schemes like dynamic pricing in peak periods, while DSM uses no customer-specific historical or real-time data. Table 8.1 categorizes the identified concepts in a temporal context and shows that future data is completely neglected in the attempt to match energy demand and supply.

Today's advanced information technologies and IT trends could lead to a better match between energy demand and supply by predicting future energy consumption. One aspect of such as system would be the use of historical data, which might or might not be personalized with smart grid technologies that identify existing energy usage patterns, such as a household's average cooking time or an individual's average sleeping time. The required data is already available at a granular level through data collection with smart grid technologies. Scientists at the University of Applied Sciences in Münster were even able to conclude which movie a household is watching by analyzing the transmitted data from smart meters (Greveler et al. 2011).

The interconnection of smart grid technologies and other technological concepts, such as the Internet of Things, could influence the short-term forecasts of energy demands. The Internet of Things is a concept in which each real-world object is connected to its environment via sensor networks and wireless communication devices. The dependencies of real-world objects could be automatically

**Table 8.1** Considered data types and concept

| Considered data | Historical | Real-time | Future |
|---|---|---|---|
| Concepts | DSM, SGT | SGT | – |

**Table 8.2** Considered data types and concepts #2

| Considered data | Historical | Real-time | Future |
|---|---|---|---|
| Concepts | DSM, SGT | SGT | Usage pattern detection, internet of things technologies |

analyzed and the upcoming energy demand predicted and reported to the energy provider. A typical scenario could be the combination of geographical information about the resident of a house and the previous usage patterns of electrical appliances. For example, the resident typically goes home from work at 6:00 p.m., and the smart meter detects a typical "cooking-pattern" on weekdays at 7:00 p.m., so the power supplier is notified about the likely upcoming energy consumption.

Table 8.2 shows that future energy consumption could be predicted by detecting use patterns, leveraging historical data, and incorporating new technologies, such as the Internet of Things. However, several obstacles must be overcome in order to enable the future data prediction. First, the Internet of Things technologies must take the step from the visionary or at least roughly implemented concepts to a mature, globe-spanning application. To accomplish this goal, ICT hardware, software, and reference architectures must be developed further, which might be of interest to IS scholars, computer scientists, and electrical engineers.

Second, intelligent algorithms that can detect historical usage patterns and predict future energy demands must be developed. The application of new technologies, such as in-memory computing, could be leveraged in order to improve the demand forecast even more.

Third, all the collected data, independent of the temporal condition, must be treated carefully in order to address privacy and security issues and come up with the most effective, reliable, and trustworthy concept.

## 8.4 Conclusion

This paper contributes to the emerging body of research on the integration of supply and demand data by means of information systems in order to increase energy efficiency. A literature review revealed existing concepts related to the integration of energy demand and supply data. The two major concepts identified are DSM and smart grid technologies. However, both concepts consider, at best, only historical or real-time data, while future data could be taken into consideration in order to providing accurate forecasts of energy demands. The required predictive data could be collected through the sophisticated use of new information technologies, such as the Internet of Things, and by applying mature usage-pattern-analysis algorithms sped-up by in-memory technologies. However, further research needs to be done in the field of forecasting future energy demand, including development of the required algorithms, hardware, software, and reference architectures and resolving privacy and security issues. These research topics are not only for IS scholars but also for computer scientists and electrical engineers.

## References


Andrew, J. W. (2010). The future is now: The new face of demand side management. *Management Quarterly, 51*(1), 1–29.

Auffhammer, M., Blumstein, C., & Fowlie, M. (2007). Demand-side management and energy efficiency revisited. *Energy Journal, 29*(1), 174–185.

Bloom, S. (2010). A New Dawn for Demand-Side Management. Electrical Wholesaling, (August).

Darby, S. (2010). Smart metering: What potential for householder engagement? *Building Research & Information, 38*(5), 442–457.

Davies, S. (2010). Internet of energy. *Engineering & Technology, 5*(16), 42–45.

Farnworth, E., & Castilla-rubio, J. C. (2008). *SMART 2020: Enabling the low carbon economy in the information age*. Report by The Climate Group on Behalf of the Global eSustainability Initiative (GeSI).

Geller, H., & Attali, S. (2005). *The experience with energy efficiency policies and programmes in IEA countries: Learning from the Critics*. Paris: IEA Information Paper.

Gobble, M. M. (2011). Building the smart grid in Europe. *Research-Technology Management, 54*(2), 2–8.

Graml, T., Loock, C.-M., Baeriswyl, M., & Staake, T. (2011). Improving residential energy consumption at large using persuasive systems. In *ECIS 2011 Proceedings*. Paper 184. (pp. 1–15).

Greveler, U., Justus, B., & Löhr, D. (2011). Hintergrund und experimentelle Ergebnisse zum Thema "Smart Meter und Datenschutz", 1–5.

Hoffert, M. I., Caldeira, K., Benford, G., Criswell, D. R., Green, C., & Herzog, H. et al. (2002). Advanced technology paths to global climate stability: Energy for a greenhouse planet. *Science* (New York, N.Y.), 298(5595), 981–987.

Jansen, H., & Green, D. (2011). Building the framework to accelerate smart grid benefits. *Powergrid International, 16*(7), 46–49.

Johnson, B. Y. P. (2010). The meter data manager: A starting point for an information-driven utility. *Powergrid International, 15*(8), 1–46 (August).

Loughran, D. S., & Kulick, J. (2004). Demand side management and energy efficiency in the United States. *Energy Journal, 25*(1), 19–44.

Masters, G. M. (2004). *Renewable and efficient electric power systems*. Hoboken: Wiley.

Potter, C. W., Archambault, A., & Westrick, K. (2009). Building a smarter smart grid through better renewable energy information. *IEEE/PES Power Systems Conference and Exposition, 2009*, 1–5.

The Millennium Project. (2011). How can growing energy demands be met safely and efficiently? Retrieved from http://www.millennium-project.org/millennium/Global_Challenges/chall-13.html

Watson, R. T., Boudreau, M. -C., Chen, A. J., & Huber, M. (2008). Green is: Building sustainable business practices. *Information Systems*, 1–15. Athens, GA: Global Text Project.

Watson, R. T., Boudreau, M. -C., & Chen, A. J. (2010). Information systems and environmentally sustainable development: Energy informatics and new directions for the IS community. *MIS Quarterly, 34*(1), 23–38.

Zhang, Y. (2012). IT enabled environmentally friendly consumption: IT features addressing challenges in consumer decision making. In *ECIS 2012 Proceedings*. Paper 233. (pp. 1–11).

# Chapter 9
# How Can Scientific and Technological Breakthroughs Be Accelerated to Improve the Human Condition?

Sandro Weber

## 9.1 Introduction

Technology has often been the driving force in changing the nature of work, but not only the nature of work. From the miracle of the microchip to the divine practice of cloning, the human imagination for technical ingenuity often creates revolutions that outpace society's evolution (Barbian 2003), which can be both empowering and disturbing (Buechner and Balog 1998). The goal of challenge 14 of the Millennium Project is to extract the positive, empowering effects of technologies and use them to improve the human condition. The core of the challenge lies in the question concerning how the breakthroughs of these technologies can be accelerated by means of information systems (IS).

This article first provides an overview of the most important technologies that affect the human condition and then explains how these converging technologies are interrelated and how their use can be improved by means of IS. The results of the literature review are discussed in Sect. 9.1, and the last part of the article offers a research agenda for research fields found during the literature review.

## 9.2 Related Work

This chapter describes the six increasingly interconnected megatrends Roco and Bainbridge (2003) identified in the field of technology. These megatrends are aligned with the converging technologies and the most important technologies regarding the challenge, and their contribution to the human condition is highlighted.

The term "converging technologies" refers to the convergence of nanotechnology, biotechnology, information technology, and new technologies based on cognitive science. It brings together four previously separate domains of science



J. vom Brocke et al., *Grand Societal Challenges in Information Systems Research and Education*, SpringerBriefs in Information Systems,
DOI 10.1007/978-3-319-15027-7_9

and technology and unites non-traditional conceptions of reality with useful applications. Biotechnology—not only genetic engineering, but all of it—is evolution in action, while information technology is mind embodied in machines. Nanotechnology, in part because it concerns the domain just above that where quantum events take place, demonstrates that the world consists of mechanisms, not mysteries, and cognitive technologies drag human beings themselves under the atomic force microscope to offer the many means by which humans will augment and alter themselves in future decades. Once the technology is used to transform ourselves, it becomes more relevant to our hopes and beliefs than any ancient myth could be (Bainbridge 2007). Convergence can create intelligent systems and environments to improve everyone's quality of life and create access for people with special needs. For example, a combination of wireless technology and nanoscale sensors could allow blind people to walk alone and even to drive (Roco 2008).

### *9.2.1 Nanotechnology*

The societal dimensions of nanotechnology have been a concern since the establishment of the National Nanotechnology Initiative. The workshops on this topic were dominated by visions of applications of nanotechnology in industries from automobiles to space exploration (Scheufele and Lewenstein 2005). Nanotechnology proponents expect that it will offer solutions to key problems, especially those that developing countries face (Invernizzi and Foladori 2005). Several influential articles have claimed that nanotechnology can resolve most of the United Nations' Millennium Development Goals (Invernizzi and Foladori 2005) by making fleets of medical nanorobots smaller than a human cell that can to eliminate cancer, infections, clogged arteries, and even old age (Merkle 2001). Surgeries may be performed internally, and non-invasive procedures could be performed by programmed motes. Dozens of companies are currently developing nanotechnological ways to deliver drugs to specific sites, thus enhancing effectiveness while minimizing side effects. Another use of nanotechnology is improving chemotherapy by using nanopowders and tubes that specifically target tumour sites (McGrady et al. 2010).

In addition to developing computers much more powerful than today's and new medical capabilities that will heal and cure cases that are now viewed as hopeless, this new and very precise way of fabricating products will also eliminate the pollution from current manufacturing methods. Molecular manufacturing will make exactly what it is supposed to make, no more and no less, eliminating pollutants (Merkle 2001). Nanotechnology will be especially important for the developing world because it has the potential to perform work at a small scale (Invernizzi and Foladori 2005). While nanotechnology could also be used to do great harm (Merkle 2001), this article concentrates on its positive capabilities that provide powerful means to improve the human condition.

### 9.2.2 Biotechnology

Biotechnology can produce new materials that could facilitate daily life, such as clothing that automatically adjusts to changing temperatures and weather conditions (Roco and Bainbridge 2003). This field also promises a revolution in agriculture by providing the means to develop plant species that are resistant to diseases and pests and more tolerant of drought and temperature changes. Craig Venter is creating the first artificial life form by building a synthetic chromosome. New forms of photosynthesis could create hydrogen instead of oxygen and new microbes could be created to eat the plaque that can build up between neurons and reduce brain activity in the elderly (Glenn et al. 2011).

### 9.2.3 Information Technology

Information technology is mind embodied in machines (Bainbridge 2007). The bit-based language (0 and 1) has allowed expanding communication, visualisation, and control of intellectual power (Roco and Bainbridge 2003). Information technology is primarily a component and an enabler of the other science areas.

### 9.2.4 Cognitive Science

The term "cognitive science" indicates that the study of mind in itself is a worthy scientific pursuit. It is more of a loose affiliation of disciplines than a discipline of its own. An important pole is occupied by artificial intelligence, so the computer model of the mind is a dominant aspect of the field. The other affiliated disciplines are linguistics, neuroscience, psychology, sometimes anthropology, and the philosophy of mind. Each discipline would give a different answer to the question of what is mind or cognition (Varela et al. 1991).

Fodor's monograph, "The Modularity of Mind," (1983) has had a dramatic impact on all of the cognitive sciences. The essence of Fodor's position is that the human mind is best viewed as a distinction between the central system, which is responsible for rational thought and the "fixation of belief," and the modular input systems, one for each of the senses, that feed the central system. An input system translates a sensory stimulus into a format that the central system can recognise so it can be used in cognition: If I *smell* smoke in the kitchen, I may believe that the toast is burning; if I also *see* flames emerging from the toaster, I may believe that breakfast is going to be less palatable than usual (Karmiloff-Smith 1994).

The challenge lies in the question concerning how IS can support the development of the technologies described above, such as by establishing converging technologies informatics to use existing data better and advance communication among the scientists involved in various technologies. There are separate efforts in each emerging technology, so the synergy created by such development would support converging technology development in areas like new discoveries, design, and manufacturing, and would address common environmental, health, and ethical implications. Converging technologies informatics would also help to close the gap between the creation of new knowledge in research and policy decisions by improving information transfer and framing of new technologies (Roco 2008).

Bioinformatics is defined as research, development, and/or application of computation tools and approaches to expanding the use of biological, medical, behavioural, or health data, including tools used to acquire, store, organize, archive, analyse, or visualize such data (Cummings and Temple 2010). Computations biology, which provides additional possibilities, especially in education, is defined as the development and application of analytical and theoretical methods, mathematical modelling, and computation simulation techniques to the study of biological, behavioural, and social systems (Cummings and Temple 2010).

Nanoinformatics is a collection of multi-disciplinary approaches to cataloguing, correlating, and modelling nanomaterial properties. CaNanoLab is an early example of a nanobioinformatics portal that is dedicated to fostering the rapid dissemination of nanobiological information across the scientific community. Nanobioinformatics studies are complex because they must simultaneously deal with the large dispersion of chemical formulations of nanobiomaterials (ranging from polymer to metal oxide particles), the lack of a common language across contributing disciplines, and the lack of a low-level language that can be used across nanoparticles. A structure-based annotation and analysis of nanobioparticles could help to cross-analyse their properties (Gonzalez-Ibanez et al. 2009).

Given the range of devices and applications that may be generated and addressed, the development of novel and advanced core characterization and nanomanufacturing technologies will be a requisite strategy for realizing the potential underlying nanotechnological development. The new nanotechnologies can be used to achieve a true coalescence of nanoscience and nanotechnology, which will ultimately benefit the human condition by using the building blocks and fundamental findings of nanoscience to develop systems based on the fusion of biology, nanotechnology, and informatics, with embedded intelligence and emergent behaviour (Ho et al. 2006). Nanoinformatics can also accelerate the introduction of nano-related research and applications into clinical practice, leading to an area that could be called "translational nanoinformatics."

At the same time, DNA and RNA computing present an entirely novel paradigm for computation. Nanoinformatics and DNA-based computing are likely to change completely how information is modelled and processed in biomedicine and to impact the emerging field of nanomedicine significantly (Maojo et al. 2010). While bioinformatics is usually applied in the context of analysing DNA sequence data, nanoinformatics is applied in the context of characterizing particles and materials with applications in nanotechnology and biotechnology. A key issue in nanoinformatics is the link between information, data, results, and findings using nano-related processes and the standard clinical procedures. This task will be fundamental for the future evolution in the medical arena (De La Iglesia et al. 2009) (Fig. 9.1).

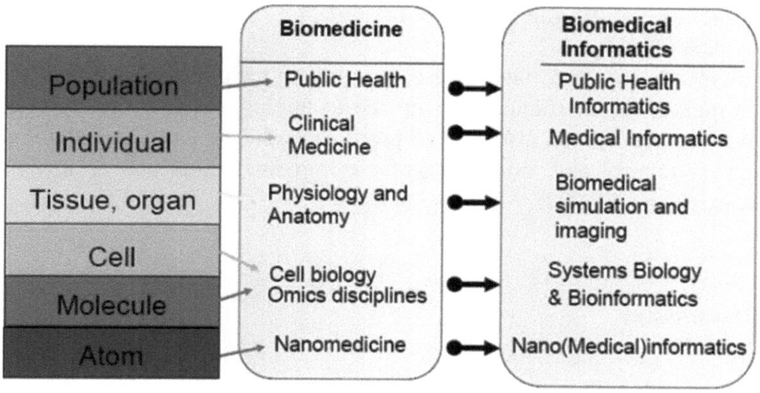


**Fig. 9.1** Biomedical informatics going down to the atom level, towards nanoinformatics (De La Iglesia et al. 2009, p. 990)

The next-generation nanoinformatics tools will focus on computational cooperation and systems integration, at which time the overflow of data will be translated into knowledge. These advances will have to move beyond data-sharing to task-sharing when a nanotechnology lab experiment touches on the computing power of a set of equipment behind a common user interface. Future progress in this field will move toward intelligent automation, complex data mining, and intuitive visualization of results (Ruping and Sherman 2004).

The brain-machine interface (BMI) is an example of the convergence of ICT and cognitive science. Recent progress in fundamental neurophysiological research suggests that the vision displayed science fiction movies, such as implanted electrodes, may become possible as the human brain is directly interfaced with computers or embedded in external devices. Information-gathering and information-processing devices could also be incorporated in the human body, but there is still a long way to go before these applications are used in real life. All current prototypes are one-directional—usually from the brain to the external environment—with no feedback loops, but bi-directional interfacing will be necessary for real-life applications like the control of paralyzed limbs or complex prosthetic devices so the brain can use its sophisticated feedback control strategies. In addition, a number of technological challenges, such as better sensory input or better understanding of neural coding of primary motor regions, must be solved before this highly invasive technique can be applied to humans. Bi-directional brain-computer interfacing (BBCI) holds great promise in the treatment of neurological and trauma patients, but before this kind of treatment can be implemented, it is necessary to identify the exact brain regions that are used for specific tasks. Real-time encoding and decoding software for brain input and output signals is also required (Malanowski and Compañó 2007).

Robotics is another convergence of ICT and cognitive science. The increased use of personal-assistance robots in the health care sector could save resources and be an effective and affordable option for the care of older citizens (Malanowski and Compañó 2007).

IS can also help to facilitate the use of emerging technologies in practice. Many decision-makers have difficulties using the technology, and IS can help to overcome this obstacle. It can also help to prove the value of the new technology and can improve control and monitoring of the implementation and accuracy of the technology (Kendall 1997).

## 9.3 Discussion

Science and technology and business policies play important roles in converging technologies. Science is reaching closer to technological applications in the emerging fields and is increasing its impact on society (Roco 2008). The convergence of nanotechnology, modern biology, the digital revolution, and cognitive sciences will bring about tremendous improvements in transformative tools, generate new products and services, enable opportunities to meet and enhance human potential and social achievements, and eventually reshape societal relationships (Hamelink 2006). Therefore, intelligent IS are required that can facilitate communication between the fields of science and the development of more complex, interrelated technologies.

Developments in systems approaches, mathematics, and computation in conjunction with emerging technologies help to clarify the natural world and scientific research as closely coupled, complex, hierarchical entities. At this unique moment of scientific and technical achievement, improvement of human performance at the individual and group levels and the development of suitable revolutionary products, primary goals for converging new technologies, become possible. Fundamentally new tools, technologies, and products will be integrated into individual and social human architecture (Roco 2004).

In the face of an ageing population, the health care system can benefit from using converging applications and products. New diagnostic and therapy procedures offer opportunities for patients and for the health care systems in general. The use of the potentials in medicine could be stimulated by an analysis of the broad spectrum of medical applicability and preventive medicine (Malanowski and Compañó 2007).

## 9.4 Research Agenda

Developments in the field of information and communication technologies may soon lose sight of the human scale, so an assessment of emerging technologies and their application in terms of the human condition is urgently needed. A useful field for such an assessment is the digitalization of health care (Hamelink 2006).

First examples of sociable technologies were toys like Tamagotchies and Sony's AIBO. Certain positive effects of the use of such robots in hospitals and homes for the elderly, such as mood improvement through interaction with a seal robot, have already been demonstrated (Wada et al. 2004). Nonetheless, a deeper theoretical understanding of emotions is necessary for the advancement of emotional robotics. The ubiquitous computing concept includes wearable computers, such as fabrics that incorporate electrical circuits that can be used for health monitoring (Beckert et al. 2007).

The first nanotechnology-based drug-delivery systems that release drugs in particular locations of the body are already commercially available (Wagner and Zweck 2005), but since research on the interaction of nanomaterials and living organisms has just begun, only limited information concerning long-term effects and the potential toxicity of nanoparticles inside the human body is available (Beckert et al. 2007).

The awareness among scientists of the concept of convergence must be improved in the future, given the uneven picture of real science and technology developments. The convergence concept works first as a political concept or a concept of research managers, but the term is so new that most researchers are not aware of convergence, even if they work in the middle of a converging discipline, busy with technology developments in the conceptual overlapping of nano, bio, info and cognitive science. Therefore, the concept of technological convergence still has a long way to go from vision to the guiding actions of scientists and finally to concrete technology development (Beckert et al. 2007).

A long-term problem regarding the introduction of enhancement technologies in a market environment is the risk that individuals must assume as new enhancement technologies are made available. The concern with protecting human subjects during clinical trials and in other experimental settings is a precursor to the difficult question of the conditions under which a proposed enhancement is considered sufficiently safe to be made available for the whole population. States and interstate bodies will need to provide some sort of welfare safety-net or insurance against the risks that individuals will undertake (and be encouraged to undertake) by subjecting themselves to enhancement regimes (Fuller 2009).

## 9.5 Conclusion

The main function of IS in the development of technologies and the interconnection between science and technology is to provide the means for disseminating information among the fields of science. IS also facilitates communication among experts by creating the common notation of information that is required to make a highly technical topic and the complex language of the specialists understandable for scientists in other fields.

## References


Bainbridge, W. S. (2007). Converging technologies and human destiny. *The Journal of medicine and philosophy, 32*(3), 197–216.

Barbian, J. (2003). High-tech times. *Training, 40*(10), 52–55.

Beckert, B., Blümel, C., & Friedewald, M. (2007). Visions and realities in converging technologies. *Innovation: The European Journal of Social Science Research, 20*(4), 375–394.

Buechner, M. M., & Balog, J. (1998). Techno sapiens. *Time International (Canada Edition), 151*(20), 8–13.

Cummings, M. P., & Temple, G. G. (2010). Broader incorporation of bioinformatics in education: opportunities and challenges. *Briefings in bioinformatics, 11*(6), 537–543.

De La Iglesia, D., Chiesa, S., Kern, J., Maojo, V., Martin-Sanchez, F., Potamias, G., et al. (2009). Nanoinformatics: new challenges for biomedical informatics at the nano level. *Studies in health technology and informatics, 150*, 987–991.

Fuller, S. (2009). Knowledge politics and new converging technologies: A social epistemological perspective. *Innovation: The European Journal of Social Science Research, 22*(1), 7–34.

Glenn, J. C., Gordon, T. J., & Florescu, E. (2011). 2011 State of the future (Challenge 14). *Security*, 2025, 1–203.

Gonzalez-Ibanez, Alvaro M., Gonzalez-Nilo, F., & Cachau, R. (2009). The collaboratory for structural nanobiology. *Biophysical Journal, 96*(3), 491.

Hamelink, C. J. (2006). Rethinking ICTs. *European Journal of Communication, 21*(3), 389–396.

Ho, D., Garcia, D., & Ho, C.-M. (2006). Nanomanufacturing and characterization modalities for bio-nano-informatics systems. *Journal of Nanoscience and Nanotechnology, 6*(4), 875–891.

Invernizzi, N., & Foladori, G. (2005). Nanotechnology and the developing world: Will nanotechnology overcome poverty or widen disparities? *Nanotechnology Law & Business, 2*(3), 294–304.

Karmiloff-Smith, A. (1994). Beyond modularity: A developmental perspective on cognitive science. *International Journal of Language & Communication Disorders, 29*(1), 95–105.

Kendall, K. E. (1997). The significance of information systems research on emerging technologies: Seven information technologies that promise to improve managerial effectiveness. *Decision Sciences, 28*(4), 775–792.

Malanowski, N., & Compañó, R. (2007). Combining ICT and cognitive science: Opportunities and risks. *Foresight, 9*(3), 18–29.

Maojo, V., Martin-Sanchez, F., Kulikowski, C., Rodriguez-Paton, A., & Fritts, M. (2010). Nanoinformatics and DNA-based computing: Catalyzing nanomedicine. *Pediatric Research, 67*(5), 481–489.

McGrady, E., Conger, S., Blanke, S., & Landry, B. J. L. (2010). Emerging technologies in healthcare: navigating risks, evaluating rewards. *Journal of Healthcare Management, 55*(5), 353–364.

Merkle, R. C. (2001). Nanotechnology: What will it mean? *Spectrum, IEEE, 38*(1), 19–21.

Roco, M. C. (2004). Science and technology integration for increased human potential and societal outcomes. Annals of the New York Academy of Sciences, The Coevolution of Human Potential and Converging Technologies, 1013, 1–16.

Roco, M. C. (2008). Possibilities for global governance of converging technologies. Journal of Nanoparticle Research, 10, 11–29.

Roco, M. C., & Bainbridge, W. S. (2003). *Converging technologies for improving human performance: Nanotechnology, biotechnology, information technology and cognitive science*. Dordrecht: Kluwer.

Ruping, K., & Sherman, B. W. (2004). Nanoinformatics : Emerging computational tools in nanoscale research. *Physics, 3*, 525–528.

Scheufele, D. A., & Lewenstein, B. V. (2005). The public and nanotechnology: How citizens make sense of emerging technologies. *Journal of Nanoparticle Research, 7*(6), 659–667.

Varela, F. J., Thompson, E., & Rosch, E. (1991). The embodied mind: Cognitive science and human experience. In (F. J. Varela, E. Thompson, & E. Rosch (Eds.)) *An International Journal of Complexity and (pp. xx, 308)*. Cambridge: MIT Press.

Wada, K., Shibata, T., Saito, T., & Tanie, K. (2004). Effects of robot-assisted activity for elderly people and nurses at a day service center. In *Proceedings of the IEEE, 92*(11), 1780–1788.

Wagner, V., & Zweck, A. (2005). *Nanobiotechnology in the medical sector—drivers for development and possible impacts*. Düsseldorf: VDI Future Technologies Consulting.

# Chapter 10
# Global Challenges for Humanity: How Can Ethical Considerations Become More Routinely Incorporated into Global Decisions?

**Nebojsa Milic and Krzysztof Oleszkiewicz**

## 10.1 Introduction

Incorporating ethical principles into global decision-making and implementation is not an easy or monolithic task. Different cultures perceive ethics from different perspectives, making it challenging for international bodies and organizations make ethics-related decisions and set ethics-related goals.

The Oxford Dictionary defines ethics as "a set of moral principles, especially ones relating to or affirming a specified group, field, or form of conduct". Roots of the term are based on Latin word *ethice*, which was derived from Greek word *ēthikē* (the science of morals). As civilization progressed, the understanding of ethics has evolved. In the modern era ethics is often related to duty (Waller 2005) and to scientific progress (Liszka 2011).

Today, ethics is globally and tacitly negotiated among cultures, religions, and generations of the human society on social networks (SN). SNs have become an irreplaceable part of everyday life for a large percentage of the world's population, so it is rational to expect that the age structure of SNs corresponds closely with the age structure of the analysed society. However, not all parts of society participate equally in SN, as research shows that "digital production inequality suggests that elite voices still dominate in the new digital commons" (Schradie 2011).

This research paper seeks to denote the state of the art and the future of incorporating ethical principles into worldwide decisions. In so doing the literature review provides a detailed overview of published work in IS on this topic, and explores the potential role of information systems (IS) in this area.

The paper is structured as follows: first, we review the academics' and practitioners' literature and analyse existing SN and IS-mediated platforms that have emerged. Then a summary of this review is used to derive future fields



J. vom Brocke et al., *Grand Societal Challenges in Information Systems Research and Education*, SpringerBriefs in Information Systems,
DOI 10.1007/978-3-319-15027-7_10

of research that might contribute to solving the challenge by leveraging IS. Finally, the paper discusses the literature review and concludes with questions aimed toward further augmentation and promotion of this only lightly investigated topic.

## 10.2 Research Method

The paper follows a systematic literature review process (vom Brocke et al. 2009) with the aim of creating a valid foundation for future research. The criteria applied in the initial literature search required that the papers' publication dates be after 2005, and keywords were searched in the titles, abstracts, and (where applicable) investigated areas of the papers based in the disciplines of IS, economics, sociology, and politics. The literature review was conducted from the end of October to the beginning of December 2011. The journals searched were those in the Senior Scholars Consortium (2011) "Basket of 8".

The literature research focused on four fields of interest that correlate with IS:

(1) **Global ethical considerations**: What are the current issues that are discussed globally and have a global effect? What is the consensus for how the ethical dimension is regarded? How are global decisions being made and implemented?
(2) **Beneficial movements supported by IS**: How are ISs contributing to increasing the transparency efforts of non-governmental organizations? What effect is observable?
(3) **Effect of public opinion on global decisions**: Do ISs help to inject public opinion into decision processes that have global effects? What is necessary to influence those processes?
(4) **Ideal place for ethical discussions**: What is the ideal environment in which ethical issues can be considered and archived in the most effective way? How such an environment be achieved?

As a consequence of the low number of results, an additional search was conducted using loosened criteria. First, we conducted a full text search, with most of the queries based on the combinations of the words *decision*, *ethic**, *global*, *information system**, and *routine*. Furthermore, year constraints were completely removed, as well as the discipline exclusions. The articles from these search results were selected by analysing their introductions, abstracts, and conclusions. The relevant research papers are grouped into four categories (Table 10.1), which are used as the basis for the upcoming discussion of the results.

**Table 10.1** Qualitative classification of relevant research papers

| Category | Literature |
|---|---|
| IS augmenting global decisions | Brey, P. (2007). Is Information Ethics Culture-Relative? International Journal of Technology and Human Interaction, 2–24 |
| | Davis, T., & Forester-Miller, H. (1996). A Practitioner' s Guide to Ethical Decision Making |
| | Miscione, G., Staring, K., Ostmo, L., & Fossum, C. (2008). The Shifting Legitimation of an Information System: Local, Global and Large Scale. GlobDev 2008, Paper 7. AIS |
| IS aligned with ethical principles | Altman, C. (2004). Converging Technologies: The Future of the Global Information Society. San Francisco |
| | Ess, C. (2006). Cybernetic pluralism in an emerging global information. Ethics and Information Technology, Volume: 8, Issue: 4, 215–226 |
| | Maner, W. (1996). Unique Ethical Problems in Information technology. Science and Engineering Ethics, 137–154 |
| | YanFang, C. (2005). Ethical Factors in Decision Making for Information System Management. Department of SCIS, Baruch College and Graduate Center CUNY |
| IS as a broker for social interaction | Floridi, L. (2008). Network Ethics: Information and Business Ethics in a Networked Society. Journal of Business Ethics, Volume 90, Supplement 4, 649–659 |
| | Nye, J. S. (2004). Power in a global information age. New York: Taylor & Francis |
| | Pippa, N. (2001). Digital Divide: Civic Engagement, Information Poverty, and the Internet Worldwide. New York: Taylor & Francis |
| IS as supporter of global activism | Floridi, L. (2007). Global Information Ethics: The Importance of Being Environmentally Earnest. International Journal of Technology and Human Interaction, 1–11 |
| | Nye, J. S. (2004). Power in a global information age. New York: Taylor & Francis |
| | Sorj, B. (2006). Internet, Public Sphere and Political Marketing: Between the Promotion of Communication and Moralist Solipsism. Rio de Janeiro: The Edelstein Center for Social Research |

## 10.3 Research Results

(1) Information systems augmenting global decisions

Scholars have gained insight into the mechanisms of global decision-making that are related to the field of IS. There is no universal IS that has been defined and developed solely for the purposes of global decision-making, although many existing ISs can improve the environment for global decisions. Miscione et al. (2008) note that "social development in the context of globalization is related to the use of ICT and determined by the ability to establish synergistic interactions between technological innovations and human values." However, existing mechanisms of decision-making on the global level should incorporate tools that take into account that "cross-cultural normative ethics cannot be practiced without a thorough understanding of the prevailing moral system in the culture that is being addressed" (Brey 2007). Scholars suggest that all ethically relevant decisions can be enhanced through a framework for ethical decision-making as "a useful method for exploring ethical dilemmas and identifying ethical courses of action" (Davis and Forester-Miller 1996).

(2) Information systems aligned with ethical principles

Although early insight into current research states that "there is no general consensus with respect to the complete framework of ethical decision making in IS" (YanFang 2005), it possible to find beneficial works that relate, although not directly, to this paper's topic of interest. As a foundation, ethical pluralism must be recognized (Ess 2006). However, since Western and Asian traditions define the majority of the stakeholders, and since those two moral centres have shaped and influenced the overall understanding and implementation of ethics around the globe, if IS is to be enabled to cooperate and somewhat improve global ethics and decision-making processes (by making them more ethical in a routine sense), analogies based on these two traditions must be used for upcoming ethical decisions and considerations.

However, scholars raise the question concerning whether the use of IS is beneficial or malignant in its ethical effects, as they "have the power to transform civilization into potential utopia—or dystopia—depending on society's ability to confront the questions with maturity and tolerance" (Altman 2004).

(3) Information systems as a broker for social interaction

As important as the use of IS in the area of decision-making and ethical support is the role of IS in mediating the discussion that reshapes existing ethical norms. Academia has already distinguished specific spheres that can be mediated by IS. Researchers have investigated various forms of newly emerging cultures and found numerous ways to derive overall trends and opinions from very different discussion places. Participants in those discussions are usually so dynamic and free that research has even coined a new term to describe those changes, "fragmegration", which expresses the idea that "both integration toward larger identities and fragmentation into smaller communities can occur at the same time" (Nye 2004). IS-mediated discussions and communications are widely recognized as intriguing and fruitful spheres for further research endeavours. The sphere of social communication has been defined as the *infosphere*, and research has recognized its importance of it by stating that the "infosphere is a common space [that] needs to be preserved and improved to the advantage of all" (Floridi 2008).

(4) Global activism supported by information systems

Global activism, which is supported by IS and technology, is not unlike other forms of protests against social, political, or economic deficits on the streets and in front of town halls, but researchers have found a need for "effort to foster all those informational conditions that facilitate participation, dialogue, negotiation and consensus-building practices among people, across cultures and through generations" (Floridi 2007). Although IS have fostered and supported global activism, there are fears that "political campaigns on the Internet may increase the active participation of the people, the moralist-solipsist[1] trend, fostered by the new medium, could lead to viewing political institutions as being ever more illegitimate and to lowering the quality of the democratic debate" (Sorj 2006).

## 10.4 Practical Examples—Digest of the Real World

### *10.4.1 The Chunked Structure of Owners of the Global Network*

Various organisations and stakeholders are responsible for organising the Internet technically and organizationally. The W3-Consortium is an institution that maintains standards, private companies, Internet service providers, and research facilities that provide technical infrastructure, routers, and backbones, without which the Internet would not be possible. On the other hand, country-based splitting of subnets allows countries like China to maintain information walls that filter information and the myriad contributors that provide endless amounts of information. This divided structure offers great potential for creativity and balance as well as for disorder and vandalism (Capurro 2000); in any case, it definitely hinders a standardised approach to contributions to global decisions that is owned by one stakeholder.

### *10.4.2 How Social Media Changes Minds*

Social media has ignited vast amounts of political and ethical influence. For example, a significant beneficial change that was gained through the usage of SN is the expression of opinion. Specific networks[2] are used to influence others and even to create news (The Economist 2011). Furthermore, networks are now being used as a significant factor in freedom. Starting in the middle of 2010 Facebook became a dominant information source in almost all democratically developed countries (The Economist 2010), and the Arab Spring revolutionary wave was initially started through SN. The level of SN participation in this social movement was so great that the US State Department even asked Twitter to delay its scheduled network upgrade because it could limit Iranians' ability to express their concerns about elections results (Pleming 2009). As opposing arguments it could be stated that critics state the opinion expression and creativity is a myth in its basis when it comes to the freedom and opinion expression (Flanagan 2011).

---

[1] Moralism–solipsism is a theory that holds that the self can know nothing but its own modifications and that it is all that exists (Merriam-Webster).

[2] http://Storify.com is an example of this specific SN resource.

The next section introduces several platforms that provide a step toward achieving the goal of routinely incorporating ethical issues in global decision-making.

### *10.4.3 Ushahidi—Platform for Instant Collaboration*

An example of how individuals can participate in global decisions is the African crowdsourcing platform Ushahidi, which emerged after the 2007 presidential election and the following riots in Kenya. According to David Kobia, founder of the platform and open source software, Ushahidi is an interactive mapping tool that enables individuals to record events on a map. Ushahidi takes those elements of location and time and puts them into a functional and usable platform (Vericat 2010). Anyone can contribute information via SMS, MMS, e-mail, and other channels. This information is published in real time on the webpage and presented graphically in a map of the region for everyone to use and respond to. The resulting platform enables the "crowd" to collaborate instantly; it is an incubator for the creation of a "hive mind" among the participating crowd.

The benefit of a platform like Ushahidi rises with the number of participating individuals, so there must be a reason that mobilises lots of people, such as natural disasters or political turmoil, two major areas where Ushahidi is used. Such platforms add to the traditional information flow of singular contributors like the Red Cross or the parties that are involved in a conflict by creating a platform for everyone.

Various problems arise in applying this platform to ethical considerations. Ethical positions tend to be the result of a long-lasting process that involves continuous participation. Its uneven level of participation also impedes a process that incorporates ethical considerations in a routine way. However, when ethical questions arise whose effect must be managed instantly, Ushahidi might provide a place where individuals can raise their voices and participate in global decisions. Still, the direct link to the process of global decision-making is missing and difficult to establish.

### *10.4.4 Edge.org—Experts' Collaboration*

The belief behind the idea of http://edge.org, stated by John Brockman, owner of the website http://edge.org is that "a finite number of people throughout history have invented, rather than discovered, the world. The words of the world are the life of the world, and a certain number of people have uttered those words" (Zaleski 2000). One question is published on the http://edge.org page each year, and experts from all over the world post their answers, thus contributing to human knowledge. The published questions touch specific areas that affect various fields of knowledge and are relevant and express a current trend. The question asked in

2010 was "How is the internet changing the way you think?" The results were responses—usually essays—from more than 170 experts and various articles in newspapers.

Edge.org shows a way for a complex, broadly defined question to be answered by experts from a variety of perspectives. Every expert tries to look "beyond the edge" and to contribute from his or her field's perspective. This approach might also be feasible in creating ethical viewpoints that can be incorporated into global decisions. As the result is the common conclusion of all participating experts, it may be more likely to be accepted in global decision-making than are ethical considerations from other, perhaps more biased sources. Edge.org also presents its results in various media, increasing its effect on public opinion and global decision-making. For groups with no relationship to global decision-makers, Edge.org offers a way to create publicity in the media in support of their ethical points of view.

Through the contributions of experts to solving current questions of interest and the resulting interest in the media, http://edge.org shows how ethical considerations can find their way into global decisions. Attention in the media raises the importance of an ethical issue that facilitates its incorporation into global decisions. Similar to Ushahidi, http://edge.com has no direct link to any global decision process because its intention is to accumulate viewpoints and contribute to knowledge.

### *10.4.5 Avaaz.org—Political Campaign Incubator*

In the same way Ushahidi provides a crowd-sourcing platform for instant collaboration and information interchange among peers, the international non-governmental organization Avaaz brings people together to influence global problems. The website http://avaaz.org is a platform that provides each visitor the ability to participate online on campaigns for or petitions against various issues. The large number of participants at http://avaaz.org is increasing its effect, as it has increased media attention. With this approach http://avaaz.org influenced a new anti-corruption law in Brazil, a move by Britain to create a marine-conservation zone in the Indian Ocean, and a proposal to allow more hunting of whales (The Economist 2010). In short, Avaaz reduces the effort involved in activism from that of attending demonstrations to that of a simple click. [An often-mentioned criticism of such platforms, is "clicktivism" (A town crier in the global village 2010)]. However, Avaaz does not only provide the web-based platform but also organises demonstrations, such as when Ricken Patel, the co-founder of Avaaz, called on Avaaz' members to put pressure on Brazilian and Turkish politicians to intercede with Iran.

This reduction in the complexity of participation is one of Avaaz' strengths: its potential to collect peers from all over the world to participate collectively in a common goal. As for Ushahidi, an increase in participants also increases the awareness of public media and global decision-makers.

Avaaz' structure as a campaigning platform is not designed for consensus-finding in controversial ethical questions but, rather, allowing ethical viewpoints to be expressed as a campaign or petition. While the decision process concerning what campaigns are created is not fully transparent or influenceable by everyone, the platform uses campaigns and the related media support to influence global decision-making directly, an aspect of the site that is missing in Ushahidi and http://Edge.org.

## 10.5 Discussion

There is a reciprocity between global ethics and culture on the one side and IS in their current shape on the other. IS are grounded in ethical and cultural values but they also influence these values [see Brey (2007) and Capurro et al. (2007)]. However, a literature review that analysed papers primarily from information science, management, decision science, economics, sociology, philosophy, and political science revealed no firm stance on or definitions of such interrelation.

When this research occurred, the academic community had no clear interest in expanding the Millennium Project's Global Challenges with additional disciplines. The majority of academic endeavours within the sphere of the Global Challenges was being conducted from the stances of political science or sociology. Ethics and similar themes were tied with the philosophical corpus of the papers, while global decision mechanism and IS co-existed in the cross-sections of information systems, economics, management, and decision theory.

In that context IS is a dimension that both involves and is affected by parts of each of the existing dimensions. For example, the mass communication theory, "Spiral of Silence" (Noelle-Neumann 1991), describes how the majority affects individuals' expressions of opinion because they fear punishment if they express unpopular views. Mass communication in IS today heavily affects and is affected by this Spiral of Silence, as the Internet is the platform on which to express unpopular opinions but is also a place for radical agitation and heavy manipulation efforts.

## 10.6 Research Agenda

Several questions emerged from the literature review and discussion of IS and ethics:

### *10.6.1 How to Implement? Routines for Highly Volatile Issues Like Ethics*

The introduction and the literature review showed that the term "ethics" has numerous attributes that prevent standardised processes from being implemented in order to incorporate ethics in global decision-making. The issue of ethics is highly volatile and variable, especially in global decisions. As the example of the crowd-sourcing platform Ushahidi (see Sect. 10.4.3) shows, ethical issues can mobilise large numbers of people in a short time, and in the same way people lose interest at the same speed. Moreover, ethics are intangible, they differ regionally, and parties involved in decision-making are often unable to reach consensus about them.

One way to meet the challenge of incorporating ethical considerations routinely into global decisions is to provide a modelling notation that provides the flexibility to support ethics' intangibility and regional differences. No institution has implemented such a process, and the literature review revealed no literature that has dealt with this issue. Future research could explore this in subject an academic way in an attempt to establish a way to incorporate ethical considerations routinely in global decision-making.

While various modeling techniques exist, to represent uncertainty or creativity in processes (Seidel et al. 2010) it might be of interest to consider the volatility and instability of ethics.

What is the trigger that will start the process of incorporating ethical considerations into global decision-making? What are the relevant steps? How can participation be maintained during decision-making? These and other questions arise when traditional approaches to process modelling are applied. Similar to existing process management approaches, various roles must be defined, which have political implications. For instance, the roles of business leader, process owner, and operational manager must be defined when implementing this kind of process.

### *10.6.2 How to Measure? Effects of Routine Involvement*

There are various ways to create and maintain systems that measure process outcomes, such as measuring qualitative indicators and aligning process goals with the overall strategy (Harmon 2010). Such measurement is a key component of continuous process improvement methods like Six Sigma and the Balanced Scorecard. Recent contributions have shown how to incorporate fuzzy criteria like "value" into business process management (e.g., vom Brocke et al. 2009). The term "ethics" has a fuzzy component that interferes with a functioning measurement system in several ways, but it is more difficult to measure the quality of the single contributions or the overall output because ethical considerations come into play and blur the measurement. It might be difficult to determine valid criteria with which to evaluate the quality of the result—a kind of "seal of ethical quality" that is recognised globally.

A second challenge is the effect of global decisions that affect third parties that are not involved in the decision process. For example the discussion concerning whether to outsource factories in other countries might consider the stakeholders in these countries and the deciding company, but effects of the outsourcing might also occur in neighbouring countries, among suppliers, or to logistics. It is often difficult to foresee the whole bandwidth of external effects when making a decision, so it might be challenging to incorporate those externalities into the process evaluation.

The ambiguity of the term "ethics" impedes the design of a typical performance measurement system and the traditional approaches of business process implementation (e.g., as-is models and to-be models). The main challenge is to incorporate ethical considerations into a functioning measurement system in order to make them meaningful for global purposes.

### *10.6.3 How to Manage? Long-Term Management for Short-Term Problems*

Finally, the question of using IS to manage processes within the domains of routine global decision-making remains a good starting point for future research. For example, it could be possible to implement concepts from workflow management to streamline and optimise processes that see significant deviations. Will flexible modelling techniques help the entire process to become more manageable and easier to model? Is it possible to make routine decisions by means of IS flexible and adherent to rapidly changing environments and optional unbalancing externalities? Will the diversified instrumentarium of flexibility techniques help to solve the particular problems of: handling exceptions and unpredictable circumstances—such as the Arab Spring phenomenon, which could significantly change the ethical background of the decision-making process—by means of exception handling, late modelling, or case handling techniques. These are patterns in modelling workflow management systems that provide the required flexibility. Whether these patterns are applicable to ethical activities could be an interesting research question. Applying existing business process management strategies like Six Sigma to these modelled processes could also enable continuous process improvement.

## 10.7 Conclusion

The polyvalent planes between which this research is conducted and the modestly researched interconnections make drawing conclusions a daunting task. It is difficult to conclude whether there is potential for routine decision-making through IS without the mediation of the Internet, yet that medium could prove completely adequate to the task. The literature review conducted here does not provide a complete study of research in this field, although it does provide an overview of currently discussed topics in the frame of ethics.

Approaches to incorporating uncertainty in the design of business processes, such as in modelling, already exist, yet considerable research remains to be done in this area that could lead to frameworks that make it possible for corporations and institutions to include ethical considerations in local processes and at the same time align ethical issues on a global level using a globally accepted framework of ethical modelling of business processes. The literature has not discussed using the techniques of business process management to incorporate ethical considerations into decision-making, so the proposals stated in this paper have a somewhat fragile grounding.

Should the solutions or further research try to fit within the domains of already defined scientific disciplines, integrate existing disciplines through creation of common ground, or evolve into completely new scientific disciplines or approaches? One might think about overarching SN that could interact with all existing SN and provide global bodies (e.g., a UN organizational unit) with a tacit overview and understanding of how global ethics are developing, evolving, and being perceived.

As there is no such global body that might be used for ethical purposes, various platforms have emerged that people use for ethical considerations. Websites like Ushahidi.com and the campaigning platform Avaaz.org meet the rising demand to leverage the power of the Internet to collect the energy of the masses in pursuit of social, economic, ecological, and political goals. It is expected that the importance of these platforms will increase and that some of them will grow into globally relevant institutions that attend to ethical issues in global decision-making.

## References

A town crier in the global village. (2010). *Economist*, p. 62.

Altman, C. (2004). *Converging technologies: The future of the global information Society*. San Francisco.

Brey, P. (2007). Is information ethics culture-relative? *International Journal of Technology and Human Interaction, 3*, 2–24.

Capurro, R. (2000). Ethical challenges of the information society in the 21st century. *International Information & Library Review, 32*, 257–276.

Capurro, R., Frühbauer, J., & Hausmanninger, T. (2007). Localizing the Internet. Ethical aspects in intercultural perspective. *ICIE Series, 4*, 21–38.

Davis, T., & Forester-Miller, H. (1996). *A Practitioner's guide to ethical decision making*. Dictionary.com. (n.d.). Retrieved 12 May 2011, from Dictionary.com: http://dictionary.reference.com/browse/society.

Ess, C. (2006). Cybernetic pluralism in an emerging global information. *Ethics and Information Technology, 8*(4), 215–226.

Flanagan, B. (2011, May 18). *Facebook Revolution 'a myth', Critics Say*. Retrieved May 20, 2011, from The National. http://www.thenational.ae/business/media/facebook-revolution-a-myth-critics-say.

Floridi, L. (2007). Global information ethics: The importance of being environmentally earnest. *International Journal of Technology and Human Interaction, 3*, 1–11.

Floridi, L. (2008). Network ethics: information and business ethics in a networked society. *Journal of Business Ethics, 90*(Suppl 4), 649–659.

Harmon, P. (2010). The scope and evolution of business process management. In J. vom Brocke, & M. Rosemann, (eds.) *Handbook on Business Process Management* (vol. 1). Berlin: Springer.

Liszka, J. (2011). *What is Pragmatic Ethics*. Retrieved October 29, 2011, from Department of Philosophy, College of Liberal Arts and Sciences. http://www.philosophy.uncc.edu/mleldrid/SAAP/USC/TP25.html.

Maner, W. (1996). Unique ethical problems in information technologoy. *Science and Engineering Ethics, 2*, 137–154.

Miscione, G., Staring, K., Ostmo, L., & Fossum, C. (2008). The shifting legitimation of an information system: Local, global and large scale. *GlobDev 2008*, Paper 7. AIS.

Noelle-Neumann, E. (1991). Communication Yearbook 14. In J. A. Anderson (Ed.), *The theory of public opinion: The concept of Spiral of SIlence* (pp. 256–287). Newburry Park: Sage.

Nye, J. S. (2004). *Power in a global information age*. New York: Taylor & Francis.

Pippa, N. (2001). *Digital divide: Civic engagement, information poverty, and the Internet Worldwide*. New York: Taylor & Francis.

Pleming, S. (2009, June 16). *U.S. State Department speaks to Twitter over Iran*. Retrieved June 1, 2011, from Reuters. http://www.reuters.com/article/2009/06/16/us-iran-election-twitter-usa-idUSWBT01137420090616.

Schradie, J. (2011). *The digital production gap: The digital divide and Web 2.0 collide*. Berkeley: Department of Sociology, University of California.

Seidel, S., Muller-Wienbergen, F., & Rosemann, M. (2010). Pockets of creativity in business processes. *Communications of the Association, Pockets of Creativity in Business Processes, 27*(1).

Senior Scholars Consortium. (2011). *Senior Scholars' Basket of Journals*. Retrieved 12 March, 2011, from http://home.aisnet.org/displaycommon.cfm?an=1&subarticlenbr=346.

Sorj, B. (2006). *Internet, public sphere and political marketing: Between the promotion of communication and moralist solipsism*. Rio de Janeiro: The Edelstein Center for Social Research.

The Economist. (2010, September 29). Facebook and freedom. Retrieved June 1, 2011, from The Economist. http://www.economist.com.

The Economist. (2011, May 12). *The Economist*. Retrieved June 8, 2011, from How to use Facebook friends and influence people. http://www.economist.com/blogs/democracyinamerica/2010/09/social_networks.

Vericat, J. (2010, Fall/Winter). Open source mapping as liberation technology. *Journal of International Affairs, 64*(1) 195–201.

vom Brocke, J. M., Simons, A. M., Niehaves, B., Riemer, K., Plattfaut, R., & Cleven, A. (2009). Reconstructing the giant: On the importance of rigour in documenting the literature search process. In *17th European Conference on Information Systems* (pp. 2206–2217).

Waller, B. N. (2005). *Consider ethics, theory, readings and contemporary issues*. New York: Pearson Longman.

YanFang, C. (2005). *Ethical factors in decision making for information system management*. New York: Department of SCIS, Baruch College and Graduate Center CUNY.

Zaleski, J. (2000, 01 24). *PW talks with John Brockman* (Vol. 247, Issue 4, p. 306.). Publishers Weekly.

MIX
Papier aus verantwortungsvollen Quellen
Paper from responsible sources
FSC® C105338

If you have any concerns about our products,
you can contact us on
**ProductSafety@springernature.com**

In case Publisher is established outside the EU,
the EU authorized representative is:
**Springer Nature Customer Service Center GmbH**
**Europaplatz 3, 69115 Heidelberg, Germany**

Printed by Libri Plureos GmbH
in Hamburg, Germany